Fundamentals of the Optics of Materials

Fundamentals of the Optics of Materials
Tutorial and Problem Solving

Vladimir I. Gavrilenko
Volodymyr S. Ovechko

Jenny Stanford
Publishing

Published by

Jenny Stanford Publishing Pte. Ltd.
101 Thomson Road
#06-01, United Square
Singapore 307591

Email: editorial@jennystanford.com
Web: www.jennystanford.com

British Library Cataloguing-in-Publication Data
A catalogue record for this book is available from the British Library.

Fundamentals of the Optics of Materials: Tutorial and Problem Solving

Copyright © 2024 Jenny Stanford Publishing Pte. Ltd.

All rights reserved. This book, or parts thereof, may not be reproduced in any form or by any means, electronic or mechanical, including photocopying, recording or any information storage and retrieval system now known or to be invented, without written permission from the publisher.

For photocopying of material in this volume, please pay a copying fee through the Copyright Clearance Center, Inc., 222 Rosewood Drive, Danvers, MA 01923, USA. In this case permission to photocopy is not required from the publisher.

ISBN 978-981-4877-93-0 (Hardcover)
ISBN 978-1-003-25694-6 (eBook)

Dedicated to

Natasha, Alexander, Alena,
Joshua, Sophia, Adelyn, Elias, and Sebastian

and to

Evdokia, Oleg, Olga, Timur, Anthony, and Eli

Contents

Preface	xv

1 Basics of the Light Interaction with Matter — 1

1.1	Introduction: Description of the Electromagnetic Field in Medium	2
1.2	Linear Optical Response: Classical Picture	8
1.3	Linear Optical Response: Quantum Mechanical Picture	12
1.4	Dispersion Relations for Rare Gas: Lorentz–Lorenz Formulae	22
1.5	Review Questions and Exercises	24
	1.5.1 Complex Dielectric Susceptibility	24
	1.5.2 Dielectric Permittivity and Refractive Index	25
	1.5.3 Reflection Coefficient for Germanium	25
	1.5.4 Group Velocity	25
	1.5.5 Refractive Index of Iodine	26
	1.5.6 Refractive Index Dispersion	26

2 Nonlinear Light Interaction with Matter — 27

2.1	Introduction	27
2.2	Nonlinear Processes	28
2.3	Nonlinear Optical Processes Classification	29
2.4	Classic Theory of Nonlinear Optical Response	39
2.5	Quantum Picture: Electron Charge Density within the Perturbation Theory	41
2.6	Review Questions and Exercises	42
	2.6.1 Relationship between Linear and Nonlinear Optical Susceptibilities	42
	2.6.2 Quadratic Nonlinear Dielectric Polarization	42
	2.6.3 Second Harmonic Generation	42

viii | *Contents*

	2.6.4	Parametric Frequency Mixing	43
	2.6.5	Phase Modulation	43
	2.6.6	Relationship between Nonlinear Functions	43

3 Fundamentals of Electron Energy Structure and Optics of Molecules — **45**

3.1	Introduction		45
3.2	Electron Energy Structure of Simple Molecules		46
	3.2.1	A Homonuclear Diatomic Molecule: The Hydrogen Molecule	47
	3.2.2	Heteronuclear Diatomic Molecule	50
3.3	Linear Combination of Atomic Orbitals Method		51
	3.3.1	Electron Energy Structure of Methane Molecule	52
3.4	First Principles Methods in Computational Materials Science		55
	3.4.1	Thomas–Fermi Approximation	56
	3.4.2	First Principles Methods Based on the Density Functional Theory	59

4 Optical Absorption and Fluorescence of Materials — **63**

4.1	Light Absorption of Materials		63
4.2	Interaction of Electromagnetic Radiation with Atom		65
4.3	Light Emission and Absorption by Atom		69
4.4	Einstein Coefficients for Condensed Matter		73
4.5	Principles of the Signal Processing in Optics		76
	4.5.1	Basic Principles of Operation in Optical Spectrometry	76
	4.5.2	Optical Response Function: Inverse Problem in Optics	77
	4.5.3	Tikhonov Method of Solution Regularization	80
	4.5.4	Fourier Transform Spectrometer	82
	4.5.5	Non-harmonic Analysis in Optical Spectroscopy	85
4.6	Photoluminescence		91
4.7	Luminescence Kinetics		95
	4.7.1	Rate Measurement of Photonic Processes in Materials	95
	4.7.2	Stern–Folmer Equation	96
	4.7.3	Chemical Reactions Limited by Diffusion	97

| | | 4.7.4 Energy Transfer in Luminescence | 98 |

| | | | |

4.7.4   Energy Transfer in Luminescence 98
4.8   Measurements of Luminescence 100
4.9   Review Questions and Exercises 105
    4.9.1   Bouguer–Lambert Law 105
    4.9.2   Amplification by Optical Pumping 105
    4.9.3   Homogeneous Spectral Line Broadening by Finite Lifetime 105
    4.9.4   Inhomogeneous Spectral Line Broadening by Thermal Motion 105
    4.9.5   Einstein Coefficient 106
    4.9.6   Atomic Concentration Detection 106
    4.9.7   Luminescence in Semiconductors 106

5 Atomic and Electron Energy Structure of Solids 107

5.1   Useful Terminology of Solid-State Physics 108
5.2   Atomic structure 109
5.3   Elements of Band Structure Theory 112
    5.3.1   Quantum Mechanics of Particles in Periodic Potential 112
    5.3.2   Tight-Binding Band Structure of Some Periodic Crystalline Structures 116
5.4   Review Questions and Exercises 125
    5.4.1   Amorphous State 125
    5.4.2   Bragg's Law 125
    5.4.3   Brillouin Zone 125
    5.4.4   Valence and Conduction Bands 125

6 Basics of Surface Science and Surface Optics 127

6.1   Surface Atomic Geometries and Classification 128
6.2   Electron Energy Structure of Solid Surfaces 132
    6.2.1   Modeling of Solid Surfaces 133
    6.2.2   Surface Electron Energy Structure 137
6.3   Optics of Solid Surfaces 139
    6.3.1   Probing of Solid Surfaces 140
    6.3.2   Surface-Enhanced Raman Spectroscopy 141
    6.3.3   Surface Reflectance Differential Spectroscopy 147
    6.3.4   Surface Analysis by the Second Harmonic Generation 150

x | Contents

6.3.5 Si-SiO$_2$ Interface Analysis by the Second
Harmonic Generation   155
6.4 Review Questions and Exercises   160
6.4.1 Anharmonicity on the Surface   160
6.4.2 Carrier Concentration on the Surface   160

7 Plasma Optics   **161**
7.1 Theoretical Model Based on Classical
Electrodynamics   161
7.2 Plasma in Metals   163
7.3 Surface Plasma   164
7.4 Surface Plasmon Polaritons   166
7.5 Application of Plasma Resonance Spectroscopy   173
7.6 Practical Examples   173
7.6.1 Absorption Coefficient of Free Electrons   174
7.6.2 Plasma Frequency Dependence on Material
Parameters   175
7.6.3 Reflectance at Plasma Frequency   175
7.6.4 Geometrical Profile for Generation of Surface
Plasmons   177
7.7 Review Questions and Exercises   178
7.7.1 Free Electron Absorption   178
7.7.2 Free Electron Scattering   178
7.7.3 Dielectric Constant of Plasma   179
7.7.4 Refractive Index of Plasma   179
7.7.5 Reflection Coefficient of Plasma   179
7.7.6 Dispersion in Plasma   179
7.7.7 Penetration Depth of Electromagnetic
Radiation   179

8 Optics of Composites, Alloys, and Artificial Materials   **181**
8.1 Effective Medium Approximations in the Optics
of Composite Materials   182
8.2 Electronic Band Structure and Optical Functions
of Alloys   184
8.2.1 Method of Pseudopotential for Electronic
Band Structure Calculations of Alloys   185
8.2.2 Calculation of Optical Functions from the
Electronic Band Structure of Alloys   187

8.3	SiGe Alloys		188
	8.3.1	Atomic Structure of SiGe Alloys	188
	8.3.2	Optical Functions of SiGe Alloys	190
8.4	$A^{III}B^{V}$, and $A^{III}B^{V}$-Based Alloys		195
	8.4.1	Gallium Arsenide	195
	8.4.2	Aluminium Gallium Arsenide	197
	8.4.3	GaP, InAlP, and Other Phosphorous-Based Alloys	197
	8.4.4	Group III Nitride-Based Alloys	198
8.5	Inorganic–Organic Composites		201
8.6	Metamaterials		208

9 Anisotropic Material Optics — 217

9.1	Basics of Anisotropic Materials		217
	9.1.1	Crystallography and Crystallophysical Coordinate Systems	218
	9.1.2	Linear Optical Anisotropy of Non-cubic Crystals	220
9.2	Jones Vectors and Jones Matrices		224
9.3	Birefringence		228
9.4	Practical Examples		233
	9.4.1	Polarization of Light by Calcite Dichroic Polarizer	233
	9.4.2	Snell's Law of Refraction in Anisotropic Crystal	234
	9.4.3	Light Beam Propagation in Calcite Crystal	236
	9.4.4	Light Transmition through the System of Three Polarizers	239
9.5	Stokes Vector and Mueller Matrices		240
9.6	Review Questions and Exercises		246
	9.6.1	Permittivity Tensor	246
	9.6.2	Electric Field Vectors Relationship	246
	9.6.3	Poynting Vector in the Iceland Spar Crystal	246
	9.6.4	Circularly Polarized Light on a Quartz Plate	247
	9.6.5	Huygens' Construction	247
	9.6.6	Birefringence of a Quartz Plate	247
	9.6.7	Refraction in a Calcite Crystal	247
	9.6.8	Stokes Vector	248

10 Optics of Organic and Biological Materials 249

10.1 Optical Properties of Molecular Systems 250
10.2 Optics of Molecular Crystals 259
10.3 Polymers 263
10.4 Biological Materials 273
    10.4.1 Surface Plasmon Resonance for Biosensing 275
    10.4.2 Fluorescence of Biomaterials 279

11 Optics of Moving Media 283

11.1 Introduction 283
11.2 Relativistic Effects in Optics 284
11.3 Light Propagation in Ether: Michelson-Morley
    Experiment 287
11.4 Practical Examples 290
    11.4.1 Frequency of Moving Light Source 290
    11.4.2 Doppler Effect in Gas 291
    11.4.3 Light Propagation in a Moving Glass Bar 292
11.5 Review Questions and Exercises 293
    11.5.1 Optical Doppler Effect 293
    11.5.2 Speed of Light in a Moving Medium 294
    11.5.3 Optical Radar 294
    11.5.4 Emission of a Rarefied Gas 294
    11.5.5 Gyroscope 294
    11.5.6 Dispersion in Moving Medium 295
    11.5.7 Optical Doppler Effect Example 295

12 Applied Optics 297

12.1 Introduction 297
12.2 Thermal Optics 297
    12.2.1 Thermal Lensing 299
    12.2.2 Laser Heat Treatment of Metals 303
12.3 Waveguides: Fiber-optics 306
12.4 Electro-optics 312
    12.4.1 Principles of Electro-optics 312
    12.4.2 Primary and Secondary Electro-optical Effect 316
    12.4.3 Practical Examples 317
12.5 Magneto-optics 323
    12.5.1 Principles of Magneto-optics 324
    12.5.2 Practical Examples 325

12.6 Acousto-optics		328
	12.6.1 Basics of Acousto-optics	329
	12.6.2 Practical Examples: Optical Difraction Wave Amplitude for $Q \ll 1$	331
	12.6.3 Practical Examples: Optical Diffraction Wave Amplitude for $Q \gg 1$	332

Appendix A Selected Physical Constants   337

Appendix B Optical Units Conversions   339

Appendix C The Vector Triple Product   343

Appendix D Mueller versus Jones Vectors and Matrices   345

Bibliography   349

Index   363

Preface

The optical properties of materials are of key importance for a variety of applications. Rapid developments in materials engineering, recent progress in nanotechnology, and discoveries of new artificial materials (such as fullerene, graphene, and metamaterials) have revolutionized electronics, energy production, medicine, and other features of everyday life in the 21st century.

This book presents the rapidly developing field of materials optics. It focuses on a broad audience requiring basic knowledge of physics, chemistry, and optics, normally at the university entry level. It also provides a more advanced description of selected topics (in the appendices) to provide better orientation in the field to those interested in further reading the subject. The first general chapters of the book correspond to standard university courses for bachelor's and master's degrees in physics. Other chapters focus on modern developments in materials optics, such as nano-plasmonics, nano-photonics, and optical properties of nano-sized materials, and are for those familiar with the basic elements of quantum mechanics.

The book aims to provide a systematic description of the basic theory and numerical simulation methods in optics of different materials, ranging from gases, liquids, and inorganic solids to organic materials, atoms, molecules, and polymers, and including elements of biological materials optics. Special attention is given to developing a clear understanding of the basic processes and the mechanisms that govern the optical response of materials as well as the skills to solve problems related to these processes. For this, the book is written self-consistently with all chapters linked together and every chapter contains a theoretical introduction followed by an extended list of problems relevant to the topic. About one-third of the problems contain solutions with detailed descriptions and explanations. However, solutions to most of the problems presented in the book are given in a separate supplemental manual. This textbook and its solution manual form the complete set of this teaching material.

The book contains chapters describing the basics of the interactions of light with matter (atoms, molecules, solids), including fundamentals of electron energy structure as well as a detailed description of the physical mechanisms that govern the linear and nonlinear optical response of matter (Chapters 1 to 5). It comprises more specialized chapters addressing recent developments in the relevant field, such as optics of solid surfaces (Chapter 6); plasma optics (Chapter 7); optics of composites, alloys, and metamaterials (Chapter 8); optics of anisotropic materials (Chapter 9); and optics of organic and biological materials (Chapter 10). Relativistic effects in optics are addressed in Chapter 11. Chapter 12 overviews the basics of optical engineering, focusing on typical and widely used applications of materials optics and presenting numerous examples and problems. The book also contains appendices with important reference materials, a subject index, and an extended list of publications. We hope the book will be helpful for university and college students and instructors as well as a broad audience interested in improving their knowledge of the subject.

Vladimir I. Gavrilenko
Williamsburg, Virginia, USA

Vladimir S. Ovechko
Kyiv, Ukraine

Chapter 1

Basics of the Light Interaction with Matter

Interaction of light with materials could be understood as a reaction of atoms, molecules, or solids with the external electromagnetic radiation. This reaction is described in terms of different kinds of excitations, e.g., appearance of additional induced polarization due to electron transitions between available energy states, generation of lattice vibrations, and electron–hole pairs. This chapter presents an introduction to the basics of electron energy structure and the theoretical basics of optical functions of materials and focuses on the physical nature of phenomena.

For tutorial purposes this chapter considers a few relevant topics of the electron energy structure in materials in order to illustrate their specific features in optics. Within this chapter we consider only the basics of electron energy structure, its modeling and applications for optics. For further reading more specialized literature is recommended, e.g. [Martin (2005); Gavrilenko (2020)].

After a brief overview of the electron energy structure, the classical theory of light and the optical functions of materials (dielectric permittivity, polarization functions) are considered within both the classical and quantum mechanical theories. This provides a bridge

Fundamentals of the Optics of Materials: Tutorial and Problem Solving
Vladimir I. Gavrilenko and Volodymyr S. Ovechko
Copyright © 2024 Jenny Stanford Publishing Pte. Ltd.
ISBN 978-981-4877-93-0 (Hardcover), 978-1-003-25694-6 (eBook)
www.jennystanford.com

2 | Basics of the Light Interaction with Matter

between the classical theory of optics and the modern state-of-the-art approaches in the optics of materials from first principles.

1.1 Introduction: Description of the Electromagnetic Field in Medium

Within the electromagnetic (EM) Maxwell's theory, a signature of the medium is embodied in the relationship between the polarization vector function ($P(r, t)$) (or in magnetic medium in magnetization vector function, $M(r, t)$), on the one hand, and the electric ($E(r, t)$) (or magnetic, $H(r, t)$) field of external EM radiation, on the other [Born and Wolf (1999); Jackson (1998); Zahn (1979); Vysotskii *et al.* (2011); Ovechko and Sheka (2006); Ovechko (2017)]. These are known as *constitutive relations* (or materials equations) [Born and Wolf (1999); Saleh and Teich (2007)].

First we consider nondispersive, homogeneous, and isotropic medium. In nonmagnetic materials (which are mostly addressed in this book) the constitutive relation describes a relationship between P and E according to the equation

$$P = \chi E, \tag{1.1}$$

where χ is the *dielectric susceptibility* of the medium. According to EM theory (see e.g. [Born and Wolf (1999); Saleh and Teich (2007); Zahn (1979)]) the electric field in a medium is described in terms of the *electric flux* density (or the *electric displacement*) D according to

$$D = \varepsilon E = E + 4\pi P. \tag{1.2}$$

Equations (1.1) and (1.2) define the *dielectric permittivity* function of the medium:

$$\varepsilon = 1 + 4\pi\chi. \tag{1.3}$$

The dielectric permittivity function in the static case is also called the *dielectric constant*. In the same way the magnetic field in a medium

(**B**) can be written as

$$B = \mu H, \tag{1.4}$$

where μ is the *magnetic permeability* of the medium. The Maxwell's equations, which are the general laws governing electromagnetic fields interacting with matter, are given by

$$\nabla \times H = \frac{1}{c}\frac{\partial D}{\partial t} + \frac{4\pi}{c}j, \tag{1.5}$$

$$\nabla \times E = -\frac{1}{c}\frac{\partial B}{\partial t}, \tag{1.6}$$

$$\nabla \cdot D = 4\pi\rho, \tag{1.7}$$

$$\nabla \cdot B = 0, \tag{1.8}$$

where the symbols $\times$ and $\cdot$ denote the *vector* and *scalar* products [Zahn (1979)], respectively, j is the electric current density, and c is the phase speed of light in vacuum. The *continuity equation* states that the divergence of the current density is equal to the negative rate of change of the charge density:

$$\nabla \cdot j = -\frac{\partial \rho}{\partial t}. \tag{1.9}$$

The continuity equation can either be regarded as an empirical law expressing (local) *charge conservation* or be derived as a consequence of two of Maxwell's equations, (1.6) and (1.7). The functions of ρ and j can be expanded in terms of polarization and magnetization functions. For nonmagnetic media the last contributions can be neglected. Within the electric dipole approximation the values of ρ and j are given by

$$\rho = \rho_0 - \nabla \cdot P,$$

$$j = j_0 + \frac{\partial P}{\partial t}, \tag{1.10}$$

where ρ_0 and j_0 are free electric charge and current densities, respectively. In a free medium there are no free charges and current,

4 | *Basics of the Light Interaction with Matter*

i.e., $\rho_0 = 0, j_0 = 0$. Consequently Eq. (1.7) becomes

$$\nabla \cdot E + 4\pi \nabla \cdot P = 0. \tag{1.11}$$

Equations (1.5) and (1.6) can be combined into the wave equation

$$\nabla^2 E - \nabla(\nabla \cdot E) - \frac{1}{v^2}\frac{\partial^2 E}{\partial t^2} - \frac{4\pi}{v^2}\frac{\partial^2 P}{\partial t^2} = 0, \tag{1.12}$$

where v denotes the speed of electromagnetic waves (light) in a medium

$$v = \frac{c}{\sqrt{\varepsilon\mu}} = \frac{c}{\tilde{n}}. \tag{1.13}$$

The following vector identity was used to derive equation (1.12) [Zahn (1979)]

$$\nabla \times (\nabla \times E) = \nabla(\nabla \cdot E) - \nabla^2 E. \tag{1.14}$$

The quantity $\tilde{n}$ in equation (1.13) defined as

$$\tilde{n} = \sqrt{\varepsilon\mu} = n + i\kappa, \tag{1.15}$$

is a complex *index of refraction*, with real n and imaginary κ parts of $\tilde{n}$ called *refraction* and *extinction* coefficients, respectively. Note that the complex optical functions of $\tilde{n}$ (Eq. (1.15)) define the *phase velocity* of electromagnetic waves [Born and Wolf (1999)].

Vector potential A and scalar potential U of electromagnatic field are introduced in the following way:

$$E = -\nabla U - \frac{1}{c}\frac{\partial A}{\partial t}, \tag{1.16}$$
$$B = \nabla \times A. \tag{1.17}$$

We are free to change both A and U by a *gauge transformation*:

$$A' = A + \nabla\Lambda, \quad U' = U - \frac{\partial\Lambda}{\partial t} \tag{1.18}$$

where Λ is a scalar function. This transformation leaves the fields invariant, however, the form of the dynamical equations will be changed. Here we will use the *Coulomb gauge* (that is also called velocity, momentum, or radiation gauge) that is defined by:

$$\nabla \cdot \boldsymbol{A} = 0. \tag{1.19}$$

Vector potential obeys the wave equation, that in free space is given by:

$$\nabla^2 \boldsymbol{A} = \frac{1}{c^2}\frac{\partial^2 \boldsymbol{A}}{\partial t^2}, \tag{1.20}$$

with the solution in the following form:

$$\boldsymbol{A}(\boldsymbol{r},t) = A\hat{\boldsymbol{e}}\frac{1}{2}\left[e^{i(\boldsymbol{k}\cdot\boldsymbol{r}-\omega t)} + e^{-i(\boldsymbol{k}\cdot\boldsymbol{r}-\omega t)}\right] = A\hat{\boldsymbol{e}}\cos(\boldsymbol{k}\cdot\boldsymbol{r}-\omega t), \tag{1.21}$$

where for the wave vector we have $\hat{\boldsymbol{e}} \cdot \boldsymbol{k} = 0$, $k^2 = \omega^2/c^2$. For a linearly polarized field, the polarization vector $\hat{\boldsymbol{e}}$ is real. However, for a circularly and elliptically polarized fields $\hat{\boldsymbol{e}}$ is complex, e.g. for a circularly polarized field it is given by $\hat{\boldsymbol{e}} = (\boldsymbol{x} \pm i\boldsymbol{y})/2$. The $+$ and $-$ signs correspond to positive and negative helicity, respectively i.e. they correspond to left and right hand circular polarization, respectively and are traditionally used in optics. The electric and magnetic fields are given by:

$$\boldsymbol{E}(\boldsymbol{r},t) = -i\frac{\omega}{c}A\hat{\boldsymbol{e}}\frac{1}{2}\left[e^{i(\boldsymbol{k}\cdot\boldsymbol{r}-\omega t)} - e^{-i(\boldsymbol{k}\cdot\boldsymbol{r}-\omega t)}\right]$$
$$= kA\hat{\boldsymbol{e}}\sin(\boldsymbol{k}\cdot\boldsymbol{r}-\omega t) \tag{1.22}$$

$$\boldsymbol{B}(\boldsymbol{r},t) = -ik(\boldsymbol{k} \times \hat{\boldsymbol{e}})A\frac{1}{2}\left[e^{i(\boldsymbol{k}\cdot\boldsymbol{r}-\omega t)} - e^{-i(\boldsymbol{k}\cdot\boldsymbol{r}-\omega t)}\right]$$
$$= k(\boldsymbol{k} \times \hat{\boldsymbol{e}})A\sin(\boldsymbol{k}\cdot\boldsymbol{r}-\omega t) \tag{1.23}$$

The space- and time-dependent polarization function of $\boldsymbol{P}(\boldsymbol{r},t)$ (see Eq. (1.1)) is given by:

$$\boldsymbol{P}(\boldsymbol{r},t) = \int \chi\left[\boldsymbol{r},\boldsymbol{r}',(t-t')\right]E(\boldsymbol{r}',t')d\boldsymbol{r}'dt'. \tag{1.24}$$

A similar relation for the electric displacement function (see Eq. (1.2)) reads:

$$D(r,t) = \int \varepsilon\,[r,r',(t-t')]\,E(r',t')dr'dt'.$$

(1.25)

In homogeneous space and in the absence of time-dependent perturbations these equations are simplified and can be represented as:

$$P(r,t) = \int \chi\,[(r-r'),(t-t')]\,E(r',t')dr'dt'.$$

(1.26)

$$D(r,t) = \int \varepsilon\,[(r-r'),(t-t')]\,E(r',t')dr'dt'.$$

(1.27)

In anisotropic media, the quantities ε and χ become tensors. In the linear, homogeneous, and nondispersive medium the vector components of P and D are given by

$$P_i = \sum_j \chi_{ij}E_j,$$

(1.28)

$$D_i = \sum_j \varepsilon_{ij}E_j.$$

(1.29)

For the nonlinear anisotropic medium with dispersion it is instructive to consider optical response of the medium (described e.g. by the $P(r,t)$ function) as a reaction to the monochromatic light. The electric field of the monochromatic EM wave could be represented in the form:

$$E(r,t) = E(k,\omega)e^{i(kr-\omega t)}.$$

(1.30)

with the wave vector of light (k) that defines the light propagation in the direction of s. The vector of k depends on the complex refraction coefficient $\tilde{n}$ of the medium according to:

$$k = \frac{\omega}{c}\tilde{n}s = \frac{\omega}{c}(n+i\kappa)s.$$

(1.31)

The unit vector s in equation (1.31) defines the direction of the *Poynting vector S* that represents the amount of energy which crosses per second a unit area normal to the directions of E and H. The Poynting vector is defined by the following relation [Born and Wolf (1999)]:

$$S = \frac{c}{4\pi}(E \times H).\qquad(1.32)$$

Note, that in optically anisotropic media (with the properties being directionally dependent) the function of $\tilde{n}$ is a tensor and vectors k and S are not collinear any more (as in isotropic medium or in free space). This is addressed more in details in Chapter 9.
The vector components of the polarization function can be expanded in the frequency domain, according to the power of the applied E field showing harmonics generations [Boyd (1992); Shen (2003)]:

$$P_i(\omega) = p_i^{(0)} + \sum_j \chi_{ij}^{(1)}(\omega)E_j(\omega)$$

$$+ \sum_{jk} \chi_{ijk}^{(2)}(-2\omega,\omega,\omega)E_j(\omega)E_k(\omega) + \dots \qquad(1.33)$$

$$+ \sum_{jkl} \chi_{ijkl}^{(3)}(-3\omega,\omega,\omega,\omega)E_j(\omega)E_k(\omega)E_l(\omega) + \dots$$

The notation of photon processes related to the optical susceptibility function in the last equation (see e.g. the second-order $\chi_{ijk}^{(2)}(-2\omega,\omega,\omega)$ component of the full susceptibility function) means optical response to the two incoming photons at the frequency ω with the j and k components of the E field resulting in one photon of outgoing radiation at the frequency 2ω with the i component of the E field. Strictly speaking, the second-order process contributes to the static optical constant (which is called *optical rectification*), and the third-order process contributes to the optical response at the frequency ω modifying the linear optical functions. These phenomena however are not considered in this book and the interested readership can refer to the more specialized literature in field (see e.g. [Boyd (1992); Shen (2003); Sipe and Boyd (2002)]).

Optical functions of materials introduced and discussed in this book (see Eqs. (1.15) and (8.10)) relate to each other. These

Basics of the Light Interaction with Matter

relationships could be derived from the equations given in this section and for the nonmagnetic materials are summarized below:

$$
\begin{cases}
\varepsilon(\omega) = & \varepsilon_1(\omega) + i\varepsilon_2(\omega), \\
\chi(\omega) = & \chi'(\omega) + i\chi''(\omega), \\
\tilde{n}(\omega) = & n(\omega) + i\kappa(\omega), \\
\varepsilon = & 1 + 4\pi\chi, \\
\tilde{n} = & \sqrt{\varepsilon}, \\
\varepsilon_1 = & n^2 - \kappa^2, \\
\varepsilon_2 = & 2n\kappa, \\
n = & \dfrac{1}{\sqrt{2}}\sqrt{\sqrt{\varepsilon_1^2 + \varepsilon_2^2} + \varepsilon_1}, \\
\kappa = & \dfrac{1}{\sqrt{2}}\sqrt{\sqrt{\varepsilon_1^2 + \varepsilon_2^2} - \varepsilon_1}.
\end{cases}
\tag{1.34}
$$

1.2 Linear Optical Response: Classical Picture

Within the classical microscopic theory the optical medium is described though an array of harmonic oscillators [Saleh and Teich (2007); Shen (2003); Fox (2003)]. The starting point of most optical response theories of media is the postulate that initially all materials are electrically neutral with homogeneous distribution of charge. Incoming light breaks the local neutrality resulting in a mutual displacement of negative and positive charges of atoms (induced charge inhomogeneity). If electrons are bounded to the nucleus, the charge displacement generates dipoles. The full optical polarization function of the medium (given by equation (1.1)) is represented as an array of N independent (non-interacting) light-induced dipoles:

$$
P(r) = N\alpha(r)E(r, t) = -Ner.
\tag{1.35}
$$

Here we assumed that the medium consist of N atoms in the volume unit, with one electron per each of these atoms. Equation (1.35) defines the optical microscopic *polarizability* function $\alpha(r)$. We will consider further the one-dimensional case for simplicity (replacing **P(r)** by **P(x)**). Note that the assumption made in the postulate represented by equation (1.35) (mutual independence of the dipoles) is an essential simplification of the phenomenon. The light-induced dipoles do interact with each other, modifying the resulting polarization function. This contribution is called the *local field* (LF) effect in optics of materials. A description of the polarization function within the quantum theory including the LF effect is described in details in more specialized literature [Gavrilenko (2020)].

The external driving force is determined through the monochromatic light field (see Eq. (1.30)) and can be given in the following form:

$$F_d(t) = -eReE(t) = -eE_0 \cos(\omega t) = -\frac{e}{2}E_0 \left(e^{i\omega t} + e^{-i\omega t}\right). \quad (1.36)$$

A system of bound electrons includes a restoring force ($F = -kx$, where k is the force constant) that causes an electron harmonic vibration. The dependence of F on the electron displacement follows from the coordinate dependence of the potential energy $U(x)$ of electrons in an atom. A typical shape of the $U(x)$ function is shown in Fig. 1.1.

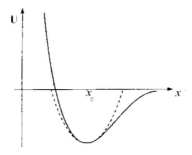

Figure 1.1 A typical coordinate dependence of the electron potential energy. By x_0 the equilibrium position of electrons in an atom is indicated. The dashed line shows the parabolic function.

Around equilibrium the function of $U(x)$ can be expanded in a Taylor series

$$U(x - x_0) = U(x_0) + \frac{dU}{dx}(x - x_0) + \frac{1}{2!}\frac{d^2 U}{dx^2}(x - x_0)^2$$
$$+ \frac{1}{3!}\frac{d^3 U}{dx^3}(x - x_0)^3 + \dots \tag{1.37}$$

In equilibrium (at $x = x_0$) the second term in equation (1.37) vanishes. The restoring force (by taking $x_0 = 0$) is given by

$$F(x) = -kx - mBx^2 - mCx^3 + \dots, \tag{1.38}$$

where x is the displacement of the electron from its equilibrium position in the atom, the constant k is frequently called the *force constant* of an atomic bond and B and C stand for anharmonic effects. Note that $|F| \approx |E| \approx \sqrt{I}$, where I is the light intensity proportional to the number of incoming photons per time and area units. An increase in I (up to laser intensities) leads to an increas in the amplitude of vibrations, thereby making important contributions to the anharmonic (nonlinear) effects. Nonlinear optical effects are a direct consequence of the nonparabolicity of the potential energy function $U(x)$ around the equilibrium, as illustrated in Fig. 1.1.

First we consider the linear optical response that corresponds to low light intensities and small average amplitudes of electron vibrations. The equation of motion of the electron under the effect of the driving light force field given by equation (1.36) follows from Newton's second law as

$$m\frac{d^2 x}{dt^2} + m\gamma\frac{dx}{dt} + kx = -\frac{e}{2}E_0\left(e^{i\omega t} + e^{-i\omega t}\right), \tag{1.39}$$

where γ is the damping coefficient (like the *friction* in mechanics) related to the *optical losses* and e and m are electron charge and mass, respectively. The trial solution can be taken in the form

$$x = Ae^{i\omega t} + c.c., \tag{1.40}$$

where *c.c.* refers to the complex conjugate. Substituting equation (1.40) into equation (1.39) and separately equating the coefficients at $e^{i\omega t}$ and $e^{-i\omega t}$ yields

$$A = -\frac{eE_0}{2m}\frac{1}{\omega_0^2 - \omega^2 + i\gamma\omega}, \tag{1.41}$$

where $\omega_0 = \sqrt{\frac{k}{m}}$ stands for the resonant frequency of harmonic vibrations. Combining equation (1.35), (1.40), and (1.41), the polarization function of this system is given by

$$P = \frac{Ne^2 E_0}{2m}\frac{1}{\omega_0^2 - \omega^2 + i\gamma\omega}e^{i\omega t} + c.c. \tag{1.42}$$

Following the definition of equation (1.33) the linear susceptibility function is now given by

$$P^{(1)} = \frac{1}{2}\left[\chi^{(1)}(\omega)e^{i\omega t} + c.c.\right]E_0,$$

$$\chi^{(1)}(\omega) = N\frac{e^2}{m}\frac{1}{\omega_0^2 - \omega^2 + i\gamma\omega}. \tag{1.43}$$

Note that the $\chi^{(1)}(\omega)$ function defined by equation (1.43) is complex. Separating into real and imaginary parts (and suppressing the index$^{(1)}$) the linear susceptibility function can be written as

$$\chi(\omega) = \chi'(\omega) + i\chi''(\omega),$$

$$\chi'(\omega) = N\frac{e^2}{m}\frac{\omega_0^2 - \omega^2}{\left(\omega_0^2 - \omega^2\right)^2 + (\gamma\omega)^2}, \tag{1.44}$$

$$\chi''(\omega) = N\frac{e^2}{m}\frac{\gamma\omega}{\left(\omega_0^2 - \omega^2\right)^2 + (\gamma\omega)^2}.$$

In the absence of a restoring force (i.e. when $k = 0$, see equation (1.39)) the electrons are free (plasma) and follow external electric field. In this case equation (1.43) reduces to

$$\chi^{(1)}(\omega) = N\frac{e^2}{m}\frac{1}{i\gamma\omega - \omega^2} \tag{1.45}$$

Basics of the Light Interaction with Matter

and if one neglects the losses in plasma (i.e. when $\gamma = 0$) the optical functions are given by

$$\chi^{(1)}(\omega) = -\frac{Ne^2}{m\omega^2},$$

$$\varepsilon(\omega) = 1 - \left(\frac{\omega_p}{\omega}\right)^2 \tag{1.46}$$

where

$$\omega_p^2 = 4\pi\frac{Ne^2}{m} \tag{1.47}$$

is known as *plasma frequency* [Yu and Cardona (2010); Saleh and Teich (2007)]. Equation (1.46) represent the *Drude* model of media. This model, which describes dispersion of the optical function in the case of the optical excitations of the free electron system, has been shown to be very useful for applications in the optics of metals [Liebsch (1997)]. This model is referred in this book while considering optical plasmon excitations in materials (see Chapter 7).

1.3 Linear Optical Response: Quantum Mechanical Picture

The optics of materials can be understood as a reaction of atoms, molecules, or solids with external electromagnetic radiation. This reaction is described in terms of different kind of excitations: appearance of additional induced charge redistribution due to, e.g., electron transitions between available energy states, generation of lattice vibrations, electron–hole pairs, etc. The starting point for optical modeling from first principles is to built up a parameter-free theoretical model of the material that realistically reproduces its basic properties [Gavrilenko (2009)]. Using the full potential of quantum theory and elementary particles physics, one builds up the atomic structure corresponding to the minimum of the free energy of the system, as follows from thermodynamics. The basics of optical response can be fairly understood by neglecting the effects of temperature, i.e., by modeling at zero temperature. In this case free energy is equal to the total energy of the system. In order to study optics at a finite temperature one needs to calculate entropy, which substantially complicates the problem. Therefore the first step in first-principles modeling of materials is the creation of an atomic structure that at the minimum of the total energy (the *ground state*) realistically reproduces well-known measured properties (e.g., known atomic geometry parameters, bulk modulus for solids). After that, complete information about electronic states and their wavefunctions is required as a next step for a quantitative analysis of the optical response of materials. Accurate description of optics requires inclusion of spin-related (relativistic) effects in the band structure model [Yu and Cardona (2010)]. In other words, before studying excitations (optics) by microscopic modeling, one needs first to determine the electron energy structure of the system in equilibrium (ground state).

In modern materials science the ground state is realistically predicted by the density functional theory, DFT. The basic ideas of DFT are explained in Section 3.4.2. For advanced reading one can refer to the books and original papers given in the bibliography (for references see Section 3.4.2 and [Gavrilenko (2009)]). In this section modeling of optical functions of materials is described assuming that eigenenergies and eigenfunctions of the system in equilibrium have already been obtained (e.g., from DFT calculations).

As stated before, in equilibrium all electronic materials consisting of charged particles (electrons and nucleus) are neutral. Light illumination (or more generally, application of an external electromagnetic field) deforms electron density. This perturbation can be described in terms of induced charge density (*longitudinal* response) or induced current (*transversal* response). In a wide spectral range, up to vacuum ultraviolet, both approaches are equivalent. Below, we consider the *longitudinal* response picture. The equilibrium electron charge density is defined through the density operator (using definition of Trace, Tr, as the sum of the diagonal elements):

$$n_{eq}(\boldsymbol{r}) = Tr[\rho_0, \delta(\boldsymbol{r} - \boldsymbol{r}_0)], \tag{1.48}$$

where subscript eq stands for *equilibrium*. In an external optical field, when the light quanta strike electrically neutral atoms, the equilibrium is broken through the deformation of electron clouds.

Time-dependent changes of the electron charge density can be represented as a Taylor expansion. The number of the terms to be included into the Taylor sum for the induced part of the charge is determined by the excitation intensity:

$$n(\boldsymbol{r}, t) = n_{eq}(\boldsymbol{r}) + n_{ind}(\boldsymbol{r}, t) = Tr[\rho, \delta]$$
$$= Tr[\rho_0, \delta] + eTr[\rho^{(1)}, \delta] + eTr[\rho^{(2)}, \delta] + \dots \qquad (1.49)$$

The first-, second-, and higher-order corrections $(\rho^{(1)}, \rho^{(2)},$ etc.) to the density operator (ρ_0) are determined from standard perturbation theory. Here we briefly consider the linear response only. Detailed evaluation of optical functions within the perturbation theory is considered in Section 2.5.

Dynamic optical response is described through the time-dependent density operator, assuming that in an external electromagnetic field the perturbation is harmonic. The unperturbed density operator in matrix presentation has the form

$$\rho_{ss'}^{(0)} = f(E_s)\delta_{ss'}. \qquad (1.50)$$

Without illumination if the system is periodic (at least in one dimension) the density operator could be defined in energy representation on a set of Bloch functions accordingly to [Davydov (1980)]:

$$\rho|s\rangle = \rho_0|\boldsymbol{k}, l\rangle = f(E_{\boldsymbol{k},l})|\boldsymbol{k}, l\rangle, \qquad (1.51)$$

where equilibrium Fermi distribution function is given by:

$$f(E_s) = \left(e^{\frac{F-E_s}{kT}} - 1\right)^{-1}, \qquad (1.52)$$

The Bloch functions

$$|s\rangle = |\boldsymbol{k}, l\rangle = u_{\boldsymbol{k},l}(\boldsymbol{k})e^{i\boldsymbol{k}\boldsymbol{r}}, \qquad (1.53)$$

are solutions of undisturbed Schrödinger equation with periodic potential:

$$H_0|\mathbf{k},l\rangle = \left[-\frac{1}{2}\nabla^2 + V_0(\mathbf{r})\right]|\mathbf{k},l\rangle = E_{\mathbf{k},l}|\mathbf{k},l\rangle, \tag{1.54}$$

In external optical field when the light quanta strike electrically neutral atoms the equilibrium is broken through the deformation of electron clouds. Time-dependent changes of the electron charge density could be represented as Taylor expansion that is given by Eq. (1.49).

The first and higher order corrections to the density operator are determined from the standard perturbation theory:

$$i\hbar\frac{d\rho}{dt} = [H,\rho] = (H\rho - \rho H), \tag{1.55}$$

with

$$H = H_0 + V^{(1)} + ...,$$
$$\rho = \rho_0 + \rho^{(1)} + \rho^{(2)} + ... \tag{1.56}$$

In equation (1.55) for simplicity we neglected effect of the energy dissipation, which could be included through the relaxation time. Plugging (1.56) in (1.55) and choosing terms of the same order on both left and right sides of the equation of motion for density operator, the equation (1.55) splits into series of equations for zero, first, second, etc. orders of perturbations, respectively:

$$i\hbar\dot{\rho}_0 = [H_0,\rho_0],$$
$$i\hbar\dot{\rho}^{(1)} = \left[H_0,\rho^{(1)}\right] + \left[V^{(1)},\rho_0\right] \tag{1.57}$$
$$i\hbar\dot{\rho}^{(2)} = \left[H_0,\rho^{(2)}\right] + \left[V^{(1)},\rho^{(1)}\right]$$

Dynamic optical response is described through time-dependent density operator. In external electromagnetic field the perturbation

Basics of the Light Interaction with Matter

is harmonic, i.e.:

$$\rho(t) = \rho(0)e^{i\omega t},$$
$$i\hbar\dot{\rho}(\omega) = -\hbar\omega\rho(\omega) \tag{1.58}$$

It is convenient now to switch to the matrix representation in equations (1.58) by projecting the relevant quantities on a set of Bloch functions equation (1.53). To this end one should multiply each and every term in equation (1.58) on the function equation (1.53); the complex conjugate of equation (1.53) is then multiplied on the left and right sides of relevant equation. Through integration over the entire space and by taking into account orthonormality conditions for Bloch functions, this leads to the following expression for the first order terms:

$$-\hbar\omega\rho_{ss'}^{(1)} = (E_s - E_{s'})\rho_{ss'}^{(1)} + \sum_t V_{st}^{(1)}\rho_{ts'}^0 - \sum_p \rho_{sp}^0 V_{ps'}^{(1)}, \tag{1.59}$$

Density operator defined as equation (1.51) in matrix presentation has the form:

$$\rho_{ss'}^{(0)} = f(E_s)\delta_{ss'}, \tag{1.60}$$

The equation (1.59) is now transformed into

$$-\hbar\omega\rho_{ss'}^{(1)} = (E_s - E_{s'})\rho_{ss'}^{(1)} + [f(E_{s'}) - f(E_s)]\,V_{ss'}, \tag{1.61}$$

At zero temperature, optical excitations occur between completely filled and empty states with Fermi functions equal either 1 or 0, respectively. Consequently, the first-order perturbation of the density operator describing linear optical response has the form

$$\rho_{ss'}^{(1)}(\omega) = \frac{f(E_{s'}) - f(E_s)}{E_s - E_{s'} - \hbar\omega} V_{ss'} = (E_s - E_{s'} - \hbar\omega)^{-1} V_{ss'}|_{T=0}. \tag{1.62}$$

Equation (1.62) can be now used to obtain the induced charge from equation (1.49) within the first-order perturbation, the linear response.

We use here the plane wave representation. This approach is very convenient for evaluation of optical functions, and it is widely used in literature. The wavefunction is given by

$$\psi_{n,k}(r) = \frac{1}{\sqrt{\Omega}} \sum_{G} d_{n,k}(G) e^{i(q+G)r}, \tag{1.63}$$

where $d_{n,k}$ are the expansion coefficients of the wavefunction characterized by the wave vector k and related to the nth electron energy state, G is the reciprocal lattice vector, and q is the wave vector of light.

The evaluation of the induced charge within representation (1.53) can be performed using functions given by equation (8.4).

In a system with a periodicity the perturbation potential is given by:

$$V(r,t) = \sum_{qG} \int_{-\infty}^{\infty} V(q+G,\omega) e^{i(q+G)r} d\omega, \tag{1.64}$$

where G is reciprocal lattice vector. For Fourier transform of the potential:

$$V(r,t) = \int_{-\infty}^{\infty} V(r,\omega) e^{i\omega t}, \tag{1.65}$$

the expansion of the potential is given by:

$$V(r,\omega) = \sum_{qG} V(q \mid G,\omega) e^{i(q+G)r}, \tag{1.66}$$

In a periodic system with all equivalent atoms separated by R_i one has for the charge:

$$n_{eq}(r_0 + R_i) = n_{eq}(r_0), \tag{1.67}$$

with

$$n(r) = \sum_{q} e^{iqr} n(q) = \sum_{q} e^{iqr} \sum_{G} \delta_{qG} n(G), \tag{1.68}$$

Basics of the Light Interaction with Matter

For induced charge density in (1.49) we have:

$$n_{ind}(\boldsymbol{r}, \omega) = \sum_{qG} n_{ind}(\boldsymbol{q} + \boldsymbol{G}, \omega)e^{i(\boldsymbol{q}+\boldsymbol{G})\boldsymbol{r}}, \tag{1.69}$$

Where Fourier transform of the induced charge is given by Fourier integral by:

$$n_{ind}(\boldsymbol{q} + \boldsymbol{G}, \omega) = \int n_{ind}(\boldsymbol{r}, \omega)e^{-i(\boldsymbol{q}+\boldsymbol{G})\boldsymbol{r}}d^3\boldsymbol{r}, \tag{1.70}$$

Linear part of the induced charge in (1.70) follows from (1.49):

$$n_{ind}(\boldsymbol{q} + \boldsymbol{G}, \omega) = e \int Tr\left[\rho^{(1)}(\omega), \delta(\boldsymbol{r}' - \boldsymbol{r})\right] e^{-i(\boldsymbol{q}+\boldsymbol{G})\boldsymbol{r}'} d^3\boldsymbol{r}'$$

$$= eTr\left[\rho^{(1)}(\omega), e^{-i(\boldsymbol{q}+\boldsymbol{G})\boldsymbol{r}}\right], \tag{1.71}$$

Trace of the operator product is calculated accordingly to:

$$Tr\left(\hat{A}\hat{B}\right) = \sum_m \langle m|\hat{A}\hat{B}|m\rangle = \sum_m \sum_n A_{mn}B_{nm}, \tag{1.72}$$

Equation (1.71) projected on the plane wave basis (1.53) can be written as:

$$n_{ind} = n_{ind}(\boldsymbol{q} + \boldsymbol{G}, \omega)$$

$$= e \sum_{k+q} \sum_{k,l} \langle l', \boldsymbol{k} + \boldsymbol{q}|\rho_\omega^{(1)}|\boldsymbol{k}, l\rangle \langle l, \boldsymbol{k}|\delta(\boldsymbol{r}' - \boldsymbol{r})|\boldsymbol{k} + \boldsymbol{q}, l'\rangle e^{-i(\boldsymbol{q}+\boldsymbol{G})\boldsymbol{r}'}$$

$$= e \sum_{k+q} \sum_{k,l} \langle l', \boldsymbol{k} + \boldsymbol{q}|\rho_\omega^{(1)}|\boldsymbol{k}, l\rangle \langle l, \boldsymbol{k}| \sum_{G'} e^{i(\boldsymbol{r}-\boldsymbol{r}')\boldsymbol{G}'} e^{-i(\boldsymbol{q}+\boldsymbol{G})\boldsymbol{r}'} |\boldsymbol{k} + \boldsymbol{q}, l'\rangle$$

$$= e \sum_{k+q} \sum_{k,l} \langle l', \boldsymbol{k} + \boldsymbol{q}|\rho_\omega^{(1)}|\boldsymbol{k}, l\rangle \langle l, \boldsymbol{k}|e^{-i(\boldsymbol{q}+\boldsymbol{G})\boldsymbol{r}}|\boldsymbol{k} + \boldsymbol{q}, l'\rangle \tag{1.73}$$

In (1.73) bra-ket notation means space integration. We also used the definition of the δ-function given by:

$$\delta(\boldsymbol{r} - \boldsymbol{r}') = \sum_G e^{i(\boldsymbol{r}-\boldsymbol{r}')\boldsymbol{G}}, \tag{1.74}$$

Now equation (1.62) can be written as:

$$\langle l', k+q|\rho_\omega^{(1)}|k,l\rangle = \frac{f[E_{l'}(k+q)] - f[E_l(k)]}{E_{l'}(k+q) - E_l(k) - \omega - i\eta}\langle l', k+q|V(r,\omega)|k,l\rangle$$

$$= \frac{f[E_{l'}(k+q)] - f[E_l(k)]}{E_{l'}(k+q) - E_l(k) - \omega - i\eta}$$

$$\times \sum_{q,G} V(q+G,\omega)\langle l', k+q|e^{i(q+G)r}|k,l\rangle, \qquad (1.75)$$

Complex part of the energy in denominator of equation (1.75) is introduced to prevent unphysical divergences at resonance frequencies.

Plugging (1.75) into (1.73) we arrive in the following expression for the induced charge:

$$n_{ind}(q+G,\omega) = e \sum_{G'} P_{G,G'} V(q+G,\omega), \qquad (1.76)$$

Using notation $(k' = k+q)$ the polarization function is defined as (see (Gavrilenko, 2020) for details)

$$P_{G,G'}(\omega) = \sum_{k',k}\sum_{l',l}\langle l', k'|e^{i(q+G)r}|k,l\rangle\langle l, k|e^{i(q+G')r}|k',l'\rangle$$

$$\times \frac{f[E_{l'}(k+q)] - f[E_l(k)]}{E_{l'}(k+q) - E_l(k) - (\omega + i\eta)} \qquad (1.77)$$

Here we present simplified results obtained in the limit of $q \to 0$ (neglect of spatial dispersion) and neglecting the local field effects and the nonlocality (i.e., if $G = G' = 0$). The result is given by

$$n_{ind}(\omega) = \hat{P}V(\omega), \qquad (1.78)$$

where $V(\omega)$ is the external perturbation potential. In this case the the polarization function given by equation (1.77) is reduced to the local tensor function, where the tensor components $(\alpha, \beta = x, y, z)$ of the polarization function $\hat{P}$ are given by:

$$P_{\alpha,\beta}(\omega) = \frac{2}{\Omega N} \sum_{k}\sum_{c,v} \frac{p_{cv}(k)p_{cv}^*(k)}{[E_c(k) - E_v(k)]^2} F_{cv}(\omega, k), \qquad (1.79)$$

Basics of the Light Interaction with Matter

where indices c and v run over the conduction (empty) and valence (filled) electron states, respectively. In the plane-wave basis given by equation (1.53) Cartesian components of momentum matrix elements are given by

$$p_\alpha^{cv}(\mathbf{k}) = \sum_{\mathbf{G}'} d_{c\mathbf{k}}^*(\mathbf{G}')d_{v\mathbf{k}}(\mathbf{G}')(k_\alpha + G_\alpha). \qquad (1.80)$$

The spectral function is given by

$$F_{cv} = \left\{ \frac{1}{\omega + i\eta - E_c(\mathbf{k}) + E_v(\mathbf{k})} + \frac{1}{-\omega - i\eta - E_c(\mathbf{k}) + E_v(\mathbf{k})} \right\}. \qquad (1.81)$$

The prefactor in equation (1.79) appears after summation over volume Ω of homogeneous ambient of noninteracting N dipoles (the random phase approximation, RPA, or independent particles picture). A more general case describing the effect of the ambient in terms of dynamic interaction with surrounding light-induced dipoles (the local field effect, LF) is addressed in (Gavrilenko, 2020).

The potential energy in materials can be separated into two parts, the external and induced potentials:

$$V(\omega) = V_{\text{ext}}(\omega) + V_{\text{ind}}(\omega). \qquad (1.82)$$

Equation (1.82) can be understood as a reduction (screening) of external potential through the induced charge in materials. This process can be represented in terms of the dielectric function:

$$V_{\text{ext}}(\omega) = \hat{\varepsilon}(\omega)V(\omega). \qquad (1.83)$$

The tensor components of the dielectric function $\varepsilon(\omega)$ can be now expressed in terms of the polarization function:

$$\varepsilon_{\alpha,\beta}(\omega) = \delta_{\alpha,\beta} - 4\pi P_{\alpha,\beta} \qquad (1.84)$$

with function $P_{\alpha,\beta}$ given by equation (1.79).

Equation (1.84) defines dielectric functions, in terms of microscopic optical excitations described through optical transition matrix elements given by equation (1.80). The imaginary part of equation (1.84) corresponds to the widely used *golden rule* formula [Yu and Cardona (2010)]. In the plane wave basis (see equation (8.4)) and in the limit of $q \to 0$ the Cartesian components of momentum matrix elements are given by

$$p_\alpha^{cv}(k) = \sum_{G'} d_{ck}^*(G')d_{vk}(G')(k_\alpha + G_\alpha). \tag{1.85}$$

Equation (1.85) presents a plane wave evaluation of the momentum operator. Equation (1.84) is derived by evaluation of the induced charge dipole (and induced polarization) as the response to the external electric field.

Equation (1.84) defines the *longitudinal* dielectric function. This is in contrast to the microscopic definition of the *transversal* dielectric function that can be derived from the analysis of the induced current, which is parallel to the electric field of light and therefore perpendicular (*transversal*) to the penetration of light [Martin (2005)]. For historical reasons the definitions of *longitudinal* and *transversal* relate to the formulation of the perturbation in the quantum theory: if the perturbation is caused by charge displacements due to the momentum transformations (which are parallel to the wave propagation) the resulting function is called *longitudinal*. If the perturbation is taken as a current induced by the electric field component of light (which is perpendicular to the wave propagation) the calculated function is defined as *transversal* [Martin (2005)]. The *longitudinal* dielectric function determines the optical response of electron plasma. The *transversal* function determines optical response to a TE (transverse electric) electromagnetic wave, but for a TM (transverse magnetic) wave the contributions of both *longitudinal* and *transversal* dielectric functions are important. In the spectral regions where the wavelength of light is comparable with the characteristic dimensions of the elemental excitations (X-ray for the electronic part or visible light for excitons) the difference between the two functions is not negligible. However, for electronic excitations in the visible and nearest ultraviolet (UV) and/or infrared (IR) spectral regions, both definitions for dielectric function are equivalent [Gavrilenko (2020)].

1.4 Dispersion Relations for Rare Gas: Lorentz–Lorenz Formulae

In this part we continue to study linear optical response. As to Drude–Lorenz model molecule of the substance is described by an ensemble of harmonic oscillators. In Section 1.2 we assume that electric amplitude of the optical field $\mathbf{E}_0$ corresponds to $\mathbf{E}_a$-acting field. That is true for rare gas molecules. But we have to take into consideration polarization factor of surrounding molecules for higher density gases or condensed matter. Then

$$\mathbf{E}_a = \mathbf{E}_0 + \frac{4\pi}{3}\mathbf{P}. \tag{1.86}$$

We substitute (1.86) into (1.39). Then

$$(-\omega^2 - i\omega\gamma + \omega_0^2)\mathbf{P} = \frac{Ne^2}{m}\mathbf{E}_0 + \frac{4\pi}{3}\frac{Ne^2}{m}\mathbf{P}. \tag{1.87}$$

From (1.87) follows:

$$\mathbf{P} = \frac{Ne^2}{m}\left[\omega_0^2 - \omega^2 - i\gamma\omega - \frac{4\pi}{3}\frac{Ne^2}{m}\right]^{-1}\mathbf{E}_0. \tag{1.88}$$

By taking into account the relationship between optical fields, i.e. $\mathbf{D} = \varepsilon\mathbf{E}_0 = \mathbf{E}_0 + 4\pi\mathbf{P}$, equation (1.88) results in the following

$$\varepsilon = 1 + \frac{4\pi Ne^2}{m}\left[\omega_0^2 - \omega^2 - i\gamma\omega - \frac{4\pi}{3}\frac{Ne^2}{m}\right]^{-1}. \tag{1.89}$$

After some algebra, equation (1.89) can by written in the following form

$$\frac{\varepsilon - 1}{\varepsilon + 2} = \frac{1}{3}\frac{4\pi Ne^2/m}{\omega_0^2 - \omega^2 - i\gamma\omega}. \tag{1.90}$$

Introducing a *polarizability* function of of a single oscillator α_0 according to

$$\alpha_0(\omega) = \frac{e^2}{m} \cdot \frac{1}{\omega_0^2 - \omega^2 - i\gamma\omega} \qquad (1.91)$$

the equation (1.90) can be written in the following form

$$\frac{\varepsilon - 1}{\varepsilon + 2} = \frac{4\pi}{3} N\alpha_0. \qquad (1.92)$$

The equation (1.92) is known as the *Clausius–Mossotti* equation[1]. For a molecular system we have

$$N = N_A n_m \qquad (1.93)$$

where N_A is the *Avogadro's* number (see appendix A) and n_m is the number of moles per unit volume.

Equation (1.92) can be written now as

$$\frac{\varepsilon - 1}{\varepsilon + 2} = \frac{4\pi}{3} \alpha_m n_m, \qquad (1.94)$$

where α_m is the polarizability per mole (i.e. the *molar polarizability*).

For a substance with a variety of different oscillators the dispersion (1.90) is given by.

$$\frac{\varepsilon - 1}{\varepsilon + 2} = \frac{1}{3} \frac{4\pi e^2}{m} \sum \frac{N_i}{\omega_i^2 - \omega^2 - i\gamma_i\omega}. \qquad (1.95)$$

Numerical parameters of the material denoted by index i in equation (1.95) can be calculated by methods of quantum mechanics. In particular, the value of N_i should be replaced by $N \cdot f_i$, where f_i is called the oscillator force.

---

[1] Equation (1.92) is called sometimes in literature the Clausius–Mossotti–Lorentz–Lorenz formula which is named after four people that derived various versions of it.

24 | Basics of the Light Interaction with Matter

By separation into real and imaginary parts in equation (1.90) the function of ε is given by

$$\varepsilon = 1 + \frac{\omega_p^2(\tilde{\omega}_0^2 - \omega^2)}{(\tilde{\omega}_0^2 - \omega^2)^2 + \omega^2\gamma^2} + i\frac{\omega_p^2\gamma\omega}{(\tilde{\omega}_0^2 - \omega^2)^2 + \omega^2\gamma^2}, \qquad (1.96)$$

where

$$\omega_p^2 = \frac{4\pi Ne^2}{m}, \qquad (1.97)$$

$$\tilde{\omega}_0^2 = \omega_0^2 - \omega_p^2/3. \qquad (1.98)$$

According to the definition, $\sqrt{\varepsilon} = \tilde{n} = n + i\kappa$ (where $\tilde{n}$ is called complex refractive coefficient) the refractive index (n) and the extinction coefficient (κ) can be derived from equation (1.96).

$$\sqrt{\varepsilon} \simeq 1 + \frac{1}{2}\frac{\omega_p^2(\omega_0^2 - \omega^2)}{(\omega_0^2 - \omega^2)^2 + \omega^2\gamma^2} + i\frac{1}{2}\frac{\omega^2\gamma\omega}{(\omega_0^2 - \omega^2)^2 + \omega^2\gamma^2}, \qquad (1.99)$$

which results

$$n \simeq 1 + \frac{1}{2}\frac{\omega_p^2(\omega_0^2 - \omega^2)}{(\omega_0^2 - \omega^2)^2 + \omega^2\gamma^2}, \qquad (1.100)$$

$$\kappa \simeq \frac{1}{2}\frac{\omega_p^2\gamma\omega}{(\omega_0^2 - \omega^2)^2 + \omega^2\gamma^2}. \qquad (1.101)$$

Dispersions of refractive index $n(\omega)$ and extinction coefficient $\kappa(\omega)$ are shown in Fig. 1.2.

In Fig. 1.2 by (ω_1, ω_2) the anomalous dispersion region is indicated.

1.5 Review Questions and Exercises

1.5.1 Complex Dielectric Susceptibility

Derive the classical formula for the complex dielectric susceptibility for a condensed medium. Consider the Lorentz formula for the local light field in condensed matter, i.e. $\mathbf{E}_{loc} = \mathbf{E} + 4\pi/3\mathbf{P}$.

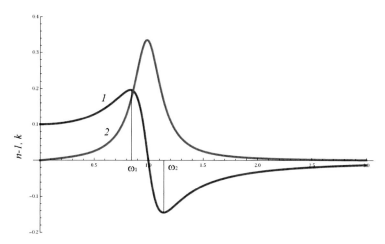

Figure 1.2 Dependencies of $n(\omega) - 1$ (1) and $\kappa(\omega)$ (2) on relative frequency, ω/ω_0.

1.5.2 Dielectric Permittivity and Refractive Index

Find a relationship between components of the complex dielectric permittivity and components of the complex refractive index. Write down the appropriate formulas for the not dense medium approximation, i.e. $|n - 1| \ll 1, \kappa \ll 1$.

1.5.3 Reflection Coefficient for Germanium

Calculate reflection coefficient, phase shift and depth of penetration of a plane optical wave with its normal incidence on a Germanium sample. Complex refractive index of Germanium is: $\tilde{n} = 3.47 - i1.40$.

1.5.4 Group Velocity

Express the group velocity V_g of an optical wave in terms of the refractive index of the medium and its derivative: (a) in infrequency domain; (b) in wavelength domain.

1.5.5 Refractive Index of Iodine

The refractive index of iodine vapor can be approximated by the following expression

$$n^2 = a + b\lambda^{-2} + d\lambda^2 \qquad (1.102)$$

where $a, b, d > 0$. Find the wavelength λ_m at which it reaches minimum. Calculate and compare group (V_g) and phase (V_p) velocities.

1.5.6 Refractive Index Dispersion

Derive approximate formulas for the refractive index dispersion $n(\omega)$ for two cases: a) transparent dielectric, i.e. $\omega \ll \omega_0$ and b) X-ray range, i.e. $\omega \gg \omega_0$.

Chapter 2

Nonlinear Light Interaction with Matter

2.1 Introduction

If a high-intensity beam of light illuminates a specimen, it will respond in a nonlinear manner, i.e. higher optical harmonics will be generated in addition to the linear optical response (see Chapter 1). Consider first the classical picture of optical response in materials given in Chapter 1 and potential energy of the bounded electrons in materials (see Fig. 1.1).

As the result of the material reaction on external excitation, the bounded electrons are displaced from their equilibrium locations. In equation (1.38) the value of x stands for the displacement of the electron from its equilibrium position in the atom, the quantity of k stands for the *force constant* of an atomic bond and B and C stand for anharmonic effects. Within the classical picture we have $|F| \approx |E| \approx \sqrt{I}$, where I is the light intensity proportional to the number of incoming photons. An increase in I (up to laser intensities) leads to an increase of the amplitude of vibrations, thereby making important contributions to the anharmonic (nonlinear) effects [Ovechko *et al.* (2015); Dmytruk and Ovechko (1995); Molebny *et al.* (1974); Ovechko and Myhashko (2018)]. Therefore, the nonlinear optical effects in classical optics of materials are direct consequence of the

---

Fundamentals of the Optics of Materials: Tutorial and Problem Solving
Vladimir I. Gavrilenko and Volodymyr S. Ovechko
Copyright © 2024 Jenny Stanford Publishing Pte. Ltd.
ISBN 978-981-4877-93-0 (Hardcover), 978-1-003-25694-6 (eBook)
www.jennystanford.com

28 | *Nonlinear Light Interaction with Matter*

nonparabolicity of the potential energy function $U(x)$ around the equilibrium, as illustrated in Fig. 1.1.

2.2 Nonlinear Processes

Optical wave propagation in nonmagnetic medium within classical electrodynamics is defined by the following waves equation:

$$\nabla^2 \boldsymbol{E} - \frac{1}{c^2}\frac{\partial^2 \boldsymbol{D}}{\partial t^2} = 0, \qquad (2.1)$$

where $\boldsymbol{D} = \boldsymbol{E} + 4\pi\boldsymbol{P}$, and polarization vector $\boldsymbol{P} = \boldsymbol{P}^L + \boldsymbol{P}^{NL}$ includes both linear $\boldsymbol{P}^L$ and nonlinear $\boldsymbol{P}^{NL}$ parts. Equation (2.1) has been written for homogeneous medium without volume charges. The polarization vector relates to the electric field $\boldsymbol{E}$ by a functional given by:

$$P_i^L = \int \chi_{ij}^{(1)}(t')E_j(t - t')dt' \qquad (2.2)$$

where $i,j = x,y,z$, the time dependent function of $\chi^{(1)}$ stands for linear susceptibility, and repeating index (see j in Eq. (2.2)) means summation.

Nonlinear polarization $\boldsymbol{P}^{NL}$ can be evaluated over a small parameter determined by the ratio, $|\boldsymbol{E}|/|\boldsymbol{E}_{inner}|$, where $\boldsymbol{E}_{inner}$ is an electric field inside a medium (i.e. gas, liquid or solid state). The function of $\boldsymbol{P}^{NL}$ is given by:

$$P_i^{NL} = \int\int dt'dt'' \chi_{ijk}^{(2)}(t',t'')E_j(t - t')E_k(t - t' - t'') + \dots \quad (2.3)$$

Equation (2.3) is truncated by the so-called *quadratic nonlinearity*, i.e. by $\chi_{ijk}^{(2)}$ function; repeating indexes mean summation.

Function related to the next order in evaluation of the polarization function is $\chi_{ijkl}^{(3)}(t',t'',t''')$. For a medium without dispersion equations (2.2) and (2.3) are given by:

$$P_i^L = \chi_{ij}^{(1)}E_j,$$

$$P_i^{NL} = \chi_{ijk}^{(2)}E_jE_k + \chi_{ijkl}^{(3)}E_jE_kE_l + \dots \qquad (2.4)$$

2.3 Nonlinear Optical Processes Classification

Last equation in (2.4) can be used for nonlinear processes classification. There are two types of nonlinear processes:

1. processes that are determined by phase correlations between interacting electromagnetic waves
2. processes that are determined by a modification of the intermediate state.

To the first group of nonlinear processes belong frequency multiplication, sum frequency generation, and parametric amplification. To the second group belong multi-photon absorption, Raman scattering, etc. Other nonlinear optical effects include different types of elementary excitation in medium such as optical phonons, acoustic phonons, magnons, plasmons, etc.

For nonlinear processes analysis by consideration of the dielectric polarization widely used is a method of *Feynman diagrams*. Consider first the Feynman diagram technique for linear dielectric susceptibility $\chi_{ij}(\omega,\omega)$ as an example. Linear optical processes related to absorption and emission of photons described by $\chi^{(1)}$ are represented by two Feynman diagrams shown in Fig. 2.1.

Optical transition diagram corresponding to emission and absorption of photons in linear optical processes is shown in Fig. 2.2.

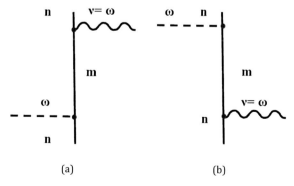

Figure 2.1 Feynman diagrams describing absorption (a) and emission of photons (b) in linear optical processes.

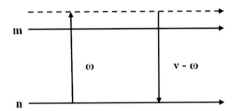

Figure 2.2 Optical transition diagram describing emission and absorption of photons in linear optical processes.

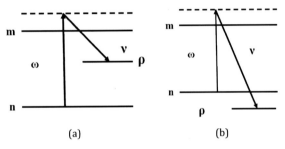

Figure 2.3 Optical transition diagrams representing spontaneous Raman scattering with $\hbar\omega > \hbar\nu$ (Stokes scattering) (a) and $\hbar\omega < \hbar\nu$ (anti-Stokes scattering) (b).

Dashed lines in Fig. 2.1 represent elementary photon absorption process of external electromagnetic wave at frequency (ω). Wavy lines represent elementary emissions of photons at frequency ($\nu = \omega$). Nodes (or vertices) represent photon–particle interactions described by dipole matrix elements. Vertical lines between nodes describe electron transitions between energy states (denoted by (E_m, E_n)) that are stimulated by photons. These processes obey the energy conservation law, i.e. $\hbar\omega = \hbar\omega_{mn} = E_m - E_n$.

In the systems where there are no final states (of atom or molecule) such that the photon–matter optical interaction process can obey the energy conservation law the spontaneous Raman scattering occurs that is represented by the diagrams shown in Fig. 2.3 and Fig. 2.4:

where m denotes intermediate virtual state, ρ indicates final energy value corresponding to $\nu = \omega - \omega_{pn}$ with $\nu > \omega$ (Stokes frequency) or $\nu < \omega$ (anti-Stokes frequency).

Consider now the nonlinear optical processes in materials that are described by the second-order dielectric susceptibility function, $\chi^{(2)}_{ijk}(\nu, \omega, \omega')$. Complete set of Feynman diagrams describing the second-order optical processes in nonlinear medium includes eight

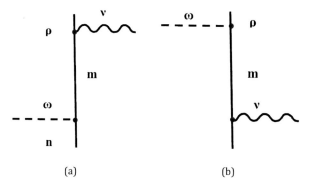

Figure 2.4 Feynman diagrams representing spontaneous Raman scattering with $\hbar\omega > \hbar\nu$ (Stokes scattering) (a) and $\hbar\omega < \hbar\nu$ (anti-Stokes scattering) (b).

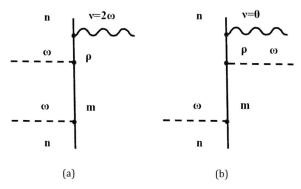

Figure 2.5 Feynman diagrams describing second harmonic generation (a) and optical rectification (b).

diagrams [Shen (2003)]. Here we discuss only two second-order nonlinear optical processes that are represented by the Feynman diagrams shown in Fig. 2.5.

If $\omega = \omega'$, $\nu = 2\omega$, the function of $\chi^{(2)}_{ijk}(2\omega, \omega, \omega)$ describes the second harmonic generation (SHG). SHG is also called in literature frequency doubling. This is a nonlinear optical process, in which photons interacting with a nonlinear material are effectively combined to generate new photons having twice the frequency of initial photons. A typical configuration for frequency doubling is the following: an infrared input beam at wavelength $\lambda = 1064$ nm generates a green wave at wavelength $\lambda = 532$ nm during its path through a nonlinear crystal. The SHG process is represented by Feynman diagram shown in Fig. 2.5(a).

The nonlinear optical process in wich $\nu = 0$ corresponds to $\chi^{(2)}(0, \omega, -\omega)$ and is shown in Fig. 2.5(b). It is called *optical rectification*. Optical rectification, is a non-linear optical process that consists of the generation of a quasi-DC polarization in a non-linear medium by passing an intense beam of light[1]. Typical application of optical rectification phenomena is a generation and detection of short pulses of terahertz (THz) frequency electromagnetic radiation [Nuss and Orenstein (1998)].

Higher order nonlinear optical processes are more complicated and include more quasiparticles and photons. For example, the third order processes are described by the third order nonlinear susceptibility function $\chi^{(3)}_{ijkl}(\nu, \omega, \omega'\omega'')$. If $\omega = \omega = \omega, \nu = 3\omega$ then $\chi^{(3)}_{ijkl}(3\omega, \omega, \omega'\omega'')$ determines third harmonic generation represented by Feynman diagram and optical transitions depicted in Fig. 2.6.

Two photons absorption with spontaneous emission of two photons is illustrated in Fig. 2.7.

Let us consider now examples of solving the wave equations for nonlinear processes.

Example 1. Write down the wave equation for high harmonic generation in slowly varying amplitude approximation.

Solution. Wave equation for $P(r, t) = P^L(r, t) + P^{NL}(r, t)$ is

$$[\nabla \times [\nabla \times E(r, t)]] + \frac{1}{c^2}\frac{\partial^2}{\partial t^2}E(r, t) = -\frac{4\pi}{c^2}\frac{d^2}{\partial t^2}P(r, t) \qquad (2.5)$$

For plane monochromatic wave equation (2.5) reads:

$$E(r, t) = \frac{1}{2}\sum_i e_i E^{(i)} \exp(-i\omega_i t + ik_i r) + c.c. \qquad (2.6)$$

---

[1] Optical rectification is also frequently called in literature *electro-optical rectification*. This phenomenon, the generation of the DC electric field by intense optical field is reverse to the electro-optical effect i.e. changes in the optical properties of the material due to the presence of an electric field.

Fundamentals of the Optics of Materials: Tutorial and Problem Solving | 33

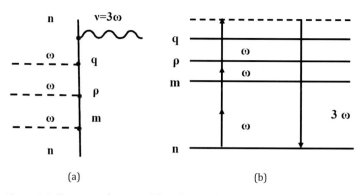

Figure 2.6 Feynman diagram (a) and optical transitions (b) corresponding to the third harmonic generation.

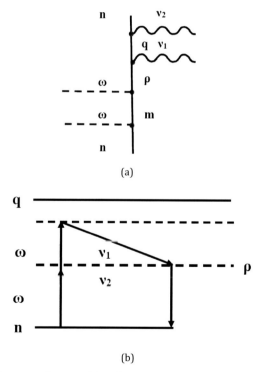

Figure 2.7 Feynman diagram (a) and optical transitions (b) corresponding to the two-photon absorption.

Nonlinear Light Interaction with Matter

In the same way the equation for polarization is given by:

$$P(r,t) = \frac{1}{2}\sum_m P^{(n)} \exp(-i\omega_m t + ik_m r) + c.c. \tag{2.7}$$

where

$$P^{NL}(r,t) = \sum_m P^{NL}(k_m, \omega_m). \tag{2.8}$$

Equation (2.5) can be now given by

$$[\nabla \times [\nabla \times E(k,\omega)]] - \frac{\omega^2}{c^2}\varepsilon E(k,\omega) = \frac{4\pi\omega^2}{c^2}P^{NL}(k_m,\omega_m). \tag{2.9}$$

Vector of nonlinear polarization $P^{NL}(k_m, \omega_m)$ is the source of an electric components at new frequency. All components are binding to each other.

Slowly varying amplitudes approximation is defined as

$$\left| \frac{\partial^2 E(z)}{\partial z^2} \right| \ll \left| k\frac{\partial E(z)}{\partial z} \right| \tag{2.10}$$

Substituting (2.7), (2.8) into (2.10) one can get a one dimensional equation for the wave propagation in the following form:

$$k^2 E - 2ik\frac{\partial E}{\partial z} - \frac{\partial^2 E}{\partial z^2} - \frac{\varepsilon\omega^2}{c^2}E = \frac{4\pi\omega^2}{c^2}(e \cdot P^{NL})\exp(i\omega t - ikz). \tag{2.11}$$

Taking into account approximation (2.10), equation (2.11) can be simplified to the following

$$\frac{\partial E}{\partial z} = i\frac{2\pi\omega}{cn}(e \cdot P^{NL})\exp(i\omega t - ikz), \tag{2.12}$$

as the result the differential equation (2.12) has lower degree. Consider now another example.

Example 2. Solve equation (2.12) for the second harmonic generation (SHG). Analyze efficiency of the SHG with the phase mismatch

$$\Delta\phi = (k_2 - 2k_1)z. \tag{2.13}$$

Solution. Let us solve equation (2.12) at the following condition:

$$| \mathbf{E}_{2\omega} | \ll | \mathbf{E}_{1\omega} | .\tag{2.14}$$

It is therefore restricted to the small SHG efficiency, approximation

$$\frac{dE_{2\omega}}{dz} = i\frac{2\pi 2\omega}{cn_{2\omega}}\left(\mathbf{e}_{2\omega} \cdot \mathbf{P}_{2\omega}^{(2)} exp(i2\omega t - ik_2 z)\right),\tag{2.15}$$

where

$$\mathbf{P}_{2\omega}^{(2)}(z) = \frac{1}{2}\mathbf{e}_2 P_{2\omega}(\mathbf{k}_2, \omega_2) \exp(-i2\omega t + ik_2 z)\tag{2.16}$$

$$P_{2\omega} = \chi^{(2)}(2\omega, \omega, \omega)E_{\omega}^2(z)\tag{2.17}$$

Then

$$\frac{dE_{2\omega}(z)}{dz} = i\frac{2\pi 2\omega}{cn_{2\omega}}\chi^{(2)}(2\omega)E_{\omega}^2 \exp[i(2k_1 - k_2)z]\tag{2.18}$$

Solution of Eq. (2.18) is given by

$$E_{2\omega}(z) = \frac{4\pi\omega}{cn_{2\omega}\Delta k}\chi^{(2)}(2\omega)E_{\omega}^2[\exp(i\Delta kz) - 1],\tag{2.19}$$

where

$$\Delta k = 2k_1 - k_2\tag{2.20}$$

is the wave mismatch.

Intensity of the second harmonic is given by

$$I_{2\omega} = \frac{cn_{2\omega}}{8\pi} | E_{2\omega} |^2 = \frac{128\pi^3\omega^2}{c^3 n_{\omega}^2 n_{2\omega}} | \chi^{(2)}(2\omega) |^2 I_1^2 \sin c^2\left(\frac{\Delta kz}{2}\right)\tag{2.21}$$

where $\sin c^2 y = (\sin y/y)^2$. The quantity $l_{coh} = 1/\Delta k$ is called the *coherent length*. The condition of maximum SH efficiency is called the *collinear synchronism* limit

$$\Delta k = 2k_1 - k_2 = 0\tag{2.22}$$

36 | *Nonlinear Light Interaction with Matter*

In general the condition (2.22) should be written in vector form. For optical materials in spectral region far away from the absorption the refraction coefficients exhibit normal dispersion i.e. $(n_{2\omega} > n_{\omega})$. In this case $l_{coh} \ll z$. The value of SH efficiency is very small. Condition (2.22) can be met in crystals with negative anisotropy i.e. $(n_e(\omega) \leq n_0(\omega))$. Thus the second harmonic wave has to be of nonordinary nature. In this case there are two types of the interactions:

$$k_e(2\omega) - 2k_0(\omega) = 0, \tag{2.23}$$

$$k_e(2\omega) - k_0(\omega) - k_e(\omega) = 0 \tag{2.24}$$

Conditions (2.23) and (2.24) correspond to the type I and type II interactions, respectively.

Another example relates to the SHG in crystals.

Example 3. Calculate angle of collinear SHG-interaction (type I) for the KDP (KH_2PO_4) crystal.

Refractive indices have the following values:

at the wavelength of $\lambda = 1060$ nm: $n_0 = 1.494$, $n_e = 1.460$;

at the second harmonic wavelength of $\lambda = 530$ nm: $n_0 = 1.513$, $n_e = 1.471$

Solution. Equation (2.23) for the type I interaction reads

$$\frac{2\omega}{c} n_e(2\omega, \theta_c) = 2\frac{\omega}{c} n_0(\omega). \tag{2.25}$$

or

$$n_e(2\omega, \theta_c) = n_0(\omega). \tag{2.26}$$

Value of $n_e(2\omega, \theta_c)$ for a single axis crystal is given by

$$\frac{\cos^2 \theta_c}{n_0^2(2\omega)} + \frac{\sin^2 \theta_c}{n_e^2(2\omega)} = \frac{1}{n_e^2(2\omega, \theta_e)} \tag{2.27}$$

After some algebra one can get the following:

$$\sin \theta_c = \left(\frac{n_0^{-2}(\omega) - n_0^{-2}(2\omega)}{n_e^{-2}(2\omega) - n_0^{-2}(2\omega)} \right)^{1/2} \tag{2.28}$$

Plugging values of refractive indices of KDP crystal into equation (2.28) we get $\sin \theta_c = 0.66$ and $\theta_c = 41^\circ 33'$.

Next example relates to the Raman scattering.

Example 4. Calculate threshold intensity of stimulated Raman scattering if absorption coefficient at combined frequency is equal to α.

Solution. Let us examine nonlinear four photons scattering $\omega_2 = \omega_2 - \omega_1 + \omega_1$, which is determined by the third order dielectric susceptibility function $\chi_{ijkl}^{(3)}(\omega_2; \omega_2, \omega_1, -\omega_1)$. For isotropic homogeneous medium and collinear waves ω_1, ω_2 with linear polarization $(e_1 \parallel e_2)$ we have:

$$e_2 \cdot P^{NL}(\omega_2) = 6\chi_{1111}^{(3)}(\omega_2, \omega_2, \omega_1, -\omega_1)E_2 \mid E_1 \mid^2 \exp(-i\omega_2 t + ik_2 z)$$

$$(2.29)$$

With the expression (2.29) the wave equation (2.12) reads

$$\frac{dE_2}{dz} = i\frac{12\pi\omega_2}{cn_2}\chi_{1111}^{(3)} \mid E_1 \mid^2 E_2 \tag{2.30}$$

and solution of equation (2.30) is given by:

$$E_2(z) = E_2(0)\exp(gz), \tag{2.31}$$

where

$$g = i\frac{12\pi\omega_2}{cn_2}\chi^{(3)} \mid E_1 \mid^2 = i\frac{96\pi^2\omega_2}{c^2 n_1 n_2}\chi^{(3)} I_1, \tag{2.32}$$

$$I_1 = \frac{cn_1}{8\pi} \mid E_1 \mid^2 \tag{2.33}$$

For the intensity I_2 we have

$$I_2(z) = I_{20}\exp(Gz) \tag{2.34}$$

Nonlinear Light Interaction with Matter

where

$$G = -\frac{384\pi^3}{cn_1n_2\lambda_2}\left(c\chi^{(3)}_{1111}\right)I_1 \tag{2.35}$$

If imaginary part of $\chi^{(3)}_{1111}$ is negative, i.e. Im $c\chi^{(3)}_{1111} < 0$, there is an exponential amplification, however, if it is positive, i.e. Im $c\chi^{(3)}_{1111} > 0$, there is an exponential relaxation.

The complex form of the third order susceptibility $\chi^{(3)}_{1111}(\omega_2;\omega_2;\omega_1; -\omega_1)$ is attributed to the two photon absorption $\omega_1 + \omega_2 \simeq \Omega$ or combined type resonance $\omega_1 - \omega_2 = \Omega$, where Ω is a frequency of molecular oscillation.

Threshold intensity I_1^{th} can be calculated from the equation $G(I_1^{th}) = \alpha$. In this case we have:

$$I_1^{th} = \alpha\frac{n_1n_2c^2}{192\pi^2\omega_2}\left(-\text{Im } c\chi^{(3)}_{1111}\right)^{-1} \tag{2.36}$$

The value of I_1^{th} is considered as a threshold level of spontaneous Stokes scattering signal.

Next example relates to a medium with the Kerr nonlinearity (see Section 12.4).

Example 5. Calculate threshold power P^{th} for laser beam self-focusing in a Kerr nonlinear medium under following conditions:

$$n(\mathbf{r}, t) = \left[\varepsilon_0 + 4\pi|E(r)|^2\text{Re}\left[\chi^{(3)}_{1111}(\omega;\omega, -\omega, \omega)\right]\right]^{1/2}$$
$$\simeq n_0 + n_2|E(r)|^2, \tag{2.37}$$

where $n_0 = \sqrt{\epsilon_0}$ and $n_2|E(r)|^2 \ll n_0$

$$n_2 = 2\pi\text{Re}\left(\chi^{(3)}_{1111}\right)/n_0. \tag{2.38}$$

Optical constants of CS_2 liquid have the following values:

$$\chi^{(3)}_{1111} = 10^{-12} \text{ cm}^3/erg, \tag{2.39}$$

$$n_0 = 1.5 \quad (\lambda = 1000 \text{ nm}). \tag{2.40}$$

Solution. As long as $\chi^{(3)}_{1111} > 0$ the following relationship is valid:

$$\Delta n(\mathbf{r}_\perp) = n_2 \mid E(\mathbf{r}) \mid^2 \quad 0. \tag{2.41}$$

Under condition of (2.41) and if $\mid E(0) \qquad \mid E(\mathbf{r}\mid)$ we have the nonlinear focusing lens. Optical power of nonlinear lens depends on the laser beam power P. Threshold P^{th} is defined by the equality of diffraction defocusing and nonlinear focusing. Diffraction angle is

$$\theta_d = \frac{\lambda}{2n_0 a}, \tag{2.42}$$

where a is the beam radius.

Nonlinear part of refraction index induces internal reflection at the refraction angle θ^{NL} i.e.

$$\theta^{NL} = \arccos \frac{n_0}{n_0 + \Delta n} \approx \left(\frac{2\Delta n}{n_0} \right)^{1/2}, \tag{2.43}$$

note that $\Delta n \ll n_0$. Condition for the self-focusing is defined as an equality of the angles: $\theta^{NL} = \theta_d$. We have:

$$\Delta n^{th} = n_2 \mid E^{th} \mid^2 \simeq n_0 \theta_d^2 / 2 \simeq \lambda^2 / 8 n_0 a^2 \tag{2.44}$$

The total power of the laser beam is given by

$$P^{th} = \frac{cn_0}{8\pi} \mid E^{th} \mid^2 \pi a^2 = \frac{c\lambda^2}{64 n_2} \tag{2.45}$$

Plugging λ, n_2, n_0 into (2.45) for CS_2 liquid results in the following value of the threshold power:

$$P^{th} = 5 \times 10^5 \text{ W}.$$

2.4 Classic Theory of Nonlinear Optical Response

At high (laser) light intensities effects of the anharmonicity of atomic bonds can not be neglected. Below we consider the effect of the inclusion of the second term in the equation for the restoring force (see Eq. (1.38)). The equation of motion of the electron under the effect of the driving force field given by Eq. (1.38) is now given by

$$m\frac{d^2x}{dt^2} + m\gamma\frac{dx}{dt} + kx + mBx^2 = -\frac{e}{2}E_0\left(e^{i\omega t} + e^{-i\omega t}\right). \tag{2.46}$$

The trial function for Eq. (2.46) includes the additional terms

$$x = x^{(1)} + x^{(2)},$$

$$x^{(1)} = A^{(1)}e^{i\omega t} + c.c., \tag{2.47}$$

$$x^{(2)} = A^{(2)}e^{i2\omega t} + c.c.$$

The technique of solving Eq. (2.46) is the same that the one used to solve the perturbation set of the quantum mechanics equations described in Section 1.3. After plugging the trial function given by Eq. (2.47) into Eq. (2.46) and by collecting terms of the same order, the Eq. (2.46) splits into two equations (terms generating contributions higher than the second order are neglected):

$$m\frac{d^2x^{(1)}}{dt^2} + m\gamma\frac{dx^{(1)}}{dt} + kx^{(1)} = -\frac{e}{2}E_0\left(e^{i\omega t} + e^{-i\omega t}\right), \tag{2.48}$$

$$m\frac{d^2x^{(2)}}{dt^2} + m\gamma\frac{dx^{(2)}}{dt} + kx^{(2)} + mB\left[x^{(1)}\right]^2 = 0. \tag{2.49}$$

The solution is given by

$$A^{(2)} = -\left(\frac{eE_0}{2m}\right)^2 \frac{1}{\left(\omega_0^2 - \omega^2 + i\gamma\omega\right)^2} \frac{B}{\omega_0^2 - 4\omega^2 + 2i\gamma\omega}. \tag{2.50}$$

Referring to Eq. (1.33) the second-order nonlinear optical suscepti- bility function is now given by

$$\chi^{(2)}(-2\omega, \omega, \omega) = \frac{N}{2}\frac{e^3}{m^2}\frac{1}{\left(\omega_0^2 - \omega^2 + i\gamma\omega\right)^2}\frac{B}{\omega_0^2 - 4\omega^2 + 2i\gamma\omega}. \tag{2.51}$$

Note that according to the definition of the polarization function given by Eq. (1.33) the quantity B in Eqs. (1.38) and (2.51) incorpo- rates a local symmetry of the system. It vanishes in centrosymmetric systems (e.g., molecules with two identical atoms, crystals such as

Si, Ge etc.). These systems do not exhibit any even-order optical response. For the centrosymmetric systems the equation of electron motion under the effect of the driving light field force is given by

$$\frac{d^2x}{dt^2} + \gamma\frac{dx}{dt} + \omega_0^2 x + Cx^3 = -\frac{e}{2}E_0\left(e^{i\omega t} + e^{-i\omega t}\right). \qquad (2.52)$$

Following the same procedure by solving Eq. (2.52), one can derive the terms contributing to the third-order harmonic susceptibility function. For example,

$$\chi^{(3)}(-3\omega, \omega, \omega, \omega) = \frac{N}{4}\frac{e^4}{m^3}\frac{1}{\left(\omega_0^2 - \omega^2 + i\gamma\omega\right)^3}\frac{C}{\omega_0^2 - 9\omega^2 + 3i\gamma\omega}. \qquad (2.53)$$

Equation (2.53) represents a third-order contribution to the nonlinear optical susceptibility function at frequency 3ω. Other terms contributing to the third-order optical susceptibility function can be obtained in the same way, however, they are not shown here (see comments on Eq. (1.33)).

2.5 Quantum Picture: Electron Charge Density within the Perturbation Theory

In this section we describe evaluation of the light filed induced charge within perturbation theory using plane wave representation for calculations of optical functions. Linear response is considered in Section 1.3. In this section we consider both linear and nonlinear optical response within the perturbation theory in more details.

Equilibrium electron charge density is defined by equation (1.48) through the density operator (using definition of Trace, Tr, as a sum of the diagonal elements).

For the linear response at zero temperature (when optical excitations occur between completely filled and empty states with Fermi functions equal either 1 or zero) we arive in equation (1.62).

For the second-order perturbation one needs to use first-order solution Eq. (1.62). Plugging it into Eq. (1.58) after some algebra it

Nonlinear Light Interaction with Matter

leads to the following expressions at $T = 0$:

$$-\hbar\omega\rho_{ss'}^{(2)} = (E_s - E_{s'})\rho_{ss'}^{(2)} + \sum_{s''} V_{ss''}^{(1)}\rho_{s''s'}^{(1)} - \sum_{s''} \rho_{ss''}^{(1)} V_{s''s'}^{(1)}, \quad (2.54)$$

or

$$\rho_{ss'}^{(2)}(\omega) = \frac{1}{E_s - E_{s'} - \hbar\omega}$$
$$\times \sum_{s''} \left[V_{ss''}^{(1)} V_{s''s'}^{(1)} \left(\frac{1}{E_{s''} - E_{s'} - \hbar\omega} - \frac{1}{E_s - E_{s''} - \hbar\omega} \right) \right]. \quad (2.55)$$

2.6 Review Questions and Exercises

2.6.1 Relationship between Linear and Nonlinear Optical Susceptibilities

Consider the classical oscillator model of matter with allowance for the quadratic nonlinear force. Find a relationship between linear $\chi(2\omega)$ and nonlinear $\chi^{(2)}(2\omega)$ optical susceptibilities for the second harmonic generation process.

2.6.2 Quadratic Nonlinear Dielectric Polarization

Derive the expression for the quadratic nonlinear dielectric polarization vector $\mathbf{P}$ using the shortened index notation known from crystallography, i.e. $xx = 1, yy = 2, zz = 3, yz = zy = 4, xz = zx = 5$, $xy = yx = 6$. Apply this notation for a KDP nonlinear crystal (KH_2OP_4, point symmetry group $42m$) for which $d_{14} = d_{25} = d_{36}$.

2.6.3 Second Harmonic Generation

Derive a system of differential equations for the second optical harmonic generation $(\omega + \omega \to 2\omega)$. Solve it for the following conditions (1) slowly varying amplitude approximation; (2) phase matching condition $(\Delta k = 2k_1 - k_2 = 0)$; (3) effective exchange between the fundamental and the second harmonic waves.

2.6.4 Parametric Frequency Mixing

Derive a system of differential equations for parametric mixing of frequencies, $\omega_1 + \omega_2 \to \omega_3$ in a quadratic nonlinear medium. Solve it for the parametric amplification mode $(A_I(0) = A_{10},\ A_2(0) = 0,\ A_3(0) = A_{30})$. Take into account the following:

- phase matching condition $(\Delta k = k_1 + k_2 - k_3 = 0)$;
- effective exchange between pump wave $A_3(z)$ and amplified wave $A_I(z)$.

2.6.5 Phase Modulation

Derive a differential equation for phase modulation of a plane optical wave in a cubic nonlinear medium in which $\mathbf{P}^{NL} = \varepsilon_2 |E(\omega)|^2 \mathbf{E}(\omega)$. Solve it and using the solution explain the effect of self-focusing.

2.6.6 Relationship between Nonlinear Functions

Derive a differential equation for two-photon absorption of plane optical wave in a cubic nonlinear medium in which $P^{(3)}(\omega_2) = i Im(\chi^{(3)}) |E(\omega_1)|^2 E_2$. Solve it in the approximation $I_{10} \gg I_{20}$, assuming that attenuation wave, I_{10} can be neglected.

Chapter 3

Fundamentals of Electron Energy Structure and Optics of Molecules

3.1 Introduction

A progress in many different areas of modern material science, chemistry, and physics has vastly accelerated by the synergy between theory and experiment. One can define a theory as one or more rules that are postulated to govern the behavior of a physical system. In science such rules are quantitative in nature and expressed in form of mathematical equations. Such quantitative nature of scientific theories allows them to be proved experimentally.

This chapter is focused on the theoretical methods in physics and optics of materials. One can define the *computational optics* as a branch of physics that uses modeling and simulation to assist in solving various problems. This approach is based on application of different methods in the theory of materials (atoms, molecules, polymers, solids) as incorporated in computational packages to calculate atomic, electronic structures and properties of materials. The computational results normally complement the information obtained by experiments, it can in some cases predict unobserved physical phenomena substantially saving time in the achieving

---

Fundamentals of the Optics of Materials: Tutorial and Problem Solving
Vladimir I. Gavrilenko and Volodymyr S. Ovechko
Copyright © 2024 Jenny Stanford Publishing Pte. Ltd.
ISBN 978-981-4877-93-0 (Hardcover), 978-1-003-25694-6 (eBook)
www.jennystanford.com

the research goals (e.g. in design of new materials compared to traditional "trial-and-error" approach).

The aim of the chapter is to provide the readership with basic understanding of modeling and simulations in optics. It starts with very simple molecules and semi-empirical methods, considers more complex systems, and presents in an introductory form the modern methods based on the first principles theoretical developments in quantum chemistry and physics.

3.2 Electron Energy Structure of Simple Molecules

In this section modeling of the electron energy structure of organic materials is described. First the diatomic molecule is considered reviewing some mathematical ideas that the reader may been exposed to in a course on quantum mechanics and in other chapters of this book.

Before getting into the quantum mechanics the idea of expansion of a quantum state in a set of orthonormal basis states is explained in terms of basics of vector algebra.

A vector (v) is a quantity that has both magnitude and direction. Assume we have a coordinate system specified by a three ortho-normal vectors i, j, k. Then a vector v is represented as follows:

$$v = v_i i + v_j j + v_k k \tag{3.1}$$

where $v_i = (v \cdot i)$, $v_j = (v \cdot j)$, and $v_k = (v \cdot k)$ are components (numbers) or projections of the vector v onto basis coordinate vectors i, j, k.

A similar idea is used in quantum mechanics representing a quantum state as a projection on a set of ortho-normal basis states [Landau and Lifshits (1980); Davydov (1976)]. A wave function $\Psi(r)$ represents the quantum state $|\Psi\rangle$ in physical Hilbert space. The $\langle\phi|$ is called *bra* and the $|\phi\rangle$ is called *ket* state. The quantum state $|\Psi\rangle$ can be expanded in a set of N ortho-normal basis states $|\phi\rangle$ according to:

$$|\Psi\rangle = \sum_{i=1}^{N} \langle\phi_i|\Psi\rangle \, |\phi_i\rangle, \qquad (3.2)$$

note that notation $\langle\phi_i|\Psi\rangle$ in Eq. (3.2) indicates integration over r. In our example it is a number and can be called a projection of $|\Psi\rangle$ on the basis function $|\phi_i\rangle$.

The expansion of Eq. (3.2) is a linear combination of the basis functions $|\phi_i\rangle$ that represents a wave function $\Psi(r)$. It is a key approach of the Linear Combination of Atomic Orbitals (LCAO) method that is widely used in computational physics and chemistry and is discussed in details in Section 3.3.

3.2.1 A Homonuclear Diatomic Molecule: The Hydrogen Molecule

Consider the H_2 molecule in its ground state [Sutton (2004); Gray (1965)]. Solving the Schrödinger equation exactly will require accounting for two electrons in this molecule, one from each hydrogen interacting Coulombically with each other. The full solution is quite complicated. Instead, here a simple molecular-orbital approach is used that gives a quantitative picture which reproduces most of the important features.

Let $|\Psi\rangle$ denote a state vector of an electron in the molecule. Assume that the molecule is formed by bringing two isolated hydrogen atoms together. Let $|1\rangle$ and $|2\rangle$ denote the electron states in the first and second atoms, respectively. For the chosen example it is naturally to take the $1s$-hydrogen ground states for the $|1\rangle$ and $|2\rangle$ with the energy of the electron in this state being E_f, i.e.

$$\begin{cases} H_1 |1\rangle = E_f |1\rangle \\ H_2 |2\rangle = E_f |2\rangle \end{cases} \qquad (3.3)$$

where H_1 and H_2 in Eq. (3.3) are the Hamiltonians for the isolated atoms 1 and 2. Next one can assume that the two basis states $|1\rangle$ and $|2\rangle$ represent the full orthonormal basis set in which one can expand the ground state $|\Psi\rangle$ of the hydrogen molecule:

$$|\Psi\rangle = c_1 |1\rangle + c_2 |2\rangle \tag{3.4}$$

Assumption of orthonormality of the basis set means that $\langle 1|2\rangle = \langle 2|1\rangle = 0$ and $\langle 1|1\rangle = \langle 2|2\rangle = 1$.

The time-independent Schrödinger equation for the molecular state of the hydrogen molecule has now the following form:

$$H |\Psi\rangle = E |\Psi\rangle \tag{3.5}$$

After substituting Eq. (3.2) for the wave function we have:

$$H(c_1 |1\rangle + c_2 |2\rangle) = E(c_1 |1\rangle + c_2 |2\rangle) \tag{3.6}$$

The Eq. (3.6) will now be solved by projection onto the basis set. This is done through multiplying the Eq. (3.6) by complex conjugated basis functions $\langle 1|$ and $\langle 2|$:

$$\begin{cases} \langle 1| H(c_1 |1\rangle + c_2 |2\rangle) = \langle 1| E_f(c_1 |1\rangle + c_2 |2\rangle) \\ \langle 2| H(c_1 |1\rangle + c_2 |2\rangle) = \langle 2| E_f(c_1 |1\rangle + c_2 |2\rangle) \end{cases} \tag{3.7}$$

Applying orthonormallity conditions for the basis function results in the following:

$$\begin{cases} E_0 c_1 + H_{12} c_2 = E c_1 \\ H_{21} c_1 + E_0 c_2 = E c_2 \end{cases} \tag{3.8}$$

Here the following notations are used $E_0 = H_{11} = \langle 1|H|1\rangle = H_{22} = \langle 2|H|2\rangle$ for the on-site Hamiltonian matrix elements, and $H_{ij} = \langle i|H|j\rangle$ for the overlap integral. The nontrivial solution of Eq. (3.8) has to be found from the condition that secular determinant given by:

$$\begin{vmatrix} E_0 - E & H_{12} \\ H_{21} & E_0 - E \end{vmatrix} \tag{3.9}$$

be zero that yields the following quadratic equation:

$$E^2 - 2E_0E + E_0^2 - \mid H_{12} \mid^2 = 0 \tag{3.10}$$

The Hamiltonian in Eqs. (3.8), (3.9) is hermitian representing by a hermitian matrix [Landau and Lifshits (1980); Davydov (1976)], which means $H_{12} = H_{21}^*$. In hydrogen atom the chosen 1s basis states are real. Introducing a notation $\beta = H_{12} = H_{21}$ the solution of Eq. (3.6) is given by:

$$\begin{cases} E_b = E_0 - \beta \\ E_a = E_0 + \beta \end{cases} \tag{3.11}$$

Substituting the eigen values of Eq. (3.11) into Eq. (3.6) one can find the corresponding eigen vectors that after applying the orthonormality condition are given by:

$$\begin{cases} \sigma_{1s} = |\Psi\rangle_b = \dfrac{1}{\sqrt{2}}(|1\rangle + |2\rangle) \\ \sigma_{1s}^* = |\Psi\rangle_a = \dfrac{1}{\sqrt{2}}(|1\rangle - |2\rangle) \end{cases} \tag{3.12}$$

Indexes b and a in Eqs. (3.11), (3.12) indicate *bonding* and *antibonding* states of the hydrogen molecule.

Schematic diagram of the electron energy levels and corresponding wave functions in hydrogen molecule is shown in Fig. 3.1. Two equivalent non-interacting hydrogen atoms have the same energy corresponding to the 1s-electron orbital in isolated atom. Creation of the hydrogen molecule resulted from an interaction between two atoms results in a characteristic change of the energy diagram according to the Eq. (3.11). One molecular orbital (σ_{1s}) corresponding to the two overlapped 1s-atomic orbitals from the interacting atoms has an energy lower than that in an isolated atom (E_0) by β (see Eq. (3.11)). This orbital can be occupied by two electrons from the interacting atoms having opposite spins, as shown in Fig. 3.1. In addition, energy structure of the hydrogen molecule represents another state characterizing by the energy higher than that in an isolated atom (E_0) by β (see Eq. (3.11)). This anti-

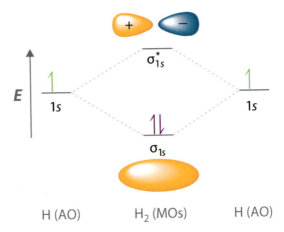

Figure 3.1 Energy diagram and schematic representation of the bonding (σ_{1s}) and antibonding (σ_{1s}^*) molecular orbitals in hydrogen molecule.

bonding molecular orbital (σ_{1s}^*) is characterized by non-overlapping 1s-atomic orbitals, as shown in Fig. 3.1.

In the ground state of the hydrogen molecule the anti-bonding molecular orbital is non-occupied. However, the σ_{1s}^* orbital can be occupied upto two electrons as a result of external excitation (e.g. by external laser radiation). In complex molecules the Highest Occupied Molecular Orbital (HOMO) are mostly bonding by nature (like σ_{1s} in hydrogen molecule) and the Lowest Unoccupied Molecular Orbital are anti-bonding (like σ_{1s}^* in hydrogen molecule).

3.2.2 Heteronuclear Diatomic Molecule

In the previous section it has been shown how in diatomic molecule the bonding and anti-bonding states arose. Now the consideration is generalized to a hetero-nuclear molecule [Sutton (2004)]. The overlap integral will be still taken as a parameter β according to the notation $\beta = H_{12} = H_{21}$. However the on-site Hamiltonian matrix elements on the A and B atoms are now different, and the wave function of the molecule can be writen in terms of the atomic states $|A\rangle$ and $|B\rangle$ as follows:

$$|\Psi\rangle = c_A |A\rangle + c_B |B\rangle. \tag{3.13}$$

The time-independent Schrödinger equation is now given by:

$$\begin{cases} (E_A - E)c_A + \beta c_B = 0 \\ \beta c_A + (E_B - E)c_B = 0. \end{cases} \tag{3.14}$$

It is instructive to compare now Eq. (3.14) with Eq. (3.8). Solution of the secular equation in this case leads to the following values for the bonding and anti-bonding eigen energies:

$$\begin{cases} E_b = \epsilon - \sqrt{(\Delta^2 + \beta^2)} \\ E_a = \epsilon + \sqrt{(\Delta^2 + \beta^2)} \end{cases} \tag{3.15}$$

where $\epsilon = (E_A + E_B)/2$ is the average on-site energy and $\Delta = (E_A - E_B)/2$. If Δ tends to zero the Eq. (3.15) results in Eq. (3.11) for the homo-nuclear molecule. Looking at the results obtained for the hetero-nuclear molecule one can state that the difference in the on-site energies leads to an increase of the splitting between the bonding and anti-bonding states.

Plugging solutions of Eq. (3.15) into the Schrödinger equation Eq. (3.14) can give a solution for the wave function.

The analysis of the charge density (i.e. the value of $|\Psi|^2$) leads to interesting conclusions about properties of *electronegativity* [Sutton (2004)] and optical response of materials [Gavrilenko (2020)].

3.3 Linear Combination of Atomic Orbitals Method

In this section the Linear Combination of Atomic Orbital (LCAO) method is discussed. The LCAO method plays a very important role in modeling electron energy structure of molecules. It is also called the tight binding approach since the basic idea of this method is the assumption that the electrons are tightly bound to their nuclei as in the atoms [Yu and Cardona (2010); Harrison (1989); Gray (1965); Bechstedt (2003)]. The valence electrons covalently bounded in atoms, molecules and solids are concentrated mainly in the bonds retaining most of their characters and characteristics

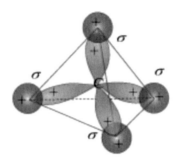

Figure 3.2 Spatial structure of the CH$_4$ molecule. The sp^3-hybrid orbitals of the carbon atom create σ-bonds to the four hydrogen atoms building ideal tetrahedral configuration.

in atoms. This justified numerous successful applications of the LCAO method to electron energy study of fundamental properties as well as optics of organic molecules, polymers, and nanomaterials [Gavrilenko (2020)].

3.3.1 Electron Energy Structure of Methane Molecule

It is instructive to consider application of the LCAO for modeling of relatively simple molecules. In this section the CH$_4$ molecule is considered [Gray (1965)].

Structure of the methane molecule is shown in Fig. 3.2. The electron bonds are tetrahedrally coordinated in space creating a spatial angle of 109.5°. The four hybrid orbitals can be formed from the s, p_x, p_y, p_z atomic orbitals of C atom creating the following linear combinations:

$$\begin{cases} |h_1\rangle = \frac{1}{2}(|s\rangle + |p_x\rangle + |p_y\rangle + |p_z\rangle) \\ |h_2\rangle = \frac{1}{2}(|s\rangle + |p_x\rangle - |p_y\rangle - |p_z\rangle) \\ |h_3\rangle = \frac{1}{2}(|s\rangle - |p_x\rangle + |p_y\rangle - |p_z\rangle) \\ |h_4\rangle = \frac{1}{2}(|s\rangle - |p_x\rangle - |p_y\rangle + |p_z\rangle) \end{cases} \quad (3.16)$$

The four sp^3-hybrid orbitals of the carbon atom represented by Eq. (3.16) form a tetrahedral as shown in Fig. 3.2. The hybrid orbitals of the carbon atom create σ-bonds to the four hydrogen atoms. The coefficient $(\frac{1}{2})$ in Eq. (3.16) stands for normalization.

Now we will follow a similar procedure as that described in Section 3.2.1 and construct a linear combination for the trial $|\Psi\rangle$ function from the eight basis functions: four of the C-atom $(|h_1\rangle, |h_2\rangle, |h_3\rangle, |h_4\rangle)$, given by Eq. (3.16)) and the four $|s\rangle$-functions of the hydrogen atoms (see Fig. 3.2). We have:

$$|\Psi\rangle = c_1 |h_1\rangle + c_2 |h_2\rangle + c_3 |h_3\rangle + c_4 |h_4\rangle$$
$$+ c_5 |s_1\rangle + c_6 |s_2\rangle + c_7 |s_3\rangle + c_8 |s_4\rangle \qquad (3.17)$$

Note that despite all $|s\rangle$ functions of the hydrogen atoms are identical their geometrical locations are different (indicated by the index in Eq. (3.17)). This will be accounted by an appropriate geometrical factor in the secular equation.

Plugging Eq. (3.17) into Schrödinger equation Eq. (3.5) we calculate the Hamiltonian matrix following a similar procedure to that described in Sections 3.2.1 and 3.2.2, multiplying by complex conjugate of every basis functions. It is important to note that we will use condition of orthonormality and completely neglect the basis functions overlaps. This method corresponds to the *Completely Neglected Differential Overlaps* method (CNDO) in computational chemistry.[1] The Hamilton matrix of the CH_4 molecule is now given by:

$$\begin{vmatrix} \epsilon & b & b & b & \alpha & \beta & \beta & \beta \\ & \epsilon & b & b & \beta & \alpha & \beta & \beta \\ & & \epsilon & b & \beta & \beta & \alpha & \beta \\ & & & \epsilon & \beta & \beta & \beta & \alpha \\ & & & & \epsilon & d & d & d \\ & & & & & \epsilon & d & d \\ & & & & & & \epsilon & d \\ & & & & & & & \epsilon \end{vmatrix} \qquad (3.18)$$

---

[1] For discussions of the overlaps affecting calculated eigen energy values (extended *Hückel* approximation) the reader is referred to more specialized literature [Harrison (1989); Gray (1965); McQuarrie and Simon (1997)].

Since the Hamiltonian matrix is hermitian (complex conjugate and symmetrical with respect to the diagonal) only upper triangular part of the H-matrix is shown in Eq. (3.18). Following notations are used in Eq. (3.18):

$$\begin{cases} \alpha = \langle s_i|H|h_i \rangle \\ \beta = \langle s_i|H|h_j \rangle , i \neq j \\ b = \langle h_i|H|h_j \rangle , i \neq j \\ d = \langle s_i|H|s_j \rangle , i \neq j \end{cases} \tag{3.19}$$

where $i,j = 1,2,3,4$. In many text books on chemistry and physics one uses following notations for the interaction matrix elements (hopping integrals), $d = \langle s_i^H|H|s_j^H \rangle = (s^H s^H \sigma)$, $\nu = \langle s^H|H|s^C \rangle = (s^H s^C \sigma)$, $\mu = \langle s^H|H|p_i^C \rangle = (s^H p_i^C \sigma)$, with σ indicating a type of the bond [Gray (1965)]. Based on Eq. (3.16) the on-site energies calculated on hybrid and atomic orbitals relate to each other as follows, $\epsilon_h = \frac{1}{4}(\epsilon_s + 3\epsilon_p)$.

The eigen energies for the CH_4 molecule calculated from Eq. (3.5) with the H-matrix given by Eq. (3.18) are given by [Gavrilenko (1993)]:

$$\begin{cases} E_{a_1^\pm} = \frac{1}{2}(\epsilon_s + \epsilon_s^H + 3d) \pm \sqrt{\frac{1}{4}(\epsilon_s - \epsilon_s^H - 3d)^2 + 4\nu^2} \\ E_{t_2^\pm} = \frac{1}{2}(\epsilon_p + \epsilon_s^H - d) \pm \sqrt{\frac{1}{4}(\epsilon_p - \epsilon_s^H + d)^2 + \frac{4}{3}\mu^2} \end{cases} \tag{3.20}$$

In Eq. (3.20) the eigen energies of molecular orbitals in methane having indexes $(a_1^\pm)$ and $(t_2^\pm)$ are denoted followed notations of the tetrahedral symmetry point group representations [McQuarrie and Simon (1997); Harrison (1989); Gavrilenko (1993); Bechstedt (2003)]. There are therefore four molecular orbitals in CH_4 molecule: two occupied orbitals (bonding orbitals indicated by index $(^+)$) and two unoccupied orbitals (antibonding orbitals indicated by index $(^-)$). Schematically the energy diagram representing by equation (3.20) is shown in Fig. 3.3.

Four electron of the carbon atom $(|s\rangle , |p_x\rangle , |p_y\rangle , |p_z\rangle)$ and four electrons of the hydrogen atoms $(4\,|s\rangle)$ initially located on the relevant atomic orbitals are redistributed between two bonding

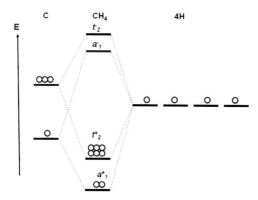

Figure 3.3 Energy diagram of the CH$_4$ molecular orbitals given by Eq. (3.20). Electrons occupied atomic orbitals in the isolated C-atom and in four isolated H-atoms are shown by circles in the left and right parts of the diagram, respectively. The electrons occupied bonding CH$_4$ molecular orbitals are shown by circles in the middle. Energy scale is indicated by the arrow.

molecular orbitals of the CH$_4$ molecule ($|a_1^+\rangle$, $|t_2^+\rangle$), as indicated in Fig. 3.3 by circles.

Many chemical and spectroscopic properties of the CH$_4$ can be described by the *frontier orbitals*, the HOMO ($|t_2^+\rangle$) and the LUMO ($|a_1^-\rangle$) orbitals.

3.4 First Principles Methods in Computational Materials Science

As stated throughout this book the total energy is a key function describing basic physical and chemical properties of materials: the ground state. It consists from both kinetic (describing motion) and potential energy parts. To make the theoretical model realistic it is very important to incorporate all most important contributions to the both parts of the total energy. In view of the large number of the particles involved in the model, this is very challenging for the first principles theory. Different approximations are applied in

Fundamentals of Electron Energy Structure and Optics of Molecules

order to achieve a trade-off between complexity and accuracy. Very successful in realistic modeling of the ground state is the *density functional theory*, DFT. In this section we present basic ideas of the DFT considering first the Thomas–Fermi approximation that is very instructive in helping to understand the DFT.

3.4.1 Thomas–Fermi Approximation

Initially Thomas and Fermi (TF) in the 1920s suggested describing atoms as uniformly distributed electrons (negative charged clouds) around nuclei in a six dimensional phase space (momentum and coordinates). This is enormous simplification of the actual many body problem. It is instructive to consider basic ideas of the TF approximation before starting with a more accurate theory: the DFT. The basic ideas and results of the TF model in application for atoms are provided here.

Following the TF approach the total energy of the system could be presented as a function (functional) of electron density [Parr and Yang (1989)]. Each h^3 of the momentum space volume (h is the Planck constant) is occupied by two electrons and the electrons are moving in effective potential field that is determined by nuclear charge and by assumed uniform distribution of electrons. Density of ΔN electrons in real space within a cube (nanoparticle) with a side l is given by:

$$\rho(\mathbf{r}) = \frac{\Delta N}{v} = \frac{\Delta N}{l^3} \tag{3.21}$$

The electron energy levels in this three-dimensional infinite well are given by:

$$E = \frac{h^2}{8ml^2}(n_x^2 + n_y^2 + n_z^2) = \frac{h^2}{8ml^2}\tilde{R}^2, \quad n_x, n_y, n_z = 1, 2, 3, ..., \tag{3.22}$$

Radius $R = \tilde{R}_{max}$ of the sphere in the space (n_x, n_y, n_z) covering all occupied states determines the maximum energy of electrons: the Fermi energy, F. The number of energy levels within this maximum value at zero temperature is given by:

$$N_F = \frac{1}{2^3}\frac{4\pi R^3}{3} = \frac{\pi}{6}\left(\frac{8ml^2 F}{h^2}\right)^{3/2} \tag{3.23}$$

Density of states is defined as:

$$g(E)dE = N_F(E + dE) - N_F(E) = \frac{\pi}{4}\left(\frac{8ml^2}{h^2}\right)^{3/2} E^{1/2}dE \tag{3.24}$$

At zero temperature all energy levels below Fermi energy are occupied:

$$f(E) = \left\{ \begin{array}{l} 1,\ E \le F \\ 0,\ E > F \end{array} \right\} \tag{3.25}$$

Consequently the total energy of the electrons in one cell will be given by:

$$E = \int_0^F Ef(E)g(E)dE = \frac{4\pi}{h^3}(2m)^{3/2}l^3 \int_0^F E^{3/2}dE$$

$$= \frac{\pi}{5}\left(\frac{2l}{h}\right)^3 (2m)^{3/2}F^{5/2} \tag{3.26}$$

The Fermi energy, F could be obtained from the total number of electrons ΔN in a cell:

$$\Delta N = 2\int_0^F f(E)g(E)dE = \frac{\pi}{3}\left(\frac{2l}{h}\right)^3 (2m)^{3/2}F^{3/2} \tag{3.27}$$

The energy of the electrons in one cell is given by:

$$E = \frac{3}{10}(3\pi^2)^{2/3}\frac{l^3}{(2\pi)^2}\rho^{5/3} = C_F\frac{l^3}{(2\pi)^{5/3}},$$

$$C_F = \frac{3}{10}(3\pi^2)^{2/3} = 2.871 \tag{3.28}$$

In Eq. (3.28) we reverted to atomic units $e = h = m_0 = 1$. The electron density is a smooth function in a real space. For systems without translational symmetry it is different for different cells.

However, for the periodic systems only consideration within the unit cell is required since all unit cells are equivalent. Now adding the contributions from all cells with energies within F we obtain:

$$T_{TF}[\rho] = C_F \int \rho^{5/3}(\boldsymbol{r}) d^3\boldsymbol{r} \tag{3.29}$$

Equation (3.29) represents the well-known Thomas–Fermi kinetic energy functional which is a function of the local electron density. The functional (3.29) could be applied to electrons in atoms encountering most important idea of the modern DFT, the local density approximation (LDA) [Martin (2005)]. Adding to Eq. (3.29) classical electrostatic energies of electron–nucleus attraction and electron–electron repulsion we arrive in energy functional of Thomas–Fermi theory of atoms:

$$E_{TF}[\rho(\boldsymbol{r})] = C_F \int \rho^{5/3}(\boldsymbol{r}) d^3\boldsymbol{r} - Z \int \frac{\rho(\boldsymbol{r})}{r} d^3\boldsymbol{r}$$
$$+ \frac{1}{2} \int \frac{\rho(\boldsymbol{r}_1)\rho(\boldsymbol{r}_2)}{|\boldsymbol{r}_1 - \boldsymbol{r}_2|} d^3\boldsymbol{r}_1 d^3\boldsymbol{r}_2 \tag{3.30}$$

Note that nucleus charge, Z, is measured in atomic units. Energy of the ground state and electron density can be found by minimizing the functional (3.30) with the constrain condition:

$$N = \int \rho(\boldsymbol{r}) d^3\boldsymbol{r} \tag{3.31}$$

The electron density in Eq. (3.30) has to be calculated in conjunction with Eq. (3.31) from the following equation for chemical potential, defined as the variational derivative according to:

$$\mu_{TF} = \frac{\delta E_{TF}[\rho]}{\delta\rho(\boldsymbol{r})} = \frac{5}{3}C_F\rho^{2/3}(\boldsymbol{r}) - \frac{Z}{r} + \frac{1}{2} \int \frac{\rho(\boldsymbol{r}_2)}{|\boldsymbol{r}_1 - \boldsymbol{r}_2|} d^3\boldsymbol{r}_2 \tag{3.32}$$

The Thomas–Fermi model provides reasonably good predictions for atoms. It has been used before to study potential fields and charge density in metals and the equation of states of elements [Feynman et al. (1949)]. However this method is considered rather crude for more complex systems because it does not incorporate the actual orbital structure of electrons. In view of the modern DFT theory the

Thomas–Fermi method could be considered as an approximation to the more accurate theory.

3.4.2 First Principles Methods Based on the Density Functional Theory

Density functional theory (DFT) is a quantum mechanical computational modeling approach widely used in physics, chemistry, and materials science to study the electronic structure (first of all the ground state) of many-body systems, in particular atoms, molecules, and the condensed phases. With this theory, the properties of a many-electron system can be determined by using functionals, i.e. functions of another function, which in this case is the spatially dependent electron density. Hence the name density functional theory comes from the use of functionals of the electron density [Martin (2005)]. DFT is among the most popular and versatile methods available in condensed-matter physics, computational physics, and computational chemistry.

The DFT has been used for calculations in chemistry and solid state physics since the 1970s. However, only after the 1990s, the approximations used in the theory were greatly refined to better model the exchange and correlation interactions. As the result, this method became very popular for research, mostly because the computational costs are relatively low when compared to traditional methods, such as Hartree–Fock theory and its descendants based on the complex many electron wave function nevertheless providing quite a good comparison to experimental data.

This section focuses on the local density approximation (LDA) method within the DFT using *ab initio* pseudopotentials. For the systems like large molecules (as well as solids, surfaces, etc.) much better predictions are provided by the DFT. Search for the ground state within the DFT follows the rule that *the electron density is a basic variable in the electronic problem* (the first theorem of Hohenberg and Kohn [Hohenberg and Kohn (1964)]) and another rule that *the ground state can be found from the energy variational principle for the density* (the second theorem of Hohenberg and Kohn [Kohn (1999)]).

According to the DFT the total energy could be written as:

$$E[\rho] = T[\rho] + U[\rho] + E_{XC}[\rho], \tag{3.33}$$

where T is the kinetic energy of the system of non-interacting particles, U is the electrostatic energy due to Coulomb interactions. The most important part in the DFT is E_{XC}, the exchange and correlation (XC) energy which includes all many-body contributions to the total energy. The charge density is determined by the wave functions, which for practical computations could be constructed from single orbitals, ϕ_j (e.g. antisymmetrized product—the Slater determinant, atomic or Gaussian orbitals, linear combinations of plane waves, etc.). Charge density is given by:

$$\rho(r) = \sum_j |\phi_j(r)|^2, \tag{3.34}$$

where the sum is taken over all occupied j-orbitals. In the spin resolved case there will be orbitals occupied with spin-up and spin-down electrons. Their sum gives total charge density and their difference gives the spin density. In terms of the electron orbitals the energy components are given in atomic units as:

$$T = -\frac{1}{2} \int \sum_j \phi_j^*(r) |\nabla^2| \phi_j(r) d^3 r \tag{3.35}$$

$$U = -\sum_j^n \sum_\alpha^N \int \phi_j^*(r) \left| \frac{Z_\alpha}{(R_\alpha - r)} \right| \phi_j(r) d^3 r$$

$$+ \frac{1}{2} \sum_{ij} \int \phi_i^*(r_1) \phi_j^*(r_2) \frac{1}{(r_1 - r_2)} \phi_i(r_1) \phi_j(r_2) d^3 r_1 d^3 r_2$$

$$+ \sum_\alpha^N \sum_{\beta < \alpha}^N \frac{Z_\alpha - Z_\beta}{|R_\alpha - R_\beta|} \tag{3.36}$$

The first term in potential energy (3.36) stands for the electron–nucleus attraction, the second term describes for electron–electron repulsion, and the third term represents nucleus–nucleus repulsion. In Eq. (3.36) Z_α, refers to the charge on nucleus α of N-atom system.

The third term in Eq. (3.33) describes the exchange and correlation energy. Rather simple for computations but surprisingly good approximation is the local density approximation, LDA, which assumes that the charge density varies slowly on atomic scale, i.e. the effect of other electrons on given (*local*) electron density is described as a uniform electron gas. The XC energy can be obtained by integrating with the uniform gas model (see e.g. [Ceperley and Adler (1980)]):

$$E_{XC} \cong \int \rho(\mathbf{r}) \tilde{E}_{XC}[\rho(\mathbf{r})] d^3(\mathbf{r}), \qquad (3.37)$$

where $\tilde{E}_{XC}[\rho(\mathbf{r})]$ is XC energy per particle in a uniform electron gas. For many systems a good approximation provides an analytic expression for $\tilde{E}_{XC}[\rho(\mathbf{r})]$ suggested in [Perdew and Wang (1992)]. In practical calculations through minimization of the total energy Eq. (3.33) one determines self-consistently the electron density and the actual XC part. A variational minimization procedure leads to a set of coupled equations proposed by Kohn and Sham [Kohn and Sham (1965)]:

$$\left[-\frac{1}{2} \nabla^2 - V_N + V_e + \mu_{XC}(\rho) \right] \phi_j = E_j \phi_j, \qquad (3.38)$$

with

$$\mu_{XC} = \frac{\partial}{\partial \rho} (\rho E_{XC}). \qquad (3.39)$$

Solution of the Kohn–Sham equation provides with equilibrium geometry and the ground-state energy of the system. However, eigen functions and eigen energies of the Kohn–Sham equation cannot be interpreted as the *quasiparticle* quantities needed for optics. The definition *quasiparticle* refers to a particle-like entity arising in certain systems of interacting particles. If a single particle moves through the system, surrounded by a cloud of other interacting particles, the entire entity moves along somewhat like a free particle (but slightly different). The quasiparticle concept is one of the most important in materials science, because it is one of the few known ways of simplifying the quantum mechanical many-body problem describing excitation state, and is applicable to an extremely wide range of many-body systems.

Calculation of the ground state from the Kohn-Sham equation does not result automatically in correct prediction of excitation energies required for optics. For example, in non-metallic systems the predicted value of the energy difference (energy gap) between Highest Occupied Molecular Orbital (HOMO) and Lowest Unoccupied Molecular Orbital (LUMO) in most cases is underestimated (*gap problem*). Special corrections (quasi-particle, QP corrections) are required to get more accurate excitation energies [Onida *et al.* (2002)]. Without corrections in complex molecules, semiconductors and insulators the local density approximation, LDA, substantially underestimates the HOMO–LUMO energy interval (or the forbidden gaps in solids) values. In this chapter we overview basics of the DFT method that was used for different applications, avoiding however analysis of theoretical details. For advance reading on the DFT one can recommend original papers [Hohenberg and Kohn (1964); Kohn (1999); Ceperley and Adler (1980)], reviews [Onida *et al.* (2002)], and monographs [Martin (2005); Parr and Yang (1989)].

Chapter 4

Optical Absorption and Fluorescence of Materials

Methods of optical spectroscopy are widely used to study optical properties of materials. This area includes different methods of both linear and nonlinear optics. This chapter is focused on linear optical absorption and luminescence. The chapter includes extended tutorial part addressing basics of the light emission and absorption mechanisms that will help students and young researchers to better understand the subject. Nonlinear optical spectroscopy is considered in Chapter 2. The phenomena of optical absorption and luminescence are widely discussed in literature [Benten *et al.* (2005); Glinka *et al.* (2002); Kolasinski (2008); Kulzer and Orrit (2004); Yoffe (2001); Ovechko and Kharchenko (2013); Ovechko (2012); Dmytruk and Ovechko (2003)].

4.1 Light Absorption of Materials

This section presents a brief overview of basic physics that governs absorption and luminescence of light in materials. As shown in Chapter 1 the monochromatic electromagnetic wave is given in

Fundamentals of the Optics of Materials: Tutorial and Problem Solving
Vladimir I. Gavrilenko and Volodymyr S. Ovechko
Copyright © 2024 Jenny Stanford Publishing Pte. Ltd.
ISBN 978-981-4877-93-0 (Hardcover), 978-1-003-25694-6 (eBook)
www.jennystanford.com

64 | *Optical Absorption and Fluorescence of Materials*

vector form by equation (1.30). In one dimensional case, assuming that light propagation direction (z) is perpendicular to the surface of a transparent material, the equation (1.30) is given by:

$$E_0(z,t) = E_0(k,\omega)e^{i(kz-\omega t)}. \tag{4.1}$$

with the wave vector of light given by:

$$k = \frac{\omega}{c}n. \tag{4.2}$$

In more general case of an absorbing material both the refractive index and the wave vector of light are complex (generally defined by equation (1.31)) i.e.:

$$k = \frac{\omega}{c}\tilde{n} = \frac{\omega}{c}(n + i\kappa). \tag{4.3}$$

Combining equations (1.30), (4.1), and (4.3) we have:

$$E(z,t) = E_0(k,\omega)e^{i\omega(\frac{z}{c}n-t)}e^{-\omega\frac{\kappa}{c}z} = E_0(z,t)e^{-\omega\frac{\kappa}{c}z}. \tag{4.4}$$

Intensity of light corresponding to the wave (4.1) is determined by the square of the field (4.4) i.e. $I = \frac{c}{4\pi}\,|\,E\,|^2$ and is given by:

$$I(z,t) = \frac{c}{4\pi}\,|\,E_0(z,t)\,|^2\,e^{-2\frac{\kappa}{c}\omega z} = I_0(z,t)e^{-\alpha z}. \tag{4.5}$$

Equation (4.5) is frequently considered as a definition of the light absorption coefficient α that is given by:

$$\alpha(\omega) = 2\frac{\omega}{c}\kappa(\omega). \tag{4.6}$$

This expression indicates that the absorption coefficient $\alpha(\omega)$ is proportional to $\kappa(\omega)$, i.e. to the imaginary part of the complex index of refraction $\tilde{n}$, (see equation (Eq. (1.15)). The function of $\kappa(\omega)$ (the extinction coefficient) is usually associated with power loss. Frequency dependence of the absorption coefficient is determined by various physical processes which govern optical properties of materials. Spectroscopy of $\alpha(\omega)$ (as well as of other optical functions) is an important branch of the materials science.

Fundamentals of the Optics of Materials: Tutorial and Problem Solving | 65

This section presents a classic picture of the optical absorption of materials. In order to make a next step and to get a quantum mechanical picture of light interaction with materials it is instructive to start with the interaction of an electromagnetic wave with an atom.

4.2 Interaction of Electromagnetic Radiation with Atom

The Hamiltonian of an isolated particle in atomic units is given by:

$$\hat{h}_0(\mathbf{r}) = -\frac{1}{2}\nabla^2 + V(\mathbf{r}) = \frac{\boldsymbol{p}^2}{2} + V(\mathbf{r}) \tag{4.7}$$

where the definition of the momentum of a free particle $\boldsymbol{p} = -i\nabla$ is used. If an atom contains N electrons the Hamiltonian is given by:

$$H_0 = \frac{1}{2}\sum_{j=1}^{N} \boldsymbol{p}_j^2 + \sum_{j=1}^{N} V(\mathbf{r}_j) \tag{4.8}$$

As the result of the interaction of an electromagnetic wave with a particle the momentum will change. Note, that according to the definition, the electric field strength $\boldsymbol{E}$ is a force per charge and by definition, the vector potential $\boldsymbol{A}$ is a change of the momentum in field per charge (see Eq. (1.16)). Therefore Hamiltonian of a particle in electromagnetic field is given by:

$$\hat{h}(\mathbf{r}) = \frac{1}{2}(\boldsymbol{p} + \boldsymbol{A})^2 + V(\mathbf{r}) \tag{4.9}$$

The Hamiltonian (4.8) of an atom in field is given by:

$$H = \frac{1}{2}\sum_{j=1}^{N} \left[\boldsymbol{p}_j + \boldsymbol{A}(\mathbf{r}_j)\right]^2 + \sum_{j=1}^{N} V(\mathbf{r}_j) \tag{4.10}$$

where spin interaction in Eq. (4.10) is neglected. Within the Coulomb gauge (i.e. $\nabla\boldsymbol{A} = 0$, see Eq. (1.19)) we have:

$$\nabla(AU) = \boldsymbol{A}\cdot(\nabla U) + (\nabla\boldsymbol{A})U = \boldsymbol{A}\cdot(\nabla U) \tag{4.11}$$

66 | *Optical Absorption and Fluorescence of Materials*

Considering the momentum operator ($\mathbf{p} = -i\mathbf{\nabla}$), the operators $\mathbf{A}$ and $\mathbf{p}$ commute:

$$\mathbf{pA} = \mathbf{Ap} \tag{4.12}$$

Expending and rearranging Eq. (4.10), we have:

$$H = \frac{1}{2}\sum_{j=1}^{N} \mathbf{p}_j^2 + \sum_{j=1}^{N} V(\mathbf{r}_j) + \sum_{j=1}^{N}\left[\mathbf{p}_j \mathbf{A}(\mathbf{r}_j)\right] + \frac{1}{2}\sum_{j=1}^{N} \mathbf{A}_j^2(\mathbf{r}_j) \tag{4.13}$$

$$= H_0 + H_{int} + H^{(2)} \tag{4.14}$$

where H_0 stands for the Hamiltonian of the unperturbed atom. Using Eq. (1.21) the interaction of a particle with the field is given by:

$$H_{int} = \mathbf{pA} = A(\hat{\mathbf{e}} \cdot \mathbf{p})\cos(\mathbf{k} \cdot \mathbf{r} - \omega t) \tag{4.15}$$

Consider matrix element of H_{int} in two level system describing electron transition between two states $|a\rangle$ and $|b\rangle$. Using Eqs. (1.21) and (4.12) the matrix element of the interaction is given by:

$$H_{int} = \frac{A}{2}\hat{\mathbf{e}}\left(\langle b|\,\mathbf{p}e^{i\mathbf{kr}}\,|a\rangle\,e^{-i\omega t} + \langle b|\,\mathbf{p}e^{-i\mathbf{kr}}\,|a\rangle\,e^{i\omega t}\right)$$

$$= \frac{1}{2}\left(H_{ab}e^{-i\omega t} + H_{ab}^*e^{i\omega t}\right), \tag{4.16}$$

where interaction matrix element between states $|a\rangle$ and $|b\rangle$ is defined as:

$$H_{ab} = A\hat{\mathbf{e}}\,\langle b|\,\mathbf{p}e^{i\mathbf{kr}}\,|a\rangle. \tag{4.17}$$

Within the spectral region upto vacuum ultraviolet the wavelength of the radiation considered here for the optical transitions is much larger than atomic dimensions, i.e. $\mathbf{k} \cdot \mathbf{r} = 2\pi r/\lambda \ll 1$. The scale of the electromagnetic interaction with matter is set by the fine structure constant[1] $\alpha \approx 1/137$ [Kinoshita (1996)]. Consequently $e^{i\mathbf{kr}} \approx 1$ that

---

[1] The fine structure constant α is one of the most fundamental constants in physics. It is a dimensionless number defined by $\alpha = \mu_0 c e^2/2h$, where $\mu_0 = 4\pi \times 10^{-7}\ Hm^{-1}$ is the permeability of the vacuum, c is the speed of light in vacuum, e is the electric charge of the electron, and h is the Planck constant.

results in the following:

$$H_{ab} = A\hat{e}\langle b|\,\boldsymbol{p}\,|a\rangle = -i\frac{E}{\omega}\hat{e}\,\langle b|\,\boldsymbol{p}\,|a\rangle\,, \qquad (4.18)$$

where Eq. (1.22) has been used. Equation (4.18) defines the *dipole approximation* in the optics of materials for the reason that is explained below.

Matrix element of the interaction (Eq. (4.18)) can be expressed in terms of the matrix element of coordinate $\boldsymbol{r}$. Consider a commutator of the coordinate and momentum operators:

$$[\boldsymbol{r},\boldsymbol{p}] \equiv \boldsymbol{rp} - \boldsymbol{pr}, \qquad (4.19)$$

using $\boldsymbol{p} = -i\nabla$ and a probe function $f(\boldsymbol{r})$ we have:

$$(\boldsymbol{rp} - \boldsymbol{pr})f(\boldsymbol{r}) = -i\left[\boldsymbol{r}\nabla f(\boldsymbol{r}) - \nabla(\boldsymbol{r}f(\boldsymbol{r}))\right] = if(\boldsymbol{r}), \qquad (4.20)$$

The Hamiltonian operator contains $\boldsymbol{p}^2$ operator (see Eq. (4.7)). The commutator for coordinate and the $\boldsymbol{p}^2$ operator is given by:

$$\left[\boldsymbol{r},\boldsymbol{p}^2\right] = \boldsymbol{p}\left[\boldsymbol{r},\boldsymbol{p}\right] + \left[\boldsymbol{r},\boldsymbol{p}\right]\boldsymbol{p} = 2i\boldsymbol{p}. \qquad (4.21)$$

Consequently there are following operator commutation relations:

$$[\boldsymbol{r},\boldsymbol{p}] = i \qquad (4.22)$$

$$\left[\boldsymbol{r},\boldsymbol{p}^2\right] = 2i\boldsymbol{p} \qquad (4.23)$$

$$\left[\boldsymbol{r},\hat{h}_0\right] = i\boldsymbol{p} \qquad (4.24)$$

If $|a\rangle$ and $|b\rangle$ are eigen functions of the particle operator $\hat{h}_0$ (defined by Eq. (4.7)) corresponding to the eigen energies E_a, E_b, respectively,

Optical Absorption and Fluorescence of Materials

we have the following:

$$\langle b|\,\mathbf{p}\,|a\rangle = \frac{1}{i}\,\langle b|\left[\mathbf{r},\hat{h}_0\right]|a\rangle$$

$$= \frac{1}{i}\left(\langle b|\,\mathbf{r}\,|a\rangle\,\langle a|\,\hat{h}_0\,|a\rangle - \langle b|\,\hat{h}_0\,|b\rangle\,\langle b|\,\mathbf{r}\,|a\rangle\right)$$

$$= i(E_b - E_a)\,\langle b|\,\mathbf{r}\,|a\rangle$$

$$= i\omega_{ba}\,\langle b|\,\mathbf{r}\,|a\rangle \qquad (4.25)$$

where notation for the optical transition frequency $\omega_{ba} = E_b - E_a$ and the orthonormality condition for $|a\rangle$ and $|b\rangle$ functions (i.e. $|a\rangle\,\langle a| = |b\rangle\,\langle b| = 1$) are used.

This relation can be generalized for the N-electron system. For example, for the N-electron atom (with the Hamiltonian H_0 (see Eq. (4.8) having orthonormalized N-eigen functions labeled by j) similar consideration results in the following:

$$\langle b|\,\mathbf{p}\,|a\rangle = \frac{1}{i}\,\langle b|\,[\mathbf{r},H_0]\,|a\rangle$$

$$= \frac{1}{i}\left(\sum_{j=1}^{N}\langle b|\,\mathbf{r}\,|j\rangle\,\langle j|\,H_0\,|a\rangle - \sum_{j=1}^{N}\langle b|\,H_0\,|j\rangle\,\langle j|\,\mathbf{r}\,|a\rangle\right)$$

$$= \frac{1}{i}\left(\sum_{j=1}^{N}\langle b|\,\mathbf{r}\,|j\rangle\,\delta_{ja}E_a - \sum_{j=1}^{N}\delta_{bj}E_j\,\langle j|\,\mathbf{r}\,|a\rangle\right)$$

$$= i(E_b - E_a)\,\langle b|\,\mathbf{r}\,|a\rangle$$

$$= i\omega_{ba}\,\langle b|\,\mathbf{r}\,|a\rangle = i\omega_{ba}\mathbf{r}_{ba} \qquad (4.26)$$

Note that in Gaussian system[2] the dipole moment of an electron is defined as $\mathbf{d} = -e\mathbf{r}$. Consequently the interaction Hamiltonian H_{int}

---

[2] Gaussian system of units is a system of electrical and magnetic quantities that uses the centimeter, gram, and second as it's main units. In this system electrical permittivity and magnetic permeability are dimensionless quantities and are equal to 1 for a vacuum. The Gaussian unit system was named in honor of K. Gauss, who in 1832 advanced the idea of devising an absolute system of units having the millimeter, the milligram, and the second as the main units.

(4.18) can be given in the following form:

$$H_{ab} = -i\frac{E}{\omega}\hat{e}\langle b|\,\mathbf{p}\,|a\rangle = E\frac{\omega_{ba}}{\omega}\hat{e}\langle b|\,\mathbf{r}\,|a\rangle = -\hat{e}E\langle b|\,\mathbf{d}\,|a\rangle \quad (4.27)$$

Thus in *dipole approximation* the interaction Hamiltonian is given by:

$$H_{ab} = -\hat{e}E\langle b|\,\mathbf{d}\,|a\rangle = -\mathbf{d}_{ba}E_0 e^{i\omega t} \tag{4.28}$$

Expanding the exponential in Eq. (4.17)

$$H_{ab} = A\hat{e}\langle b|\,\mathbf{p}\left(1 + i(\mathbf{k}\cdot\mathbf{r}) - \frac{1}{2}(\mathbf{k}\cdot\mathbf{r})^2 + \cdots\right)|a\rangle \tag{4.29}$$

one can see that the dipole approximation corresponds to the neglect of all terms in the series (4.29) except the first one. The dipole approximation is widely used in linear optics in a wide spectral range. However, at very short wavelength of the electromagnetic radiation (e.g. in the X-ray region) effects of spatial dispersion are important. In addition at very high light intensities, inclusion of higher terms in the expansion (4.29) is necessary. This is a subject of nonlinear optics that is out of scope of this chapter.

4.3 Light Emission and Absorption by Atom

It is instructive to consider processes occurring in the formation of atomic spectra in terms of the Albert Einstein classification suggested in 1916. There are three basic processes that are responsible for a formation of atomic spectra, i.e. spontaneous emission, stimulated emission, and absorption. The probability of each particular process is associated with a relevant *Einstein coefficient.*

Spontaneous emission is the process in which an electron spontaneously (i.e. without any external excitation) transitions from an excited energy state to a lower energy state (e.g. its ground state) and emits a quantum of energy in the form of a photon. This process is described by the Einstein coefficient A_{21} (s^{-1}) which is a probability that an electron in state 2 with the energy E_2 will spontaneously change to the state 1 with the energy E_1, emitting a photon with an energy $\hbar\omega = E_2 - E_1$. The change of the atomic density of an atom in state 2 (i.e. a decrease of the population of state 2) per second due to spontaneous emission is given by:

$$\frac{dn_2}{dt} = -A_{21}n_2.$$ (4.30)

The same process that results in increase of the population of state 1 is given by:

$$\frac{dn_1}{dt} = A_{21}n_2.$$ (4.31)

Stimulated (or induced) emission is the process by which an electron is forced to jump from a higher energy level to a lower one as the result of an interaction with electromagnetic radiation that has a photon energy close to the energy separation between levels. This is described by the Einstein coefficient B_{21} $(J^{-1}m^3s^{-2})$ and can be considered as a *negative absorption*. The B_{21} coefficient represents a probability (per second and per unit spectral energy density of the radiation field) that an electron in state 2 with the energy E_2 will change to the state 1 with the energy E_1 emitting a photon with an energy $\hbar\omega = E_2 - E_1$. The change of the population of state 1 per second due to the induced emission is given by:

$$\frac{dn_1}{dt} = B_{21}n_2\rho(\omega),$$ (4.32)

where $\rho(\omega)$ denotes a spectral energy density of the isotropic radiation field with a frequency (ω) of the transition (Planck radiation law):

$$\rho(\omega) = \frac{1}{\pi^2}\hbar\left(\frac{\omega}{c}\right)^3 \frac{1}{e^{\hbar\omega/kT} - 1}.$$ (4.33)

Stimulated emission is a fundamental processes in physics that led to the creation of lasers[3].

Absorption is the process by which a photon is absorbed by atom, causing an electron to jump from a lower energy level to a higher one. This is described by the Einstein coefficient B_{12} $(J^{-1}m^3s^{-2})$ and

---

[3] Laser is an acronym of Light Amplification by the Stimulated Emission of Radiation.

can be regarded as a *positive absorption*, which gives the probability per unit time per unit spectral energy density of the radiation field that an electron in state 1 with energy will absorb a photon with an energy $\hbar\omega = E_2 - E_1$ and go to state 2 with energy E_2. The change in the population in state 1 per unit time due to absorption is given by:

$$\frac{dn_1}{dt} = -B_{12}n_1\rho(\omega). \tag{4.34}$$

In thermodynamic equilibrium one has a simple balance in which the net change of the number of any excited atoms is zero, i.e. it is being balanced the by loss and gain due to all processes. In equilibrium a detailed balance requires that the change in time of the number of atoms in level 1 due to the above three processes (i.e. spontaneous emission, negative, and positive absorption) is equal to zero:

$$A_{21}n_2 + B_{21}n_2\rho(\omega) - B_{12}n_1\rho(\omega) = 0. \tag{4.35}$$

The absorption coefficient defined by the equation (4.5) is related to the *absorption cross section* $(\sigma_a(\omega))$ by the following:

$$\alpha(\omega) = n_1\sigma_a(\omega) \tag{4.36}$$

At low light intensities the stimulated emission is negligible. In this case and at not large density of atoms the absorption cross section relates to the Einstein coefficient B_{12} by the following:

$$\sigma_a(\omega) = \frac{\hbar\omega}{c}B_{12}s(\omega) \tag{4.37}$$

where $s(\omega)$ stands for the normalized line profile (or line shape) function, i.e. $\int_{-\infty}^{\infty} s(\omega)d\omega = 1$. Replacing $s(\omega)$ by the delta function (see detailed explanation below) after integration over frequency results in the following:

$$\sigma_0 = \frac{\hbar}{c}\omega_{21}B_{12}, \tag{4.38}$$

where $\omega_{21} = E_2 - E_1$.

Optical Absorption and Fluorescence of Materials

Consider the response of a system initially in state $|a\rangle$ to a perturbation $H_{ab}e^{i\omega t}$ (see Eqs. (4.27), (4.28)). According to the time dependent quantum mechanics [Tannor (2007)] and within the first-order perturbation theory, the probability amplitude a_b for the state $|b\rangle$ is given by:

$$a_b(t) = -i \int_0^t H_{ba} e^{-i(\omega-\omega_{ba})t'} dt' = H_{ba} \left[\frac{e^{-i(\omega-\omega_{ba})t} - 1}{\omega - \omega_{ba}} \right] \quad (4.39)$$

The probability that an electron made a transition from $|a\rangle$ to $|b\rangle$ at time t is given by:

$$W_{a\to b} = |a_b(t)|^2 = |H_{ba}|^2 \left[\frac{\sin\left[(\omega - \omega_{ba})t/2\right]}{(\omega - \omega_{ba})/2} \right]^2 \quad (4.40)$$

Equation (4.40) in the limit of $\omega \to \omega_{ba}$ results in the following

$$W_{a\to b} \approx |H_{ba}|^2 t^2. \quad (4.41)$$

Note, that Eq. (4.39) is valid for a very short time, i.e. $t \ll H_{ba}^{-1}$ providing $W_{a\to b} \ll 1$. For such a short time an incident radiation will have a substantial spectral width ($\Delta\omega \sim 1/t$). Integration of Eq. (4.40) over entire spectrum results in the following:

$$\int_{-\infty}^{\infty} \left[\frac{\sin\left[(\omega - \omega_{ba})t/2\right]}{(\omega - \omega_{ba})/2} \right]^2 d\omega \to 2\pi t \delta(\omega - \omega_{ba}). \quad (4.42)$$

Consequently equation (4.40) is now given by

$$W_{a\to b} = |H_{ba}|^2 2\pi t \delta(\omega - \omega_{ba}) = |H_{ba}|^2 2\pi t \delta(E_b - E_a - \omega). \quad (4.43)$$

Since the probability of a transition is a function of the time, the transition rate can be given as:

$$\Gamma_{ab} = \frac{d}{dt} W_{a\to b} = |H_{ba}|^2 2\pi \delta(\omega - \omega_{ba}) = |H_{ba}|^2 2\pi \delta(E_b - E_a - \omega). \quad (4.44)$$

In molecular systems and solids the final state is composed of many states close to each other on energy scale creating a continuum (see Sections 3.2 and 5.3). In this case the density of states ($\rho(E)$) can be given by:

$$dN = \rho(E)dE, \tag{4.45}$$

where dN is the number of states within the energy range dE. The transition rate is now given by:

$$\Gamma_{ab} = 2\pi|H_{ba}|^2\rho(E_b - E_a - \omega). \tag{4.46}$$

4.4 Einstein Coefficients for Condensed Matter

A fundamental difference between spectroscopy of molecules in gas and condensed matter is interparticle interaction affecting optical spectra. It not only changes physical-chemical properties of substance but it also affects radiation electric fields (see Section 1.4). These conclusions follow from fundamental relation between Einstein coefficients. Let us focus on this problem in detail.

Main goal of optical spectroscopy is to measure characteristic optical function of a sample (such as for example absorption coefficient $\alpha(\omega)$) and to find its relation to material parameters of molecules or solids (i.e. Einstein coefficient $A(\nu), B(\nu)$).

Let us calculate absorption of optical wave in a sample having a shape of a flat plate with the complex refractive index $\hat{n} = n - i\kappa$. Absorption of optical power in the plate of thickness dx is given by

$$dW(\nu) = \mathcal{K}(\nu)S(\nu)\mathcal{Q}dx, \tag{4.47}$$

where $\mathcal{Q}$ is the beam cross section. Using expression for the *Poynting vector* $S = \mathcal{E}^2 \cdot nc/8\pi$ we have

$$dW(\nu) = \mathcal{K}(\nu)\frac{nc}{8\pi}\mathcal{E}^2\mathcal{Q}dx. \tag{4.48}$$

On the other hand equation (4.48) can be written as follows

$$dW(\nu) = Nh\nu B(\nu)\rho_{in}(\nu)\mathcal{Q}dx, \tag{4.49}$$

where $\rho_{in}(\nu)$ is the optical wave density, and N is the molecular concentration. Comparing (4.49) to (4.48) one can get

$$B(\nu) = \frac{K(\nu)c}{Nh\nu}\theta(\nu),$$ (4.50)

$$\theta(\nu) = n\frac{\mathcal{E}^2}{\mathcal{E}_{in}^2}.$$ (4.51)

Function of $\theta(\nu)$ represents a difference between average electric field $\mathcal{E}$ and local internal field $\mathcal{E}_{in}$. For a rare gas $\mathcal{E} = \mathcal{E}_{in}, n = 1$ thus $\theta(\nu) = 1$. The above formulae are now used to calculate characteristic spectroscopic parameters: the *oscillator strength, f,* and the *relaxation time, τ.*

Oscillator strength relates to the spectral Einstein coefficient $B(\nu)$ according to the following:

$$f = \frac{3mh}{\pi e^2}\int \nu B(\nu)d\nu = \frac{3mh\tilde{\nu}}{\pi e^2}B.$$ (4.52)

For a condensed matter the equation (4.52) is given by

$$f = \frac{3mc}{\pi e^2}\int \sigma(\nu)\theta(\nu)d\nu,$$ (4.53)

where $\sigma(\nu)$ is spectral absorption cross section. In order to obtain f one has to calculate both $\sigma(\nu)$ and $\theta(\nu)$. The molecular integral absorption cross section σ is given by:

$$\sigma = \frac{1}{N}\int K(\nu)\theta(\nu)d\nu = \int \sigma(\nu)\theta(\nu)d\nu,$$ (4.54)

Finally we obtain the relaxation time τ. The ideal and real diagrams representing electron transitions that define relaxation are shown in Fig. 4.1(a) and Fig. 4.1(b), respectively.

The relaxation time is defined by the following equation:

$$\frac{1}{\tau} = A_{eg} = \int A(\nu)d\nu,$$ (4.55)

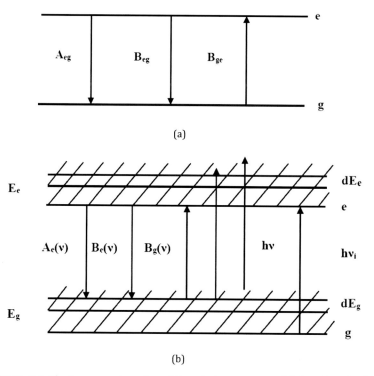

Figure 4.1 Electron quantum transitions between ground (g) and excited states (e) depicted for both ideal (a) and real cases (b).

For molecules in equilibrium states that are denoted as g and e the equation (1.55) can be written as

$$\frac{1}{\tau} = \frac{8\pi}{c^2} \cdot \frac{C_e(T)}{C_g(T)} \int n^2 \; \nu^2 \sigma(\nu) e^{h(\nu_i - \nu)/KT} d\nu. \tag{4.56}$$

Spectral factor $\exp(h(\nu_i - \nu)/KT)$ represents a difference between absorption and luminescence spectra. For the gas phase (i.e. $n = 1, \theta = 1$) and by assuming that spectral lines are narrow, the equation (4.56) can be given by

$$\frac{1}{\tau} = \frac{8\pi}{c^2} \cdot \frac{C_e(T)}{C_g(T)} \nu_i^2 \int \sigma(\nu) d\nu. \tag{4.57}$$

Using expression for f (4.53) we arrive in the following equities: $\theta(\nu) = 1, C_g(T) = C_e(T)$. Consequently equation (4.57) results in the following

$$\frac{1}{\tau} = \frac{8\pi^2}{3} \cdot \frac{e^2 \nu_i^2}{mc^3} f. \tag{4.58}$$

Equation (4.58) represents a relationship between τ of the excited state and the oscillator strength f.

4.5 Principles of the Signal Processing in Optics

This section covers the fundamental aspects of optical signal processing and analysis at an introductory level. It discusses also applied topics that helps students and professionals to bridge the gap between the current scientific and technical literature. The self-contained text reviews the essentials of signal processing, optics, and imaging as applied to the optical spectroscopy of materials.

The section reviews general principles of experimental technique frequently used to study optical absorption and luminescence. Most optical spectra are measured by an equipment that is setup following optical spectrometry principles considered below.

4.5.1 Basic Principles of Operation in Optical Spectrometry

Typical setup of an optical spectrometer is shown in Fig. 4.2. A time-dependent function of $f(t)$ in Fig. 4.2 represents an input signal, frequency dependent function of $S(\nu)$ denotes spectrum of the output signal, transformation of the $f(t)$ into the $S(\nu)$ by a spectrometer

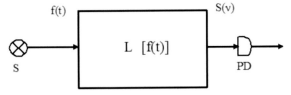

Figure 4.2 Typical spectroscopic setup, L stands for a spectrometer, S and PD denote light source and a photodetector, respectively.

is described by an operator $\hat{L}[f(t)]$, *PD* denotes a photodetector transforming an optical response into an electrical signal.

In most cases operator $\hat{L}[f(t)]$ is linear. A typical photodetector measures not an amplitude of incoming optical radiation but it's intensity given by $S(\nu) \cdot S^*(\nu) = |S(\nu)|^2$. The intensity spectrum is given by

$$G(\nu) = \lim_{T \to \infty} \frac{1}{2T} \mid \int_{-T}^{T} f(t)e^{-i2\pi\nu t}dt \mid^2 . \qquad (4.59)$$

A function reverse to $G(\nu)$ is called a *correlation* function, $\Gamma(\tau)$ is defined as:

$$\Gamma(\tau) = \lim_{T \to \infty} \frac{1}{2T} \int_{-T}^{T} f(t + \tau)f^*(t)dt. \qquad (4.60)$$

The $G(\nu)$ and $\Gamma(\tau)$ functions are related to each other through the Fourier transformations, i.e.

$$G(\nu) = \int \Gamma(\tau)e^{-i2\pi\nu\tau}d\tau, \qquad (4.61)$$

$$\Gamma(\tau) = \int G(\nu)e^{i2\pi\nu\tau}d\nu. \qquad (4.62)$$

Expressions (4.61) and (4.62) determine two different setups in spectrometry depicted in Fig. 4.3: direct measurements of optical spectra (Fig. 4.3(a)) and by a Fourier transform spectrometer (Fig. 4.3(b)), respectively. Schematic diagram shown in Fig. 4.3(a) represents direct measurements of the spectrum $G(\nu)$. The diagram shown in Fig. 4.3(b) represents measurements of the correlation function $\Gamma(\tau)$ followed by a numerical processing transforming the $\Gamma(\tau)$ into the spectrum $G(\nu)$.

4.5.2 Optical Response Function: Inverse Problem in Optics

The direct (or "normal") problem in optical physics is to predict the emission or propagation of radiation on the basis of a known

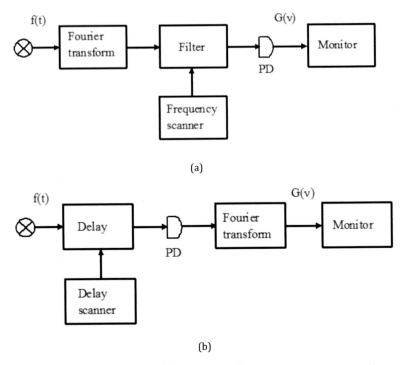

Figure 4.3 Two basic structural diagrams used in spectrometry representing direct measurements of the optical spectra (including prism, diffraction grid, or Fabry–Perot spectrometers) (a), and a schematics of a typical Fourier transform spectrometer (b).

constitution of sources. The *inverse* or indirect problem is to deduce features of sources from the detection of radiation.

An approach to the inverse optical problems can be attempted as follows: One can describe light sources by a set of space-time functions, the source functions, which are mapped by a set of (not necessarily linear) operators into a set of functions, which are the measured data.

In general case effect of the spectral instrument is described by the following integral equation

$$\int_a^b \kappa(s,t)y(t)dt = f(s), \ c \leq s \leq d, \qquad (4.63)$$

where $\kappa(s,t)$ is an instrument response function, $f(s)$ is an experimental signal.

The unknown function of $y(t)$ has to be obtained by solving the *Fredholm* integral equation of the 1st kind representing by Eq. (4.63). This is the so-called *inverse problem* in Mathematical Physics. French mathematician *Hadamard* provided classic definition of the correctness of a mathematical problem, that could be formulated in the following way:

1. the solution of the problem exists;
2. the solution of the problem is unique;
3. the solution of the problem is stable with respect to small variations (fluctuations) of the measurements $f(s)$.

One can expect that the first two conditions are satisfied for different types of the instrument function $(\kappa(s,t))$. However, the 3rd condition, as it was shown mathematically, can not be satisfied for the the integral Eq. (4.63). In this case, instead of the $y(t)$, another function is introduced:

$$z(t) = y(t) + A \cos \omega t. \tag{4.64}$$

Equation (4.63) can now be written as:

$$A \int_a^b \kappa(s,t) \cos \omega t \, dt = g(s) - f(s), \tag{4.65}$$

where function of $g(s)$ is given by

$$g(s) = \int_a^b \kappa(s,t) z(t) \, dt. \tag{4.66}$$

Value of the integral in (4.65) tends to zero at $\omega \to \infty$. In the similar way the difference between $g(s)$ and $f(s)$ vanish, (i.e. $|g - f| \to 0$) for sufficiently large ω. That means that functions of $z(t)$ and $y(t)$ differ from each other by A for any small value of $|g - f|$. Consequently solution of equation (4.63) is not stable. This situation is depicted in Fig. 4.4 where a solution of equation (4.63) includes non-physical oscillating component superimposed on the physically justified part of the homogeneously increased output. One can state that formulation of the inverse problem given by the equation (4.63)

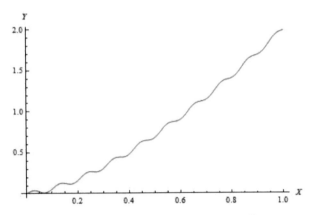

Figure 4.4 Solutions of equation (4.63) showing oscillating component superimposed on the physically justified part of the solution.

is not mathematically correct, and the oscillations shown in Fig. 4.4 is a consequence of the incorrectly formulated problem given by equation (4.63).

Consider now basic methods for correct solving of the inverse problem in spectroscopy. Asume that the optical source has a complex spectral composition. Disperse element, e.g. diffraction grade, evaluate the spectrum over the x-axis. The photodetector transforms intensity of the incoming optical radiation into electric signal, $f(x)\Delta x$. This process is described by equation (4.63), where $s = x$, $t = \nu$, $f(s)$ describe electric signal, and $y(t)$ is the required spectral distribution of the intensity.

As long as the value of Δx is not large, the electric signal contains errors. As a result there are no reliable methods to get information about the $y(t)$ function because the function of $f(s)$ contains error. In this case several effective mathematical algorithms for solving of non-correct problems have been developed. Below we present the so called *Tikhonov* method of regularization of solution.

4.5.3 Tikhonov Method of Solution Regularization

Tikhonov method of solving the Fredholm integral equation of the 1st kind, i.e. the equation (4.63), is based on an idea of suppression of the high frequency oscillations. He proposed to use the regularization operator Ω_0 given by

$$\Omega_0[y] = \int_a^b \rho_0(t)y^2(t)dt, \qquad (4.67)$$

where $\rho_0(t) > 0$ is the weight function. Instead of (4.63) the problem now is to find a minimum of the smoothing functional

$$\Phi[y] = \int_c^d \left[\int_a^b \kappa(s,t)y(t)dt - f(s) \right]^2 ds + \alpha\Omega_0[y], \qquad (4.68)$$

where α is the regularization parameter. The Euler equation that determines a minimum of $\Phi[y]$ is given by

$$\int_c^d ds\kappa(s,t) \int_a^b \kappa(s,t')y(t')dt' + \alpha\rho(t)y(t) = \int_c^d dsf(s)\kappa(s,t) \quad (4.69)$$

This is the Fredholm equation of the 2nd kind that could be written as following

$$\alpha y(n) + \int_a^b \widetilde{\kappa}(n,t)y(t)dt = \widetilde{f}(n), \qquad (4.70)$$

where functions of $\widetilde{\kappa}(n,t)$ and $\widetilde{f}(n)$ are given by

$$\widetilde{\kappa}(n,t) = \frac{1}{\rho(n)} \int_c^d ds\kappa(s,n)\kappa(s,t), \qquad (4.71)$$

$$\widetilde{f}(n) = \frac{1}{\rho(n)} \int_c^d dsf(s)\kappa(s,n). \qquad (4.72)$$

Solution of equation (4.70) is stable in contrast to the one given by Eq. (4.63). The main goal now is to determine the control parameter α. From equation (4.68) follows the expression for the error function δf that results in the following

$$\alpha = \frac{\int_c^d [\delta f(s)]^2 ds}{\int_a^b \rho(t)y^2(t)dt}. \qquad (4.73)$$

Optical Absorption and Fluorescence of Materials

The Tikhonov method explores the self-consisting procedure: one takes a probing value of α and calculates then the initial solution of y from equation (4.70). After that the new value of α is calculated by the use of equation (4.73). This procedure is repeated until convergence of α is achieved, i.e. $\alpha \simeq 1$.

4.5.4 Fourier Transform Spectrometer

Consider now the *Fourier transform spectrometer* schematically depicted in Fig. 4.3(b). Two stage measurements is the characteristic property of this type of spectrometer. As the first step the autocorrelation function of optical signal is measured. As the second step the Fourier transformation of autocorrelation function spectrum is obtained. Most Fourier transform spectrometers are equipped with Michelson interferometer. Optical signal on the output of the interferometer is given by

$$\psi(t) = \frac{1}{2}\left[f(t) + f\left(t - \frac{2\Delta}{c}\right)\right]$$ (4.74)

Intensity of the signal is given by

$$<I> = \lim_{T \to \infty} \frac{1}{2T} \int_{-T}^{T} \psi(t)\psi^*(t)dt$$ (4.75)

Combining two last equations results in the following

$$\begin{aligned} I \quad = \frac{1}{4}[\quad & f(t)f^*(t) \quad + \quad f(t-\tau)f^*(t-\tau) \\ + \quad & f(t)f^*(t-\tau) \quad + \quad f^*(t)f(t-\tau) \quad] \end{aligned}$$ (4.76)

That could be rewritten as

$$I(\tau) \quad = \frac{1}{2} \quad f(t)f^*(t) \quad + \frac{1}{2}\text{Re}(\quad f(t)f^*(t-\tau) \quad)$$ (4.77)

The second part of equation (4.77) is proportional to the real part of the autocorrelation function. The output signal with narrow frequency bandwidth $G(k)$ is given by

$$dI(k, \Delta) = \frac{1}{2}G(k)dk\,(1 + \cos 2\pi k\Delta) \tag{4.78}$$

Summation over wave number values k results in the following

$$I(\Delta) = \frac{1}{2}\int_0^\infty G(k)(1 + \cos 2\pi k\Delta)dk \tag{4.79}$$

From (4.79) follows

$$I(0) = \int_0^\infty G(k)dk \tag{4.80}$$

and finally

$$I(\Delta) - I(0) = \frac{1}{2}\int_0^\infty G(k)\cos 2\pi k\Delta dk \tag{4.81}$$

Spectrum $G(k)$ can be obtained by applying reverse Fourier transformation of equation (4.81). As an example consider interferogram of a monochromatic optical signal. Equation (4.81) can be written in the following form

$$I(\Delta) - I(0) = \frac{1}{2}\int_0^\infty B_0\delta(k - k_0)\cos 2\pi k\Delta dk = \frac{1}{2}B_0\cos 2\pi k_0\Delta \tag{4.82}$$

Functions of $B_0\delta(k - k_0)$ and $\frac{1}{2}B_0\cos(2\pi k_0\Delta)$ are plotted in Fig. 4.5. As a conclusive step to the material of this chapter consider the spectrum resolution and signal to noise ratio (SNR). Spectral resolution of any spectrometer depends on interferometric order and delay maximum of interferometric beams. Consider an interferogram

$$F(\Delta) = \begin{cases} F(\Delta), & |\Delta| \le \Delta_0 \\ 0, & |\Delta| > \Delta_0 \end{cases} \tag{4.83}$$

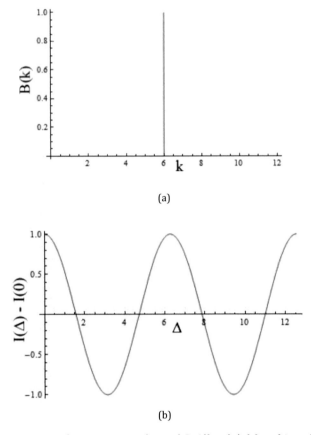

Figure 4.5 Monochromatic optical signal $B_0\delta(k-k_0)$ (a) and interferogram $\frac{1}{2}B_0\cos(2\pi k_0\Delta)$ (b), see equation (4.82).

Substitution of (4.83) info in (4.81) results in the following equation

$$G(k) = 2\int_{-\Delta_0}^{\Delta_0} (I(\Delta) - I(0))\cos 2\pi k\Delta\, d\Delta \tag{4.84}$$

For monochromatic signal instead of $B_0\Delta(k-k_0)$ we have the following function: $B_0\sin c\,(2\pi\Delta_0(k_0-k))$. The actual function range spends from $k_0 - 1/(2\Delta_0)$ to $k_0 + 1/(2\Delta_0)$. Therefore the spectral resolution $\Delta\nu$ is given by

$$\Delta\nu = \frac{c}{\Delta_0} \tag{4.85}$$

From the last equation follows the rule: the greater maximum delay Δ_0, the better spectral resolution $\Delta\nu$. The best known Fourier transform spectrometer has maximum delay of $2m$. Thus $\Delta\nu \simeq 5 \cdot 10^{-3} cm^{-1}$. Fourier transform spectrometer is multichannel instrument because it detects wide spectral interval at a time. Assume we have M channels with $\delta\nu$ spectral intervals. One-channel spectrometer detects spectral signal within the time t/M. If noise is not a quantum one, then the precision of a measurement is inversely proportional to the factor of $\sqrt{M}$.

4.5.5 Non-harmonic Analysis in Optical Spectroscopy

General theoretical approach traditionally used in modern optical spectroscopy can be characterized as an application of the slowly varying *envelope approximation* for electromagnetic waves. On the frequency scale this approach means a description of any optical signal in the form of a superposition of *narrow-band* signals. However, this approach is not good for the laser optics with femto-second pulses i.e. with a pulse duration of ($\tau \geq 10^{-15} s$).

An alternative approach in this field is a refuse on the slowly varying envelope approximation and description of ultra-short optical pulses in the form of the ortho-normalized series of new basis electromagnetic signals [Ovechko (2012)]. With other word the new method represents optical pulse as an ortho-normalized series of new *Elementary Wave Packets* (EWP). This method reliably works without any restriction on the pulse duration and at the same time incorporates all general properties of electromagnetic signals. For example, EWP series was used in the theory of *fs*-pulse propagation in a medium with dispersion [Ovechko and Myhashko (2018)]. Below we give a brief introduction to the EWP method as an alternative approach of non-harmonic analysis in optical spectroscopy.

Motion of charged particles results in a generation of electromagnetic waves with a spectrum without a zero-frequency

Optical Absorption and Fluorescence of Materials

component that can be expressed in the following way:

$$I = E_0 \int_0^\tau f(t)dt = 0, \qquad (4.86)$$

where E_0 is an electric field amplitude, τ is pulse duration, $f(t)$ represents shape of the pulse. In literature the value of I is frequently called the *unusual parameter*, see e.g. [Bessonov (1992)] and references therein. The last term means that the known objects that generate electromagnetic waves (e.g. such as atoms, molecules, electronic bunches in resonators, etc.) radiate *usual* electromagnetic pulses, for which $I = 0$. Referring to Eq. (4.86) it is important to note a natural restriction on electromagnetic wave duration that is given by

$$f(t<0) = f(t>\tau) = 0, \; f(0 \leq t \leq \tau) = f(t). \qquad (4.87)$$

Now, assuming an additional physical condition of the absence of any discontinuances in diffraction field amplitude we can write boundary conditions for the electromagnetic wave amplitude change rate at the beginning $t = 0$ and at the end $t = \tau$ of electromagnetic pulse as following

$$\frac{\partial f}{\partial t}\Big|_{t=0} = \frac{\partial f}{\partial t}\Big|_{t=\tau} = 0. \qquad (4.88)$$

At this point it is important to note that the basic harmonic functions of the Fourier transform do not satisfy conditions (4.96) to (4.88). This leads to an appearance of the multiple restrictions and even paradoxes in spectroscopic analysis.

Followed the EWP-method proposed by [Ovechko (2012)] any electromagnetic signal $f(t)$ can be expended into the following series

$$f(x) = \sum_{n=0}^\infty [A_{2n+1}U_{2n+1}(x) + B_{2n+2}U_{2n+2}(x)], \qquad (4.89)$$

where odd and even EWP-functions are given respectively by

$$U_{2n+1}(x) = \left[\frac{2(n+1)}{\pi(n+1)}\right]^{1/2}$$
$$\times \left[\frac{\sin(2n+2)]}{2(n+1)\sin[x]} - \cos[x(2n+3)]\right], \qquad (4.90)$$

$$U_{2n+2}(x) = \left[\frac{2(n+1)}{\pi(n+2)}\right]^{1/2}$$
$$\times \left[\frac{1}{2(n+1)}\left(\frac{\sin(2n+3)]}{\sin[x]} - 1\right) - \cos[x(2n+4)]\right]$$
$$(4.91)$$

where $x = \pi t/\tau, 0 \leq x \leq \pi (0 \leq t \leq \tau)$. As for orthonormalized series coefficients A_{2n+1} B_{2n+2} are evaluated by means of the following scalar products

$$A_{2n+1} = \int_0^\pi f(x)U_{2n+1}(x)dx, \qquad (4.92)$$

$$B_{2n+2} = \int_0^\pi f(x)U_{2n+2}(x)dx. \qquad (4.93)$$

EWP-functions (4.90), (4.91) are ortho-normalized ones, i.e.

$$\int_0^\pi |U_{2n+1}(x)|^2 \, dx - \int_0^\pi |U_{2n+2}(x)|^2 \, dx = 1. \qquad (4.94)$$

In Table 4.1 and Figs. 4.6(a) and 4.6(b) several odd and even EWP-functions of the lowest orders are given.

Next we present examples of the modeling and analysis of electromagnetic signals in optics and radio-physics. In most cases two types of the signals are used: the *amplitude modulated* (AM) and *phase modulated* (PM) signals. Consider first the AM-signal. In this case the EWP-analysis is performed according to the following steps:

1. analyze a shape of the signal with respect to the conformity to the conditions (4.86) to (4.88);

Optical Absorption and Fluorescence of Materials

Table 4.1 Functions of $U(x)$ describing elementary wave packets of the lowest order

	$U(x)$-*function*
$U_1(x)$	$\pi^{-1/2}(\cos x - \cos 3x)$
$U_2(x)$	$\pi^{-1/2}(\cos 2x - \cos 4x)$
$U_3(x)$	$(3\pi)^{-1/2}(\cos x + \cos 3x - 2\cos 5x)$
$U_4(x)$	$(3\pi)^{-1/2}(\cos 2x + \cos 4x - 2\cos 6x)$
$U_5(x)$	$(6\pi)^{-1/2}(\cos x + \cos 3x + \cos 5x - 3\cos 7x)$
$U_6(x)$	$(6\pi)^{-1/2}(\cos 2x + \cos 4x + \cos 6x - 3\cos 8x)$

2. calculate amplitudes A_{2n+1}, B_{2n+2} of the EWP-spectrum;

3. verify decomposition completeness of the signal with the help of Parseval's theorem;

4. reproduce resulting signal estimated using the basis of the EWP-spectrum.

First we consider signals which reasonably conform to the femto-second pulses, i.e.

$$f(x) = \frac{x}{\pi}\left(1 - \frac{x}{\pi}\right)\sin(mx), \tag{4.95}$$

where $0 \leq x \leq \pi$, $m = 2, 3...$ For this example we calculate the EWP-spectrum of a signal (4.95), for the following value of m-parameter: $m = 6$. It is easy to verify that the signal (4.95) satisfies conditions (4.86) to (4.88). We substitute (4.95) into equations (4.91) to (4.93) and find the odd part of the EWP-spectrum A_{2n+1} that is depicted in Fig. 4.7(b).

Next we apply the Parseval's theorem in order to proof the completeness of decomposition

$$\int_0^\pi |f(x)|^2\, dx = \sum_{n=0}^\infty \left(A_{2n+1}^2 + B_{2n+2}^2\right). \tag{4.96}$$

In this example the model signal of FOP (see Fig. 4.7(a)) is represented by the simple EWP-spectrum that contains 2 to 4 EWP-

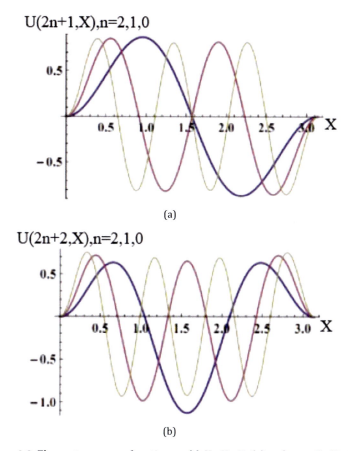

Figure 4.6 Elementary waves functions: odd, U_1, U_2, U_5 (a) and even U_2, U_4, U_6 (b), labeled in panels by index $n=$ 2, 1, 0 from left to right, respectively.

components (as shown in Fig. 4.7(b)). Considering the FM-signal it should be noted the following:

1. the EWP-spectrum of the FM-signal is more complex if compared to the spectrum of the AM-signal;
2. matching to the conditions (4.86) to (4.88) is more difficult to achieve without applying additional restrictions.

In summary, for practical applications the EWP-spectrometer should include the following components:

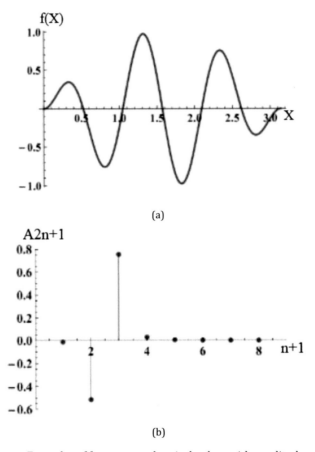

Figure 4.7 Examples of femto-second optical pulses with amplitude modulation $E(t) = E_0[1 - t/\tau)/\tau] \sin[6\pi t/\tau]$ (a), and EWP-spectrum A_{2n+1} (b).

1. a generator of the basis functions (4.90), (4.91);
2. a device responsible for a multiplication of the studied function $f(x)$ and basis functions;
3. an integrator;
4. a registration block.

However, practical applications of the electromagnetic signal analysis based on the method described in this section still require solutions of other problems ranging from microwave to an optical spectral regions [Ovechko (2012)]. One of the promising spectral method is a time-domain spectroscopy i.e. a spectral method of using ultra-wide electromagnetic band proposed by [Ovechko (2020)].

4.6 Photoluminescence

A phenomenon which involves absorption of light energy and subsequent emission of secondary light is classified as *luminescence*. Excitation of a material by absorbance of a light photon leads to a luminescent species which fluoresce or phosphorescence. In general, fluorescence is *fast* (*ns* time scale) while phosphorescence is *slow* (much longer time scale, up to hours or even days). For many technical applications, it is irrelevant whether the luminescence is fluorescence or phosphorescence. When absorption of light leads to emission, one speaks about optical excitation of luminescence or *photoluminescence* (PL). Other types of light emission are classified by the excitation. Excitation through accelerated electrons is called *cathodoluminescence* (CL). *Electroluminescence* (EL) is excited by an external electric voltage. *Chemiluminescence* is an emission of light (luminescence), as the result of a chemical reaction, and it is excited by the energy of the reaction. In this chapter only the photoluminescence (or fluorescence)[4] is considered.

The absorption of light energy, which is used to excite the luminescence in solids, takes place by either the host lattice or by intentionally doped impurities. In most cases, the light emission takes place on the impurity ions, which, when they also generate the desired emission, are called *activator* ions. In the cases when the activator ions cause a weak light absorption, a second kind of impurities can be added: *sensitizers*. The sensitizers absorb the light energy and subsequently transfer it to the activators. The energy transfer is a very important component of luminescence mechanism. The emission light color can be adjusted by choosing the proper impurity ion, without changing the host lattice in which the impurity ions are incorporated. However, only a few activator ions show emission spectra with emission at spectral positions which are

---

[4] Note that term *fluorescence* is most commonly used to refer to photoluminescence from molecular systems.

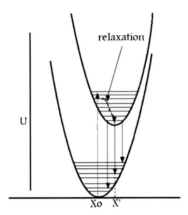

Figure 4.8 Configurational coordinate diagram of a luminescent center.

hardly influenced by their chemical environment [Dresselhaus *et al.* (2007)].

The light emission process is a release of energy in the form of photons. Initial state of a material can be represented schematically by a potential energy functional that is a function depending on coordinates of many particles surrounding the luminescent center, $U = U(x_1, x_2, x_3...)$. The U-function for a molecule (as an example) is shown in Fig. 4.8. For simplicity the initial coordinate configuration of the particles in the molecule (nucleus and electrons) is denoted X_0. The electron on a luminescent center absorbs energy and is raised to an excited state. This process is accompanied by a changes in potential and kinetic energies of particles that will result in changes of the coordinates of the particles involved in the process. On the simplified picture the new coordinates are indicated X' as depicted in Fig. 4.8.

In Fig. 4.8 a possible radiative mechanism is shown. It demonstrates that the excitation occurs at higher photon energy than the emission. The frequency downshifting mechanism of the emitted light comparing to the exciting light could be explained using the *configurational coordinate diagram* depicted in Fig. 4.8. The ion in the ground or unexcited state occupies some atomic configuration (marked as X_0 in Fig. 4.8) corresponding to the potential energy minimum. The *Franck-Condon principle* states that the atoms in solid do not change their internuclear separations during an electronic transition. The optical absorption occurs by electron transition

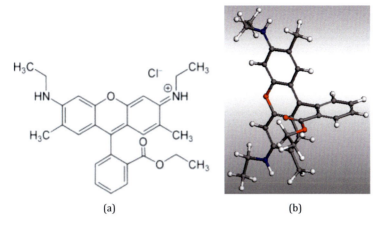

Figure 4.9 Chemical structure diagram (a) and stick-and-ball computer generator model (b) of the Rhodamine 6G dye molecule. (Adapted from [Gavrilenko and Noginov (2006)]).

from the ground energy state to an excited state within the same atomic configuration indicated by a vertical upward arrow located at X_0. When electron is excited, the ion finds a lower energy state corresponding to a lower symmetry position. After absorbing the incoming light photon the electron transfers to the higher energy excitation state within the luminescent center that mover the entire system into a nonequilibrium state. Potential energy of the excited state differs from that in the ground state and is characterized by an another energy minimum with different equilibrium atomic coordinates and interatomic distances (shown by X' in Fig. 4.8). Phonon-electron interactions then will bring the electron to a new equilibrium position: the excited electron returns to the lower energy level within the exited energy state nonradiatively through generation of phonons (see dashed line in Fig. 4.8).

The dye molecule of Rhodamine 6G (R6G) is a good example demonstrating this phenomenon. Chemical structure and computer generated stick-and-ball model of the R6G molecule are depicted in Fig. 4.9.

The R6G is a highly fluorescent rhodamine family dye. R6G as well as other Rhodamine family dyes are used extensively in biotechnology applications such as fluorescence microscopy, flow cytometry, and fluorescence correlation spectroscopy. The R6G is

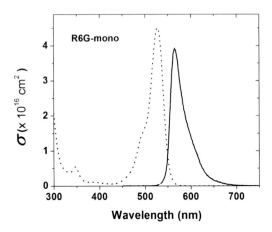

Figure 4.10 Measured absorption (dotted line) and emission (solid line) cross-section spectra of R6G dye solution in methanol (Adapted from [Gavrilenko and Noginov (2006)]).

also used as a laser dye, or gain medium, in dye lasers. It has a remarkably high photostability, high luminescence quantum yield (near 0.95), low cost, and its lasing range has close proximity to its absorption maximum. The electron energy structure of the R6G molecule has electron levels responsible for the singlet-type optical transitions well separated on the energy scale from the backbone electronic structure. This electron energy structure of the R6G molecule results in a typical optical absorption and strong luminescence lines that are well separated from the optical spectra related to the excitation of the backbone electronic orbitals [Gavrilenko and Noginov (2006)].

The optical excitation and relaxation are followed by a light emission shown by vertical arrows downwards in Fig. 4.8. According to the energy and momentum conservation principles there are several possibilities of returning to the ground state that generally may result in several luminescence lines in the optical radiation spectrum. The difference in the photon energy positions between the optical absorption and emission spectra is illustrated in Fig. 4.10.

Optical absorption caused by the singlet level excitation is located at $\lambda = 530$ nm. Energy relaxation of the optically excited electrons occurs withing the excitation state characterized by different atomic coordinates and can be shown on the potential energy diagram that is shifted in the configuration space (shown by X' in Fig. 4.8). Consequently after the relaxation to the excited state energy minimum the luminescence occurs at lower photon energy. This process is responsible for the measured R6G luminescence line near 566 nm wavelength shown in Fig. 4.10. Note, that as a result of the specific electron energy structure of R6G the non-radiative relaxation has a very low probability that results in the very high luminescence efficiency and makes the R6G dye as a very prospective material for numerous applications in applied optics and engineering.

The configurational coordinate diagrams such as the one depicted in Fig. 4.8 are widely used in the analysis of luminescence mechanisms [Dresselhaus *et al.* (2007)]. The energy absorbed by the luminescent materials which is not emitted as radiation is dissipated to the crystal lattice. For a development of highly efficient luminescent materials with numerous applications in modern optoelectronics it is crucial to suppress those radiationless processes which compete with the radiation process. Nanostructured materials provide exciting opportunities for an engineering of new luminescent materials that is reviewed in the next sections.

4.7 Luminescence Kinetics

Intensity of luminescence increases proportionally with the concentration of exited molecules. Thus anyone can measure concentration dependence in time by method of luminescence kinetics. In this chapter we present methods of measurement of photochemical and photophysical rate constants.

4.7.1 Rate Measurement of Photonic Processes in Materials

As the result of the excitation by light ($h\nu$) the molecules transit from the ground state (at concentration M) to higher energy level with a concentration (M^*)

$$M(+h\nu) \to M^* \tag{4.97}$$

Optical Absorption and Fluorescence of Materials

Rate of this process (ω) is given by

$$\omega_{ex} = b \cdot I(t) \tag{4.98}$$

where $I(t)$ is light intensity and b is a constant.

Transition of molecules back to a lower energy level is described by the following

$$M^* \rightarrow M(+h\nu'), \quad \omega_r = (k_d + k_l)[M^*], \tag{4.99}$$

where $[M^*]$ is concentration of exited molecules, k_l and k_d are rates of radiative and nonradiative relaxation, respectively. The kinetics equation for concentration changing in time can now be given by

$$\frac{d[M^*]}{dt} + (k_d + k_l)[M^*] = b \cdot I(t), \quad [M^*(0)] = 0 \tag{4.100}$$

Solution of equation (4.100) is given by

$$[M^*(t)] = \exp\left[-(k_d + k_l)t\right] \int bI(\tau) \exp\left[(k_d + k_l)\tau\right] d\tau \tag{4.101}$$

If the rate of intensity changes is much less then the value of $(k_d + k_l)$, then the equation (4.101) is given by

$$[M^*(t)] = bI_0 \exp(-\frac{t}{\tau_0}) \tag{4.102}$$

where $\tau_0 = (k_d + k_l)^{-1}$ is the lifetime of the exited state.

4.7.2 Stern–Folmer Equation

If exited molecules interact with other type of molecules then the concentration of exited molecules reduces that is described by the two processes: (1) chemical reaction, and (2) molecules relaxation. These processes are schematically represented by the following:

$$(1) \quad M^* + X \rightarrow P + Y, \quad \omega = k_{ch}[M^*][X], \tag{4.103}$$

$$(2) \quad M^* + Q \rightarrow M + Q, \quad \omega = k_r[M^*][Q], \tag{4.104}$$

where k_{ch} and k_r are chemical reaction and relaxation rates, respectively. Stationary concentration of exited molecules $[M^*]$ is defined as

$$[M^*] = \omega_0/(k_l + k_d + k_{ch}[X] + k_r[Q]),$$ (4.105)

where ω_0 is the excitation rate. If $[Q] \gg [M^*]$ the equation for luminescence relaxation can be written as

$$I_l(t) = I_0 \exp(-t/\tau),$$ (4.106)

where the life time is given by $\tau = 1/(k_l + k_d + k_{ch}[X] + k_r[Q])$, and $k_{ch} + k_r$ is the luminescence quenching rate.

4.7.3 Chemical Reactions Limited by Diffusion

Luminescence kinetics is affected by a diffusion because of comparable relaxation times of the luminescence and the diffusion. Diffusion process is defined by the 2nd *Fick's* law

$$\frac{\partial C}{\partial t} = D\nabla^2 C,$$ (4.107)

where D is a diffusion coefficient, and C stands for a concentration. Boundary conditions for the differential equation (4.107) are given by

$$C(r,0) = C_0, \quad C(\infty, t) = C_0, \quad C(R,t) = \frac{D}{k_{ch}[X]}\left(\frac{\partial^2 C}{\partial r^2}\right)_R,$$ (4.108)

where R is a radius of collision. For the case of $\sqrt{Dt} \gg R$ from equation (4.107) follows the expression for the rate of reaction $(k(t))$, i.e.:

$$k(t) = 4\pi R' D[Q](1 + R'/\sqrt{\pi Dt}),$$ (4.109)

where $R' = R/(1 + 4\pi RD/k_{ch})$.

Now equation (4.106) for the luminescence signal $I_l(t)$ can be written in the following form:

$$I_l(t) = I_0 \exp\{-\frac{t}{\tau_0} - 4\pi R'D[Q]t(1 + R'/\sqrt{\pi Dt})\}, \qquad (4.110)$$

where τ_o is defined by the equation (4.102).

Accordingly to equation (4.110) for a short time interval a faster relaxation process can be observed. In this case the relaxation rate approaches the stationary value τ_0.

4.7.4 Energy Transfer in Luminescence

Molecular excitation energy can be transferred between molecules that are closely located in space. The following energy transfer mechanisms are known:

- reabsorption—repeat absorption of luminescent light;
- nonradiative stimulated energy transfer (dipole–dipole interaction);
- nonradiative exchanging energy transfer (donor–acceptor interaction).

Energy transfer affects luminescence kinetics and distorts its time dependence.

Reabsorption. In the case of the reabsorption the luminescence kinetics depends on sample thickness, wavelength of the luminescent radiation, direction of observation. By reabsorption the concentration of excited molecules depends on time and space location of the sample. That results in an increase of the relaxation time. Time dependence of the luminescence signal deviates from the exponential type. Assume that $N(t)$ is the total number of the excited molecules. In the case of the short light pulse excitation the differential equation for $N(t)$ is given by

$$\frac{dN(t)}{dt} = -N(t)/\tau_0 + \alpha\eta N(t)/\tau_0, \qquad (4.111)$$

where

η is a quantum yield of luminescence,

α is the reabsorption probability.

A value of α depends on the time required to change a distribution of excited molecules in sample. Consequently the relaxation time of luminescence signal shows a non-exponential character. Approximate value of α (i.e. $\alpha \simeq \alpha_0$) for the initial time of light propagation is given by:

$$\alpha_0 = \frac{\int f(\nu) \int [1 - \exp(-D/(1 - P))]\, dPd\nu}{2 \int f(\nu)d\nu}, \tag{4.112}$$

where

$f(\nu)$ is spectral distribution of the luminescence signal,

$D = \varepsilon(M)d$ is an optical density of the sample,

P is a coefficient that corrects for a deviation of optical propagation length from the sample thickness d.

Consider effect of the reabsorption on the luminescence kinetics. Assume that a strong absorption luminescent radiation is emitted from the outer surface of sample, and that the luminescence is generated by initially excited molecules. Assume also that the relaxation time is close to τ_0 and that the luminescence intensity has exponential dependence. If light absorption in the sample is small we have

$$\frac{1}{I(t)}\frac{dI(t)}{dt} = \frac{1}{N(t)}\frac{dN(t)}{dt} = \frac{\alpha\eta - 1}{\tau_0} \tag{4.113}$$

For this case distribution of molecules along the sample is uniform. Consequently the luminescent relaxation follows the exponential law. However, for the high value of the reabsorption (α) and big quantum yield (η) the relaxation time τ can be much higher then τ_0.

Nonradiative transfer of energy. In this case the relaxation of the luminescence radiation is more complex. Withing the dipole–dipole mechanism the energy transfer between donor-molecule M^* and acceptor-molecule Q (i.e. the process described by the following diagram, $M^* + Q \rightarrow M + Q^*$) is defined by the *Forester's* formula and the rate constant k_t is given by

$$k_t = C\frac{\kappa^2 k_f}{N_A n^4 R^6} \int F_M(\nu)\varepsilon_Q(\nu)\frac{d\nu}{\nu^4}, \tag{4.114}$$

where
$F_M(\nu)$ is a spectrum of donor luminescence,
$\varepsilon_Q(\nu)$ is an extinction coefficient of acceptor,
k_f is a rate of donor fluorescence,
κ is a factor that accounts for dipole orientations of both donor and acceptor molecules,
n is a refractive index,
R is a distance between donor and acceptor.

In the case of random molecular distribution within the sample with uniform distribution function the emission law of luminescence is given by

$$I(t) = I_0 \exp \left(-\frac{t}{\tau_0} - \frac{\pi^{3/2} R_k^3 N_A [Q]}{750} \sqrt{\frac{t}{\tau_0}} \right), \qquad (4.115)$$

where
R_k is an effective radius. Note that $k_t = 1/\tau_0$ in Eq. (4.114).

Equation (4.115) represents luminescence kinetics in solids and liquids with hard viscosity.

4.8 Measurements of Luminescence

Luminescence is called sometime an "overtemperature emission" that is not consistent with the Planck's electromagnetic radiation. Luminescence of optical materials attracts interest of researchers for two main reasons:

1. diagnostic method of a small impurity amount;
2. process of *light amplification by stimulated emission of radiation* i.e. LASER.

Regarding the first application a key characteristic is a minimum concentration C_x that can be detected I_e i.e.

$$I_e = a I_0 C_x, \qquad (4.116)$$

where a is a proportionality coefficient, I_0 is intensity of stimulated optical signal, and C_x is concentration of a substance.

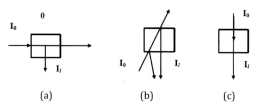

Figure 4.11 Schematics of laser pumping I_0 and luminescence I_e detecting with 90° lightening (a), 180° lightening (b), and 0° lightening (c). I_0 is laser pump, I_e is luminescence signal.

Linear dependence (4.116) fits only for small values of concentration C_x and stimulated signal I_0. For a low level of radiation the error of measurements is limited by quantum properties of light. That has made possible to defect a single atom. However, for the most of real situations the noise signal is defined by the stimulated signal scattering on the optical elements of a setup. In this case the sensitivity of luminescence method is growing upto $10^3...10^4$ particles per cm^3. Besides it has high volume and time resolution, i.e. $(\simeq 10^{-6} cm^3)$, and $(\simeq 10^{-8} s)$, respectively.

Regarding the second application of luminescence materials in LASER the main features are

1. high quantum efficiency;
2. the optimal lifetime of a stimulated atom (or molecule).

Below we will discuss typical schematics of the luminescence signal measurement.

Typical optical schematics of the luminescence measurements are shown in Fig. 4.11.

Optical schematics shown in Fig. 4.11 is applicable for samples with low absorption. Normally it is also characterized by a small scattering signal of laser pump (I_0). Schematics depicted in Fig. 4.11(b) and Fig. 4.11(c) are applicable for samples with high absorption level (for both pump and luminescence signals).

Typical laser-luminescence setup is shown in Fig. 4.12. Numbers in Fig. 4.12 denote wavelength scanning laser (1), lenses (2) and (4), sample (3), monochromator (5), and photometer (6). Laser (1) in

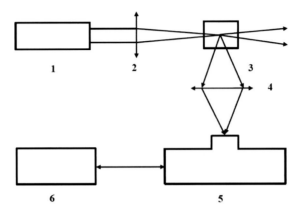

Figure 4.12 Typical experimental setup for luminescence measurements

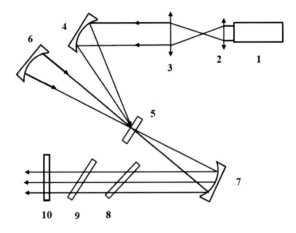

Figure 4.13 Optical schematics of a dye laser.

Fig. 4.12 is frequently a dye laser. Typical optical schematics of a dye-laser is shown in Fig. 4.13.

Typical dye-laser includes following component (see Fig. 4.13): Ar^+ - laser (1); collimator (2), (3); focusing lens (4); laser resonator (6), (7), (10); cuvette with organic dye (5); Fabry–Perot spectral filters (8), (9).

Typical characteristics of a Rhodamine 6G dye-laser are listed below:

1. Wavelength emission range: 530 to 720 nm.

2. Wavelength emission width: 0.002 nm.

3. Pulse emission power: $6 \cdot 10^3$ W.

4. Average emission power: 0.6 W.

5. Laser pulse emission duration: 20 ns.

6. Laser pulses per second: $10^4 \ s^{-1}$

7. Angle divergence: 3 angle degrees.

8. Pump: Ar^+-laser.

Condenser of optical system is denoted by (2,3) in Fig. 4.12). The main goal of the condenser is a focusing the laser-pump radiation on the small volume $10^{-3}...10^{-4}cm^3$ to $10^{-1}...10^0cm^3$. Condenser lens (4) in Fig. 4.12 has to have a high light power $D/F \simeq 0,5...1$ in order to collect luminescence signal from a maximum spacial angle (D is a diameter, F is a focus length).

Monochromator is denoted by (5) in Fig. 4.12. The main function of the monochromator is to select a specific analytic spectral line.

In some cases as a monochromator one can use a combination of an interferometric dielectric filter with an absorption filter. The first filter selects an analytic luminescence line, the second filter suppress laser pump emission. The main advantage of this combination is it's high light power.

Photometer. Photoelectric conversion of a detected light into an electric signal is typically performed by a photomultiplier or a $p-i-n$ photodiode. These devices can work in either of two modes: (1) analog signal detection or (2) photon counting. In order to improve the signal-to-noise ratio one can apply a pulse laser pump and strobe electric setup that detect only a selected time interval. This setup includes a gate circuit and an integration unit at the output. Alternative method of the signal detection is $photon\ counting$. The main advantages of the last method are: noise suppression through the pulses discrimination, reduction of a signal dependence on the photomultiplier amplification drift. The block-diagram of a pulse luminescence measurements using the photometer is shown in Fig. 4.14.

The numbers in Fig. 4.14 denote: $p-i-n$ photodiode (1); strobe generators (2), (3); photomultiplier (4); delay unit (5); strobe amplifier (6); voltage frequency converter, pulse shaper (7);

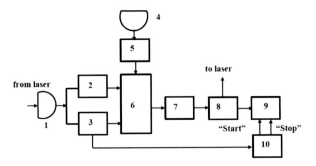

Figure 4.14 Schematics of pulse luminescence measurements.

commutator (8); reverse counters (9), (10). Photodiode (denoted by (1) in Fig. 4.14) delivers synchronization pulse. Photomultiplier signal (4), after a time delay in unit (5), enters the strobe amplifier (6). Strobe pulse duration can be varied from 3 ns to 300 ns. The time delay in unit (5) can be varied from 30 ns to 300 ns with respect to the synchronization pulse. Reverse counter (9) works as an integrator of an input signal difference. The accumulation (integration) time of measured data is adjusted by control counter (10) and can be varied from 1 s to 3 minutes. The registration setup shown in Fig. 4.14 allows to control luminescence pulses and to strobe data in digital mode.

Luminescence *quantum yield* measurement is a quite sophisticated experimental problem. It requires measurement of absorbed and emitted quanta for wide spectral range. A number of experimental problems still remain unsolved: correct accounting of scattering signal, reabsorption, refraction on the sample faces etc.

In practice the quantum yield is frequently obtained from a comparison to a well-known (reference) sample. In this case by taking into account identical measurement procedures for the two sample one has the following relationship

$$\frac{F_2}{F_1} = \frac{\eta_2}{\eta_1} \frac{D_2}{D_1}, \qquad (4.117)$$

where values of $F_{1,2}$ denote integrated signals, $\eta_{1,2}$ are the quantum yield values, $D_{1,2}$ are corresponding values of the optical densities, and indexes 1 and 2 denote measured and reference samples, respectively. By this procedure the quantum yield is obtained through a comparative analysis to a known (reference) sample.

4.9 Review Questions and Exercises

4.9.1 Bouguer–Lambert Law

Derive Bouguer–Lambert law for monochromatic optical plane wave propagation in two-level system. Assume that following parameters are known: Einstein coefficients, $B_{12}(\omega)$, $B_{21}(\omega)$; population density, N_1, N_2; spectral line shape $f(\omega)$. Find conditions for amplification (absorption) of optical beam.

4.9.2 Amplification by Optical Pumping

Show that amplification can not be achieved by optical pumping in two-level system. Two-level system is determined by the following parameters: Einstein coefficients A_{12}, B_{12}, B_{21}; population densities N_1, N_2. Write and analyze kinetic equation for population.

4.9.3 Homogeneous Spectral Line Broadening by Finite Lifetime

Determine spectral line shape for homogeneous spectral line broadening for the case when the broadening is caused by the finite lifetime of the excited state (τ).

4.9.4 Inhomogeneous Spectral Line Broadening by Thermal Motion

Determine spectral line shape for homogeneous spectral line broadening for the case when the broadening is caused by Doppler shift of the transition frequency due to the thermal motion of atoms (molecules).

4.9.5 Einstein Coefficient

Find expression for the Einstein coefficient A within the electric dipole approximation (the probability of spontaneous emission in all directions).

4.9.6 Atomic Concentration Detection

Calculate a sensitivity of the atomic concentration detection by the atomic-absorption technique. For the lowest recorded optical signal take the shot-noise level: $\Delta P_{min} = (P\hbar\omega/\Delta t\eta)^{1/2}$, where Δt is measurement time and η is photodetector quantum yield.

4.9.7 Luminescence in Semiconductors

Recombination processes due to which the electrons recombine with holes, in most cases occur through impurity centers. Two types of such centers are known: traps (N) and recombination centers (r). N and r centers are characterized by different probabilities of electron and hole capture. Write kinetic equations for the electron concentration (n) in the conduction band, hole concentration (p) at the recombination levels, and m electron concentration on the N levels. Assume W is concentration of N-centers.

Chapter 5

Atomic and Electron Energy Structure of Solids

Today, the optics of solids is expanding at an explosive rate. The advent of optical communications, television, and network interconnections across the globe has placed a heavy burden on solids and solid-state devices for signal processing and transmission. Variety of applications include optical telecommunications, computers, satellite systems, lasers, optical nanotechnology, etc.

Optical constants of materials are given in several publications [Liebsch (1997); Adachi (1999); Fox (2003); Palik (1985)]. This chapter addresses the underlying mechanisms that govern a response of solids to external electromagnetic radiation. It starts with the fundamentals of the solid-state physics, describes basics of the atomic and electron energy structure of solids. The materials of the chapter make a logical bridge to the optical properties of atoms, molecules, and other non periodic materials considered in Chapter 1 providing with an introduction to physics of the optical response from periodic materials (including elements of the band structure theory).

Fundamentals of the Optics of Materials: Tutorial and Problem Solving
Vladimir I. Gavrilenko and Volodymyr S. Ovechko
Copyright © 2024 Jenny Stanford Publishing Pte. Ltd.
ISBN 978-981-4877-93-0 (Hardcover), 978-1-003-25694-6 (eBook)
www.jennystanford.com

108 | *Atomic and Electron Energy Structure of Solids*

5.1 Useful Terminology of Solid-State Physics

Every crystal consists of a repeating pattern of objects like atoms or molecules (in molecular crystals). These objects build three dimensional array that is considered as infinite. The basis of this array is called *lattice*. Widely used mathematical approach of representing a lattice is to use *translation vector* (**T**) generating any point of the lattice:

$$\mathbf{T} = n_1\mathbf{a}_1 + n_2\mathbf{a}_2 + n_3\mathbf{a}_3 \qquad (5.1)$$

where $\mathbf{a}_j$ are non-coplanar vectors called *primitive vectors* of the crystal lattice and n_j are integer numbers. Any two lattice points (e.g. atoms) are connected by a vector given by Eq. (5.1). A group of atoms such that when repeated in space it forms the crystal is called *basis*. The primitive vectors built a cell of a lattice. The cell of the smallest volume is called the *primitive cell* (which is not the same as the *unit cell*)[1]. Either primitive or unit cell can serve as a building block for the crystal and the choice depends of specific goal by modeling and simulation of solid properties.

The *reciprocal lattice vector* **G** is defined as:

$$\mathbf{G} = m_1\mathbf{b}_1 + m_2\mathbf{b}_2 + m_3\mathbf{b}_3 \qquad (5.2)$$

where m_i are integers and primitive vectors $\mathbf{b}_1$, $\mathbf{b}_2$, and $\mathbf{b}_3$ of the reciprocal lattice are defined through the primitive crystal lattice

---

[1] In many text books a *primitive cell* is used to define the smallest *unit cell* of a crystal. These two definitions are not equivalent since the unit cell could be chosen much larger than the primitive cell. Such approach (the *super cell* method) is widely used by modeling structural imperfections while still using standard band theory of the periodic systems, e.g. by study structural (point and/or extended) defects, surfaces, interfaces, see Section 6.1 for details.

vectors $\mathbf{a}_i$ (see equation (5.1)) as [Kittel (1986)]:

$$
\begin{cases}
\mathbf{b}_1 = 2\pi \dfrac{\mathbf{a}_2 \times \mathbf{a}_3}{\mathbf{a}_1 \cdot \mathbf{a}_2 \times \mathbf{a}_3} \\[2mm]
\mathbf{b}_2 = 2\pi \dfrac{\mathbf{a}_3 \times \mathbf{a}_1}{\mathbf{a}_1 \cdot \mathbf{a}_2 \times \mathbf{a}_3} \\[2mm]
\mathbf{b}_3 = 2\pi \dfrac{\mathbf{a}_1 \times \mathbf{a}_2}{\mathbf{a}_1 \cdot \mathbf{a}_2 \times \mathbf{a}_3}
\end{cases}
\tag{5.3}
$$

Each $\mathbf{b}_i$ vector in equation (5.3) is orthogonal to the two primitive axis vectors of the lattice that mathematically is expressed as:

$$
\mathbf{b}_i \cdot \mathbf{a}_j = 2\pi \delta_{ij}
\tag{5.4}
$$

where *Kronecker delta* function has the following properties:

$$
\delta_{ij} =
\begin{cases}
0, i \neq j \\
1, i = j
\end{cases}
\tag{5.5}
$$

5.2 Atomic structure

A macroscopic bulk solid contains within a $1\ cm^{-3}$ nearly 10^{23} atoms that makes impossible to solve any classical or quantum equations of motion in a direct way. The key for a quantitative description of electronic and optical properties of crystalline solids is the fact that crystals are highly symmetrical systems and the crystal symmetry can be exploited to greatly facilitate the solution of the problem.

In Section 5.1 the *translation vector* (**T**) and the *reciprocal lattice vector* **G** of the 3D crystalline lattice are introduced. There are some conventions for specifying orientations and directions in crystals that are summarized here. The crystallographic directions are fictitious lines linking atoms, ions, or molecules. Similarly, the crystallographic planes are fictitious planes linking nodes of a crystal.

Any lattice plane is determined by three integers h, k, and l that are called the *Miller indices*. Each index denotes a plane orthogonal to a direction (hkl) in the basis of the reciprocal lattice vectors **G**. Index (100) represents a plane orthogonal to the direction h; indexes (010) and (001) represent planes orthogonal to direction k and l,

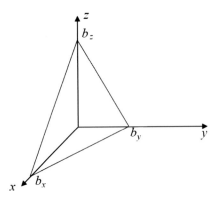

Figure 5.1 In a perfect crystal an ideal surface is determined by intersections at b_x, b_y, and b_z.

respectively. The notation $\{hkl\}$ denotes the set of all planes that are equivalent to (hkl) by the symmetry of the lattice. Similarly, the notation $[hkl]$ denotes a direction in the basis of the direct lattice vectors instead of the reciprocal lattice; the notation $<hkl>$ denotes the set of all directions that are equivalent to $[hkl]$ by symmetry.

An ideal flat surface truncates the bulk structure of a perfect crystal with intersections at b_x, b_y, and b_z as shown in Fig. 5.1. For any surface in a crystal the Miller indices can be determined by the following procedure:

- for a plane with intersections at b_x, b_y, and b_z calculate reciprocals: $\left(\frac{1}{b_x}, \frac{1}{b_y}, \frac{1}{b_z}\right)$;
- obtain Miller indices by converting all quotients to rational integers or 0.

For example:

$$b_x, b_y, b_z = 1, \infty, \infty \rightarrow \left(\frac{1}{1}, \frac{1}{\infty}, \frac{1}{\infty}\right) \rightarrow (100),$$

$$b_x, b_y, b_z = 1, 1, \infty \rightarrow \left(\frac{1}{1}, \frac{1}{1}, \frac{1}{\infty}\right) \rightarrow (110), \quad (5.6)$$

$$b_x, b_y, b_z = 1, 1, 1 \rightarrow \left(\frac{1}{1}, \frac{1}{1}, \frac{1}{1}\right) \rightarrow (111),$$

$$b_x, b_y, b_z = 1, 1, 0.5 \rightarrow \left(\frac{1}{1}, \frac{1}{1}, \frac{1}{0.5}\right) \rightarrow (112),$$

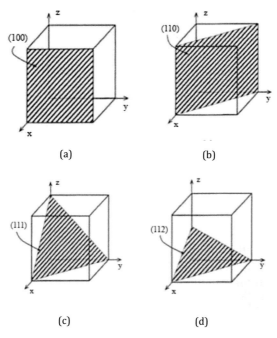

Figure 5.2 Examples of crystallographic planes in cubic crystals: (a) plane (100), (b) plane (110), (c) plane (111), and (d) plane (112).

Generally the direction [hkl] is not normal to the (hkl) plane, except in a cubic lattice. For cubic crystals (e.g. with simple cubic (sc), face centered cubic (fcc), and/or body centered cubic (bcc) lattice), the lattice vectors are orthogonal and of equal length, similar to the reciprocal lattice. Thus, in this common case, the Miller indices (hkl) and [hkl] both simply denote normals (or directions) in Cartesian coordinates.

For cubic crystals the planes considered in equations (5.6) are shown in Fig. 5.2. Spacing d between adjacent (hkl) lattice planes in cubic crystals with lattice constant a is given by:

$$d_{hkl} = \frac{a}{\sqrt{h^2 + k^2 + l^2}} \qquad (5.7)$$

Consider atomic structures of the real bulk solids. Metals are crystallized in cubic (e.g. (bcc), such as Fe, W, Cr and (fcc), Cu, Ag, Au)

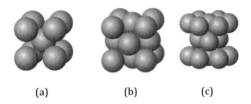

Figure 5.3 Atomic structure of selected metals: (a) body centered cubic (bcc), Fe, W, Cr; (b) face centered cubic (fcc), Cu, Ag, Au; and (c) hexagonal closed packed (hcp), Zr, Ti, Zn lattice.

or hexagonal closed packed (hcp), Zr, Ti, Zn lattice. The stick-and-ball models of these structures are shown in Fig. 5.3. One can clearly see the obvious differences in atomic packing of different surfaces.

Generally 14 different types of crystal lattice are known [Kittel (1986)]. All lattices are grouped into systems and classified according to seven types of unit cells: triclinic, monoclinic, cubic, orthorhombic, trigonal, tetragonal, and hexagonal. In Fig. 5.3 only three types of frequently occurred crystal structures are shown.

Detailed description of crystal lattices and symmetries is out of scope of this book and could be found in more specialized literature [Ashcroft and Mermin (1976); Bechstedt (2003); Kittel (1986)].

5.3 Elements of Band Structure Theory

Now we can consider electron energy structure in atomic systems having spatial periodicity: the crystals. It is useful to remind first the terminology of the condensed matter physics [Ashcroft and Mermin (1976); Kittel (1986); Harrison (1989); Singleton (2004)].

5.3.1 Quantum Mechanics of Particles in Periodic Potential

As stated above we consider here a behavior of an electron in periodic potential field within the time-independent picture (see Section 3.2). Electron energy structure of an electron in periodic potential is described by the Schrödinger equation that can be written as:

$$\left[-\frac{\hbar^2}{2m}\nabla^2 + V(\mathbf{r})\right]\Psi(\mathbf{r}) = E\Psi(\mathbf{r}) \tag{5.8}$$

Using analogy to the classical expression of the kinetic energy in terms of the momentum p:

$$E_k = \frac{1}{2}mv^2 = \frac{p^2}{2m}, \tag{5.9}$$

we use now a quantum mechanical definition of the momentum operator $\mathbf{p} = -i\hbar\nabla$ and obtain the Schrödinger equation (5.8) in the following form:

$$\left[-\frac{\hat{p}^2}{2m_0} + V(r)\right]\Psi(r) = E\Psi(r). \tag{5.10}$$

Note that $V(\mathbf{r})$ must be periodic, i.e.

$$V(\mathbf{r} + \mathbf{T}) = V(\mathbf{r}), \tag{5.11}$$

where $\mathbf{T}$ is the translation vector (5.1). As an every periodic function the potential (5.11) may be expressed as a Fourier series (see e.g. [Kittel (1986)]):

$$V(\mathbf{r}) = \sum_{\mathbf{G}} V_{\mathbf{G}} e^{i\mathbf{G}\mathbf{r}} \tag{5.12}$$

where $\mathbf{G}$ is the reciprocal lattice vector (5.2). The potential function in crystal $V(\mathbf{r})$ is the invariance under crystal translation $\mathbf{T}$ that using Eqs. (5.1) and (5.2) can be expressed by:

$$e^{i\mathbf{G}T} = e^{i2\pi(n_1m_1 + n_2m_2 + n_3m_3)} = 1. \tag{5.13}$$

According to the basic postulate of quantum mechanics a particle in the space is represented by a plane wave:

$$\phi(\mathbf{r}) = e^{i(\mathbf{k}\mathbf{r} - \omega t)} \tag{5.14}$$

Atomic and Electron Energy Structure of Solids

where the wave vector $\mathbf{k} = k_x\mathbf{x} + k_y\mathbf{y} + k_z\mathbf{z}$ is associated with the de Broglie wavelength (λ_B) as $|\mathbf{k}| = 2\pi/\lambda_B = \sqrt{k_x^2 + k_y^2 + k_z^2}$. In quantum mechanics $\mathbf{k}$ is an operator ($\mathbf{k} = -i\nabla$) related to the operator of momentum $\mathbf{p}$ (see comments to the equation (5.9)) in the following way:

$$\mathbf{p} = -i\hbar\nabla = \hbar\mathbf{k} \qquad (5.15)$$

In the periodic potential field of a crystal the wave function (5.14) is a subject to the boundary conditions that include the full symmetry of the crystal:

$$\phi(\mathbf{r} + N_j\mathbf{a}_j) = \phi(\mathbf{r}), \qquad (5.16)$$

where $j = 1, 2, 3$ and $N = N_1N_2N_3$ is the total number of primitive cells in the crystal and N_j is the number of the primitive cells in the j-th direction. The equation (5.16) is called the *Born-von Karman periodic boundary condition*.

From equation (5.16) follows:

$$e^{iN_j\mathbf{k}a_j} = 1. \qquad (5.17)$$

Comparing with the equation (5.13) and using (5.2) results:

$$\mathbf{k} = \sum_{j=1}^{3} \frac{m_j}{N_j}\mathbf{b}_j. \qquad (5.18)$$

Note that if all m_j change by one we generate a new electron state. Therefore the volume of k-space associated with one single electron state is given by:

$$\frac{\mathbf{b}_1}{N_1} \cdot \frac{\mathbf{b}_2}{N_2} \times \frac{\mathbf{b}_3}{N_3} = \frac{1}{N}\mathbf{b}_1 \cdot \mathbf{b}_2 \times \mathbf{b}_3 \qquad (5.19)$$

where $\mathbf{b}_j$ are primitive vectors of the reciprocal lattice $\mathbf{G}$ (5.2). The reciprocal lattice is a key concept in solid-state physics representing periodicity of electronic properties of solids in the k-space. The primitive cell of the reciprocal lattice is know as the first *Brillouin zone* in the *k*-space which volume is given by:

$$V_{BZ} = \mathbf{b}_1 \cdot \mathbf{b}_2 \times \mathbf{b}_3 \tag{5.20}$$

The wave-function $\Psi(\mathbf{r})$ can be represented as a sum of plane waves obeying the Born von Karman boundary conditions (5.16) in the following form:

$$\Psi(\mathbf{r}) = \sum_{\mathbf{k}} C_{\mathbf{k}} e^{i\mathbf{k}\mathbf{r}}. \tag{5.21}$$

By substitution this wavefunction into the Schrödinger equation (5.8) for a particle in periodic potential (5.12) this equation after some algebra is given by:

$$\sum_{\mathbf{k}} \frac{\hbar^2 k^2}{2m} C_{\mathbf{k}} e^{i\mathbf{k}\mathbf{r}} + \sum_{G,k} V_G C_{\mathbf{k}} e^{i(\mathbf{G}+\mathbf{k})\mathbf{r}} = E \sum_{\mathbf{k}} C_{\mathbf{k}} e^{i\mathbf{k}\mathbf{r}}. \tag{5.22}$$

After some rearrangement the Schrödinger equation (5.22) is given by:

$$\sum_{\mathbf{k}} e^{i\mathbf{k}\mathbf{r}} \left[\left(\frac{\hbar^2 k^2}{2m} - E \right) C_{\mathbf{k}} + \sum_{G} V_G C_{\mathbf{k}-G} \right] = 0 \tag{5.23}$$

The plane waves represent an orthogonal set of functions thus the coefficient of every term in (5.23) must vanish, i.e.

$$\left(\frac{\hbar^2 k^2}{2m} - E \right) C_{\mathbf{k}} + \sum_{G} V_G C_{\mathbf{k}-G} = 0 \tag{5.24}$$

It is convenient now to restrict electronic structure analysis to the first Brillouin zone only. This is achieved by a substitution $\mathbf{k} = \mathbf{q} - \mathbf{G}'$, where $\mathbf{q}$ is defined in the first Brillouin zone, and by changing the variables $\mathbf{G}'' = \mathbf{G} + \mathbf{G}'$. The equation (5.24) has now the form

$$\left(\frac{\hbar^2 (\mathbf{q} - \mathbf{G}')^2}{2m} - E \right) C_{\mathbf{q}-\mathbf{G}'} + \sum_{G''} V_{G''-G'} C_{\mathbf{q}-G''} = 0. \tag{5.25}$$

Coefficients $C_{\mathbf{k}}$ that are determined by solving the equation (5.25) are used to make up the wavefunction in equation (5.21). By substituting

$\mathbf{k} = \mathbf{q} - \mathbf{G}'$ the equation (5.21) is given now by

$$\Psi(\mathbf{r}) = \sum_{\mathbf{G}} C_{\mathbf{q}-\mathbf{G}} \cdot e^{i(\mathbf{q}-\mathbf{G})\mathbf{r}} = e^{i\mathbf{q}\mathbf{r}} u_{\mathbf{q}}(\mathbf{r}), \tag{5.26}$$

where

$$u_{\mathbf{q}}(\mathbf{r}) = \sum_{\mathbf{G}} C_{\mathbf{q}-\mathbf{G}} \cdot e^{-i\mathbf{G}\mathbf{r}}, \tag{5.27}$$

which has the periodicity of the crystal lattice (that can be proved by a substitution $\mathbf{r} \to \mathbf{r} + \mathbf{T}$). Equation (5.26) expresses the *Bloch theorem*.

The Bloch theorem states that the eigen function of a one-electron Schrödinger equation (5.8) with the periodical potential (5.11) is given by a plane wave times a function with the periodicity of the crystal lattice [Kittel (1986)].

5.3.2 Tight-Binding Band Structure of Some Periodic Crystalline Structures

In this section calculation of the electron energy structure using the *tight binding method* for crystals is described (see also applications of the LCAO method described for molecules in Section 3.2). The tight binding method has existed for many years as a convenient and transparent model for the description of electronic structure in molecules and solids. This method is described in details in original papers as well as in many text books and monographs (see e.g. [Slater and Koster (1954); Harrison (1989); Sutton (2004); Ashcroft and Mermin (1976); Yu and Cardona (2010)]) and only basic ideas are given below.

It is assumed that the potential is very large and all electrons in the crystal spend most of the time bound to ionic cores only and occasionally jumping between atoms. The atomic wavefunctions $\phi_j(\mathbf{r})$ of every j-th atom are eigen states of the Hamiltonian of a single atom (H_{at}) corresponding to the eigen energy (E_{jn}) according to the Schrödinger equation:

$$H_{at}\phi_{jn}(\mathbf{r}) = E_{jn}\phi_{jn}(\mathbf{r}), \tag{5.28}$$

where index n denotes atomic orbital and energy eigen state. It is also assumed that all atoms in the crystal are identical and close to the lattice point the crystal Hamiltonian H_0 is well described by the H_{at}. The wavefunction in the crystal can be constructed as a linear combination of the atomic wavefunctions (same as the Linear Combinations of Atomic Orbitals (LCAO) in molecule, see Section 3.2) according to:

$$\Psi_{j,\mathbf{k}}(\mathbf{r}) = \sum_{\mathbf{T}} a_{\mathbf{k},T}\phi_j(\mathbf{r} + \mathbf{T}), \tag{5.29}$$

where $\mathbf{T}$ is the vector of crystal translation. Taking account of the Bloch theorem (see equations (5.26), (5.27)), the tight binding wavefunction in crystal can be given by:

$$\Psi_{j,\mathbf{k}}(\mathbf{r}) = \frac{1}{\sqrt{N}} \sum_{\mathbf{T}} e^{i\mathbf{k}T}\phi_j(\mathbf{r} - \mathbf{T}), \tag{5.30}$$

where there are N lattice sites and the factor of $1/\sqrt{N}$ ensures that the Bloch state is normalized.

In crystal the single particle Hamiltonian is given by:

$$H = H_{at} + \Delta V \tag{5.31}$$

where ΔV encodes all the differences between the true potential in the crystal and the potential of an isolated atom. It is also assumed that $\Delta V \to 0$ at the center of each crystal atom.

The single particle states in the crystal are given by equation (5.30), where

$$H\Psi_{n\mathbf{k}}(\mathbf{r}) = E_{n\mathbf{k}}\Psi_{n\mathbf{k}}(\mathbf{r}) \tag{5.32}$$

the band index (and corresponding atomic orbital) is labeled by n and $\mathbf{k}$ is in the first Brillouin zone.

We will consider here only highest occupied (or valence) atomic orbitals in crystal. Using the equation (5.30) for the wave function in crystal, the single particle Schrödinger equation (5.32) is given by

[Singleton (2004)]:

$$H_{at} \sum_T e^{i\mathbf{k}T}\phi_j(\mathbf{r} - \mathbf{T}) + \Delta V \sum_T e^{i\mathbf{k}T}\phi_j(\mathbf{r} - \mathbf{T}) = E(\mathbf{k}) \sum_T e^{i\mathbf{k}T}\phi_j(\mathbf{r} - \mathbf{T})$$

(5.33)

Next step will be multiplication of the Eq. (5.33) by complex conjugate function of $\phi^*(\mathbf{r})$ and integration over $\mathbf{r}$ that gives:

$$\int \phi^*(\mathbf{r})H_{at} \sum_T e^{i\mathbf{k}T}\phi_j(\mathbf{r} - \mathbf{T})d^3\mathbf{r} + \int \phi^*(\mathbf{r})\Delta V \sum_T e^{i\mathbf{k}T}\phi_j(\mathbf{r} - \mathbf{T})d^3\mathbf{r}$$

$$= \int \phi^*(\mathbf{r})E(\mathbf{k}) \sum_T e^{i\mathbf{k}T}\phi_j(\mathbf{r} - \mathbf{T})d^3\mathbf{r}$$

(5.34)

Since the basis atomic functions $\phi_j(\mathbf{r} - \mathbf{T})$ are normalized, the first integral in equation (5.34) yields the atomic orbital energies (E_ϕ) and the third integral gives $E(\mathbf{k})$. Here is taken into account that value of the integral with $\mathbf{T} = 0$ is much greater than all others. By evaluation of the second integral we will account only on-site ($\mathbf{T} = 0$) and nearest neighbors values, i.e. with $\mathbf{T} = \pm a\mathbf{e}_1, \pm b\mathbf{e}_2, \pm c\mathbf{e}_3$, where a, b, c are the orthogonal lattice constants. This approximation still correctly reproduces the most essential features of the band structure since the contributions of the second, third and higher atomic neighbors of a crystal is normally weak [Harrison (1989)].

After some algebra the equation (5.34) is given by:

$$E(\mathbf{k}) = E_\phi - E_a - V_x(e^{ik_xa} + e^{-ik_xa})$$
$$- V_y(e^{ik_yb} + e^{-ik_yb})$$
$$- V_z(e^{ik_zc} + e^{-ik_zc})$$

(5.35)

or in more compact form [Singleton (2004)]:

$$E(\mathbf{k}) = E_\phi - E_a - 2V_x Cos(k_xa) - 2V_y Cos(k_yb) - 2V_z Cos(k_zc) \quad (5.36)$$

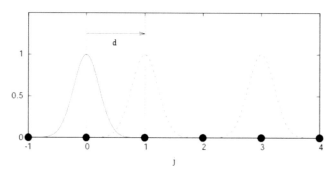

Figure 5.4 Atomic orbitals of a one-dimensional atomic chain with atoms labeled by index j and separated by a lattice constant d. The s-like atomic orbitals decay rapidly so the overlap is very small.

The following notations for on-site energies (E_a) and transfer (or overlap) integrals (V_x, V_y, V_z) are used in equations (5.35) and (5.36):

$$E_a = -\int \phi^*(\mathbf{r}) \Delta V \phi(\mathbf{r}) d^3\mathbf{r}$$

$$V_x = -\int \phi^*(\mathbf{r}) \Delta V \phi(\mathbf{r} + a_1) d^3\mathbf{r}$$

$$V_y = -\int \phi^*(\mathbf{r}) \Delta V \phi(\mathbf{r} + a_2) d^3\mathbf{r}$$

$$V_z = -\int \phi^*(\mathbf{r}) \Delta V \phi(\mathbf{r} + a_3) d^3\mathbf{r} \quad (5.37)$$

The V_x, V_y, and V_z denote *transfer integrals* that represent energies of the inter-atomic interactions describing the inter-atomic electron transfer.

It is instructive now to consider band structure of a one-dimensional crystal containing identical atoms with only s-electrons on the outer shell and with the lattice constant d. Schematically atomic orbitals of such one dimensional crystal are shown in Fig. 5.4.

The chosen 1D-crystal has only one atom in the unit cell and only s-orbitals contribute to the crystal state. There will be therefore only one electron energy band determined by the overlap integral equal to $V_x = V$. The band structure for this crystal follows from Eq. (5.36)

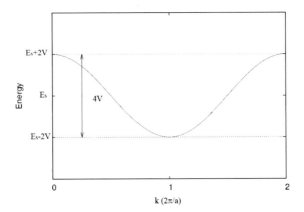

Figure 5.5 Energy dispersion of the s-band in one-dimensional crystal.

and is given by:

$$E(\mathbf{k}) = E_s - 2V\cos(kd) \tag{5.38}$$

In the one-dimensional crystal the length of the unit cell in real space is d, so the length of the unit cell in reciprocal space is $2\pi/d$.

In Fig. 5.5 the electron energy dependence on the wave vector k, the energy dispersion, is plotted inside the unit cell in reciprocal space, from $k = 0$ to $k = 2\pi/d$. At large k energy dispersion repeats. The E_s denotes the on-site energy of the s-electrons and the band width is equal to $4V$, that follows from the Eq. (5.38). Generally, the unit cell shown in Fig. 5.5 corresponds to the first Brillouin zone (BZ), but typical graphical presentation is different. The BZ length is still $2\pi/d$ but the BZ extends from $-\pi/d$ to π/d rather than 0 to $2\pi/d$.

In real materials more electrons contribute to the band structure. Typical example relevant to the subject of this book is a 2D band structure of a single layer of graphite, i.e. the *graphene*. Atomic structure of graphene is shown in Fig. 5.6.

Graphene has two carbon atoms in the unit cell defined by vectors $\mathbf{a}_1$ and $\mathbf{a}_2$ (see Fig. 5.6):

$$\begin{cases} \mathbf{a}_1 = a\left(\dfrac{1}{2}, \dfrac{\sqrt{3}}{2}, 0\right) \\ \mathbf{a}_2 = a\left(-\dfrac{1}{2}, \dfrac{\sqrt{3}}{2}, 0\right) \end{cases} \tag{5.39}$$

Fundamentals of the Optics of Materials: Tutorial and Problem Solving | 121

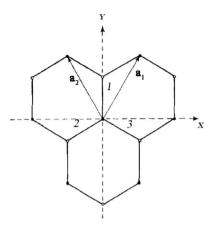

Figure 5.6 Atomic structure of a single layer of graphite, i.e. graphene.

The band structure of graphene is formed by contributions of s, p_x, p_y, and p_z electrons and electron motion has 2D character. Carbon atoms within a layer are bounded by the sp^2 hybrid atomic orbitals formed by s, p_x, and p_y electrons. The p_z electron atomic orbitals are not hybridized and are oriented perpendicular to the layer. According to the LCAO method in the tight binding theory of band structure (see Section 3.3) the atomic orbitals of the carbon atoms in graphene can be given as:

$$\begin{cases} |h_1\rangle = \dfrac{2}{\sqrt{3}} \left(\dfrac{1}{\sqrt{2}} |s\rangle + |p_y\rangle \right) \\[6pt] |h_2\rangle = \dfrac{2}{\sqrt{3}} \left(\dfrac{1}{\sqrt{2}} |s\rangle - \dfrac{\sqrt{3}}{2} |p_x\rangle - \dfrac{1}{2} |p_y\rangle \right) \\[6pt] |h_3\rangle = \dfrac{2}{\sqrt{3}} \left(\dfrac{1}{\sqrt{2}} |s\rangle + \dfrac{\sqrt{3}}{2} |p_x\rangle - \dfrac{1}{2} |p_y\rangle \right) \\[6pt] |h_4\rangle = |p_z\rangle \end{cases} \quad (5.40)$$

The hybrid orbitals $|h_1\rangle$, $|h_2\rangle$, and $|h_3\rangle$ form very strong σ-bonds that are marked by numbers in Fig. 5.6. The bond 1 binds two non-equivalent atoms of the unit cell. Atoms of the first and second atomic shells are marked by open and filled circles, respectively.

The band structure of graphite single layer within the tight binding method has been calculated first by P. R. Wallace in his pioneering work in 1947 [Wallace (1947)]. The tight binding theory

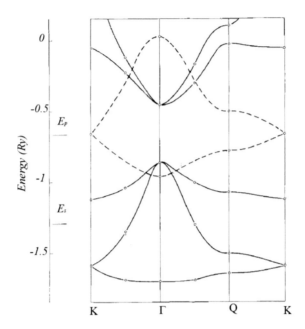

Figure 5.7 Calculated 2D band structure of graphite single layer. By solid and dashed lines σ- and π-bands are shown, respectively (redrawn from [Bassani and Parravicini (1967)]).

has been used for decades to study electron energy structure and optical properties of crystalline graphite and carbon films (see e.g. [Bassani and Parravicini (1967); Slonczewski and Weiss (1958); Doni and Parravicini (1969); Gavrilenko et al. (1990)]). Calculated energy dispersion of π- and σ-bands of a single graphite layer is shown in Fig. 5.7.

Each of two carbon atom in unit cell has four valence electrons s, p_x, p_y, and p_z. Consequently the tight-binding Hamiltonian matrix has dimension of 8×8. However, the atomic structure (represented by a single monolayer of graphite) indicates that in ideal case the p_z orbitals are perpendicular to the layer. Thus interactions between p_z and s, p_x, p_y orbitals can be neglected. Consequently, the 8×8 full $\hat{H}$-matrix can be blocked into two matrices: one matrix has dimension of 2×2 describing π-electron band structure created by the p_z orbitals (shown by the dashed lines in Fig. 5.7) and the other 6×6 matrix that describes σ- bands formed by s, p_x, and p_y electron orbitals (shown by the solid lines in Fig. 5.7). All energy levels below (above) the Fermi energy (indicated by E_p in Fig. 5.7) are filled (empty).

Strong anisotropy of interatomic bonding in graphite has very important consequences. Each carbon atom uses three of its electrons to form σ-bonds to its three close neighbors. That leaves a fourth electron in the bonding level. These electrons form the π-bonding withing a layer and become delocalized over the whole of the sheet of atoms in one layer. The π-electrons are no longer associated directly with any particular atom or pair of atoms, but are free to wander throughout the whole sheet. This semi-metalic property of graphite are indicated by a crossing of filled and empty π-bands in K-point of the band structure (see Fig. 5.7). The important consequence is that the delocalized electrons are free (like in metal) to move anywhere within the sheet—each electron is no longer fixed to a particular carbon atom. There is, however, no direct contact between the delocalized electrons in one sheet and those in the neighboring sheets. The atoms within a carbon layer are held together by very strong covalent bonds. These bond are stronger than in diamond because of the additional bonding caused by the delocalized π-electrons.

In graphite all sheets are holding together. That is the ultimate example of the van der Waals dispersion forces. The nature of these forces in graphite is the following. As the delocalized π-electrons move around in the layer, very large temporary dipoles can be induced which will set up opposite dipoles in the layers above and below, and so on throughout the whole graphite crystal.

Due to the relative weak interlayer bonding the graphite is soft and is used in pencils and as a dry lubricant for things like locks. One can think of graphite rather like a pack of cards, i.e. each card is strong, but the cards will slide over each other, or even fall off the pack altogether. When one uses a pencil, sheets are rubbed off and stick to the paper. Graphite has a lower density than diamond. This is because of the relatively large space between the sheets.

Tight binding electron band structure for crystalline bulk silicon is shown in Fig. 5.8. Optical response of material is determined by direct electron transition between filled and empty levels (see arrows in Fig. 5.8). Graphene is semimetal and there is no gap

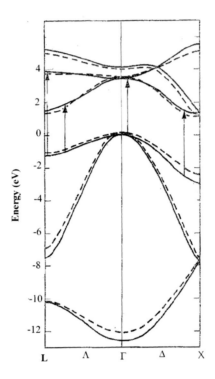

Figure 5.8 Calculated electron band structure of crystalline bulk silicon (solid lines) in comparison with a microcrystalline Si (Adapted from [Gavrilenko (1987)]).

between filled and empty bands formed by the p_z orbitals (dashed lines in Fig. 5.7) causing semimetalic properties of this material. This reflects a metallic like behavior of the π-electrons (or the p_z orbital electrons). In contrast, the bands formed by s, p_x, p_y orbitals (solid lines in Fig. 5.7) show typical semiconductor-like band structure with a gap. Silicon is a typical semiconductor having a forbidden energy gap for electrons (see Fig. 5.8). Optical transitions between valence and conduction bands allowed by the symmetry are shown by vertical arrows.

According to the band structure theory over-viewed in Section 5.3 the electron energy bands are determined by the overlap of electronic orbitals. Thus most important contributions provide nearest neighbors in atomic structure. Therefore most typical features of the band structure are observed even in poly and micro-crystalline materials. This is depicted in Fig. 5.8 by dashed lines representing a band structure calculated for microcrystalline Si in comparison with the bulk crystal [Gavrilenko (1987)]. Predicted shifts in the energy gaps in Si due to structural disorder and creation of micro grains was observed experimentally.

Now we are ready to smoothly move to the optical functions of materials and the optical response to external excitation by electromagnetic radiation that is considered in the upcoming chapters of this book.

5.4 Review Questions and Exercises

5.4.1 Amorphous State

Define amorphous state.

5.4.2 Bragg's Law

Formulate the Bragg's law.

5.4.3 Brillouin Zone

What is Brillouin zone?

5.4.4 Valence and Conduction Bands

Define valence and conduction bands in non-metals.

Chapter 6

Basics of Surface Science and Surface Optics

The goal of the surface science is to understand the fundamental principles that govern geometric and electronic structures of solid surfaces and the processes that occur on these surfaces, for example, gas-surface scattering, atomic and molecular reactions at surfaces, growth of new surface layers, etc.

Surfaces and interfaces of solids play an important role in technological device applications. Their physical and chemical characteristics are responsible for the interesting properties and making them an active area in materials research. Processes on solid surfaces play an enormously important role in modern nanotechnology. The fundamental issue is that surfaces and interfaces are (nearly) two-dimensional electronic systems. This chapter gives a brief introduction to the physics of solid surfaces as key elements of nanostructures. The following related topics are reviewed: specific atomic structures of surfaces compared to the bulk; modifications of electronic and vibrational properties upon creating a surface; and physics and chemistry that govern atoms and molecules adsorption on a surface.

Fundamentals of the Optics of Materials: Tutorial and Problem Solving
Vladimir I. Gavrilenko and Volodymyr S. Ovechko
Copyright © 2024 Jenny Stanford Publishing Pte. Ltd.
ISBN 978-981-4877-93-0 (Hardcover), 978-1-003-25694-6 (eBook)
www.jennystanford.com

Different phenomena are observed at solid surfaces that do not have a counterpart in bulk materials. The reason for this is that all bulk processes can be accounted for the universal description of electronic states due to the high level of symmetry, however, at surfaces the symmetry is broken, and the electronic and crystal structure are redefined.

The solid surface can be prepared by cutting a crystal in a certain orientation. During a creation of a surface in covalent crystals, at least one bond per atom will be cut upon cleavage, which is called *dangling bond*. The unsaturated dangling bonds make the surface unstable and are responsible for an increase in the surface free energy. A reduction in the number of dangling bonds minimizes this energy and is the driving force causing the surface atomic relaxation and reconstruction. With the atomic displacement in surface relaxation and reconstruction the surface free energy can be minimized. The atoms will find new positions which reduce the number of the dangling bonds.

As the result of the relaxation the atoms at the surface are displaced from their bulk positions, but there is no change in the surface periodicity or symmetry. The atomic reconstruction of a surface, on the other hand, involves a change in the surface unit cell, compared to an ideal or bulk truncated surface. This will result in a change in the periodicity and symmetry at the surface.

6.1 Surface Atomic Geometries and Classification

Important concept of the surface science is the *surface energy* that could be defined as the work required to build an area of a particular surface. Assuming that a piece of bulk solid with an energy E_1 is cut into two parts having a surface area of A and an energy of E_2 each than the surface energy γ is given by:

$$\gamma = \frac{E_2 - E_1}{2A}.$$

(6.1)

Surface free energy, γ^{Srf}, is a thermodynamic potential that can be defined for a thin crystalline film (or a slab) having two identical

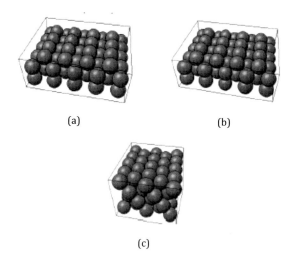

Figure 6.1 Surface atomic structures of the (fcc) lattice truncated at: (a) (100) face, (b) (110) face, and (c) (111) face.

surfaces with areas A in terms of the chemical potential μ of the surface constitute components in the following way:

$$\gamma^{Srf} = \frac{E^{Slab} - \sum_i \mu_i N_i}{2A}, \quad (6.2)$$

where E^{Slab} is the total energy of a slab, μ_i is the chemical potential of the surface constitute component, N_i is the number of the i-kind of atoms per unit cell, and factor 2 accounts for the two surfaces of the slab. Equations (6.1) and (6.2) indicate that γ and γ^{Srf} are measured in units of energy over area, e.g. J/m^2 or eV/m^2 etc.

Modern technology allows crystalline film growth with an orientation along different crystallographic directions. Consider a monocrystalline film of a bulk solid with the face centered cubic structure (see Fig. 5.3(b)) grown along [100], [110], and [111] directions. Atomic packing and planar atomic structures are very different for different planes as depicted in Fig. 6.1.

Different surfaces of a crystal have different surface free energies, depending on their orientations. The most stable surface is the one, which exhibits the lowest surface free energy. All surfaces are energetically unfavourable in that they have a positive free energy

Figure 6.2 Diamond-like bulk atomic structure.

of formation. The reason is that the formation of new surfaces by cleavage of a solid leads to the broken bonds between atoms on either side of the cleavage plane in order to split the solid and create the surfaces. Breaking bonds requires work to be done on the system, consequently the surface free energy contribution to the total free energy of a system is positive. Therefore, general thermodynamic law requires minimization of the surface free energy that can be achieved by atomic rearrangement (relaxation) within few atomic surface monolayers and/or by an adsorption of atoms or molecules on a surface after exposure to the relevant ambient.

It is instructive to consider as an example an atomic relaxation and reconstruction of the silicon (100) surface. The Si monocrystalline films grown at (100) are widely used for application in modern hi-tech electronic industry. Bulk silicon (as well as diamond and germanium) is crystallized in the diamond-like atomic structure shown in Fig. 6.2.

A side view along the [110] direction on stick-and-ball model of the slab with the (100) surface is given in Fig. 6.3(a).

Despite the bulk atomic structure of group IV elements such as diamond, Si, and Ge, is the same as shown in Fig. 6.2, the reconstructed (100) surfaces of these materials are different. The diamond (100) surface shows (2 × 1) pattern of symmetric dimers (see Fig. 6.3(b)) [Bechstedt (2003); Gavrilenko (1993)].

The Si (001) surface has received particular attention because most of silicon devices are grown on this substrate. However, in contrast to diamond (and many fcc-structured metals) the Si (100) surface shows (2 × 1) pattern but of asymmetric (buckled) dimers

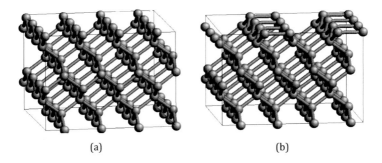

(a) (b)

Figure 6.3 Surface atomic structures of the cleaved (unrelaxed) diamond (100) surface (a) and relaxed diamond (100) surface (b) showing (2 × 1) surface pattern.

that are energetically more favorable than symmetric which is confirmed both experimentally and theoretically [Bechstedt (2003)].

Classification of the surface pattern structures requires explanation. Truncation of a bulk crystal and creation of surfaces strongly affects all electron kinetic properties on the surface (the surface area typically includes several atomic mono-layers). The system is not periodic in one dimension any more thus the electronic processes are described not by the 3D translation and reciprocal lattice vectors (given by Eqs. (5.1) and (5.2)) but by 2D vectors given by:

$$\mathbf{T} = n_1\mathbf{a}_1 + n_2\mathbf{a}_2 \tag{6.3}$$

$$\mathbf{G} = m_1\mathbf{g}_1 + m_2\mathbf{g}_2, \tag{6.4}$$

where m_i, n_i are integers and primitive vectors $\mathbf{a}_i, \mathbf{g}_i$ ($i = 1, 2$) are now 2D crystal lattice vectors of a direct and reciprocal space, respectively.

Due to different surface related processes (reconstruction, atom or molecular adsorption) the 2D periodicity on the surface will differ from that in bulk. Relationship of the surface structure and bulk is described by special notation. Here the widely used *Wood's notation* is described. The matrix notation is given in more specialized literature [Bechstedt (2003); Kolasinski (2008)].

Consider as an example the (111) surface of silver with the fcc bulk structure shown in Fig. 6.1(c). The unreconstructed surface is a periodic structure described by a primitive cell vectors $\mathbf{a}_1$ and $\mathbf{a}_2$ shown in Fig. 6.4.

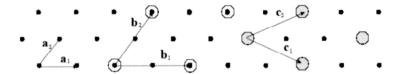

Figure 6.4 Primitive unit cells on the (111) surface of the fcc structure.

Note that length of the $\mathbf{a}_1$ and $\mathbf{a}_2$ vectors relate to the fcc bulk lattice constant a as: $|\mathbf{a}_1| = |\mathbf{a}_1| = a/\sqrt{2}$. Due to the atomic reconstruction or to the adsorption on the surface the periodicity on the surface will change describing by the bigger primitive cell created by vectors $\mathbf{b}_1$, $\mathbf{b}_2$ or $\mathbf{c}_1$, $\mathbf{c}_2$ (see Fig. 6.4). The new periodic surface structures are described by a different notation (Wood's notation) that by referring to the primitive cells shown in Fig. 6.4 are given by:

$$\left(\frac{b_1}{a_1} \times \frac{b_2}{a_2}\right) R\phi, \tag{6.5}$$

$$\left(\frac{c_1}{a_1} \times \frac{c_2}{a_2}\right) R\phi \tag{6.6}$$

The ϕ is a rotation angle with respect to the bulk primitive cell on the surface. After simple trigonometric calculations of the relevant lengths these relationships result in the following: (2×2) and $(\sqrt{3} \times \sqrt{3})R30°$ (angle of $0°$ is frequently not given in the notation).

The $(\sqrt{3} \times \sqrt{3})R30°$ surface structure is created on a silver (111) surface after adsorption of water molecules. The computer generated stick-and-ball models of the silver-water system are depicted in Fig. 6.5 [Gavrilenko et al. (2010)].

Theoretical study of the atomic structure of Ag (111) surface after adsorption of water molecules indicate a clear hexagonal structure of ice (see Fig. 6.5(a)) [Gavrilenko et al. (2010)]. The surface super cells corresponding to the $(\sqrt{3} \times \sqrt{3})R30°$ notation are shown by thin dashed lines in Figs. 6.5(a) and 6.5(b). The surface cells are created by the $\mathbf{c}_1$, $\mathbf{c}_2$ vectors shown in Fig. 6.4.

6.2 Electron Energy Structure of Solid Surfaces

This section is focused on specific features in electron energy structure and optical responses of materials that are directly caused by electronic excitation at solid surfaces and interfaces. Since the

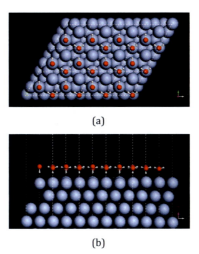

Figure 6.5 Top (a) and side view (b) of the (111) Ag surface with a one monolayer of the adsorbed water molecules calculated by [Gavrilenko *et al.* (2010)].

surface is the termination of a bulk crystal, the surface atoms have less number of neighbors than bulk atoms; i.e. part of the chemical bonds corresponding to the bulk-crystal structure are broken at the surface. Atomic reconstruction and/or atom and molecule adsorption on a surface cause substantial modification of the electron energy structure. Even in the ideal case with the surface atoms at bulk-like positions (which is called *truncated bulk*) the translation symmetry only exists in the directions within the plane of the surface describing by the translation and reciprocal lattice vectors given by equations (6.3) and (6.4), respectively.

6.2.1 Modeling of Solid Surfaces

Perpendicular to the surface the lattice periodicity breaks down and the mathematical formalism becomes more complicated. In order to apply the standard band structure theory of the 3D crystals (see Section (5.3)) the super-cell method is widely used for solid surfaces. Central idea of this method is an introduction of a new periodicity in the direction perpendicular to a surface (or interface) that must

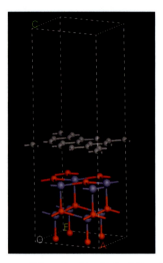

Figure 6.6 Surface super cell (marked by dashed lines) of a graphene on the ZnO surface.

include a solid slab representing a bulk and a vacuum layer thus modeling truncation of a crystal.

An example of a super cell design for a system that includes a physisorbed graphene (one monolayer of graphite) on a ZnO surface is shown in Fig. 6.6. The wurtzite ZnO bulk crystal has four atoms in the unit cell: O-Zn-O-Zn. For an illustration only in Fig. 6.6, a limiting case of a one unit cell thick slab is depicted.

The whole system is relaxed that minimized its total energy. Back surface is oxygen atom terminated. On the front surface the oxygen atoms are remarkably down-shifted as a result of the surface atomic reconstruction (see Fig. 6.6).

The following criteria are applied by the surface super cell design. The vacuum layer should also be large enough in order to prevent spurious interaction between front and back surfaces. The crystal slab should be thick enough in order to reliably reproduce bulk electron energy structure. This is normally verified by the calculated energy parameters convergence with the the slab thickness. In many cases the slab of 10 (and more) monolayers thickness is required in order to reproduce the bulk electron energy parameters [Bechstedt (2003)]. This is demonstrated in Table 6.1

Fundamentals of the Optics of Materials: Tutorial and Problem Solving | **135**

Table 6.1 Electron affinity (χ) and static dielectric permittivity values (ε) calculated by [Ilchenko *et al.* (2013)] for Si nano-slabs having different thickness, d, and number N of monolayers. Experimental values are given in the bottom

N	d (Å)	χ (eV)	ε	E_g (eV)
4	4.057	2.33	8.94	2.18
5	6.76	2.78	9.91	1.74
8	9.47	3.44	10.7	1.5
12	14.87	3.88	11.51	1.26
16	20.28	3.91	11.68	1.15
24	31.1	4.05	11.9	1.11
bulk (exp.)	∞	4.05	13.9	1.10

where calculated data of bulk parameters for the Si slabs of different thicknesses are given.

In very thin nano-films the electron energy structure is different from bulk due to the quantum confinement [Gavrilenko (2020)]. Therefore in surface science by modeling and simulation of different surface related electronic processes a care is taken to an elimination (or reduction) of the quantum effects in the model slab. On one hand the slab should be thick enough to reproduce well bulk properties, however, numerical complexity increases rapidly with the slab thickness. In the Si slabs the energy and optical parameters such as electron affinity (χ) and static dielectric constant (ε) values show convergence to bulk values with thickness (d) as shown in Table 6.1. Static dielectric function, ε, has been calculated within the Random Phase Approximation. It is converged with the slab thickness but not to the bulk value due to the neglect of many body effects in the theory used by [Ilchenko *et al.* (2013)] (see [Gavrilenko (2020)] for details). On the other hand, electron affinity, χ, converged within 5 percent to the experimental bulk value for the slab of 16 monolayer thickness, that justifies application of the local density approximation used by [Ilchenko *et al.* (2013)] for modeling of this type of phenomena.

The data given in Table 6.1 show that the super cells for realistic theoretical modeling of solid surfaces may include upto few dozens of atoms. By modeling of surfaces and interfaces with a slab super cell the artificial charging of the opposite slab surfaces should also be taken into account [Bechstedt (2003)]. For non-polar surfaces and slab thickness of more than eight monolayers this effect is small. For the system shown in Fig. 6.7 containing 16 monolayers and having two non-equivalent mono-hydride ($(2 \times 1)(001)$) and di-hydride ($(1 \times 1)(001)$) surfaces this effect is negligible.

Note, that LDA calculated values of $E_g = 0.5$ eV in bulk Si underestimates the bulk value of 1.10 eV. The calculated gap values were corrected in [Ilchenko *et al.* (2013)] for a scissors shift of 0.6 eV in order to match the bulk experimental value (the procedure frequently used in literature, see e.g. [Bechstedt (2003)]). The data given in Table 6.1 show clear exponential convergence to bulk values. After fitting an analytic (exponential) function to the values given in Table 6.1 the set of simple equations allowing corrections of relevant nano-material parameter with respect to the bulk values was obtained [Ilchenko *et al.* (2013)]:

$$E_g(d) = 1.1 + 2.28e^{-d/5.37}$$
$$\chi(d) = 4.05 - 3.52e^{-d/5.89} \quad\quad (6.7)$$
$$\varepsilon(d) = 11.96 + 5.86e^{-d/6.23}$$

The data given in Table 6.1 and Eqs. 6.7 indicate that materials parameters of the Si slab show well convergence to bulk values for slabs with $d \geq 16\text{Å}$.

However, for nano-structures having characteristic dimensions with small equivalent thickness, $d \leq 10\text{Å}$ neglect of quantum effects may result in substantial (up to 50 percent) variations of χ and ε. This should be kept in mind by applying the standard approach based on classic electrodynamics, e.g. the STM image modeling [Gavrilenko (2020); Bechstedt (2003)].

In Fig. 6.7 behavior of electrostatic and full potential electron energy on the Si-vacuum interface is depicted. As demonstrated above the Si slab with the thickness of $d = 16\text{Å}$ is a realistic models that reproduces the work function of bulk Si within 5 percent accuracy thus the model could be applied to study different phenomena related, e.g. to an electron emission, etc.

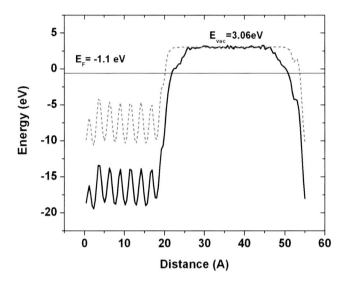

Figure 6.7 Full one-electron potential (solid line) and averaged electrostatic potential (dashed line) for a silicon (2×1) (001) surface. A Si slab contains 16 atomic monolayers (adapted from [Ilchenko et al. (2013)]).

6.2.2 Surface Electron Energy Structure

Electron energy structure of solid surfaces is typically calculated within a standard 3D band structure theory using the super cell method described in Section 6.1. In most cases the super cell used as a model for the solid surface has lower symmetry than the bulk unit cell. For example, bulk diamond crystal has a cubic symmetry but the super cell built as an atomic model for the C(001) surface widely used for application, i.e. C(001)(2 × 1) has a tetrahedral symmetry as shown in Fig. 6.3.

Crystal truncation and brake of atomic bonds cause appearance of the dangling bonds on the surface. Atomic reconstruction results in the change (reduction) of a symmetry on the surface. Consequently the electron energy structure of the surface will have essentially new electronic states in addition to the bulk states. This is demonstrated in Fig. 6.8.

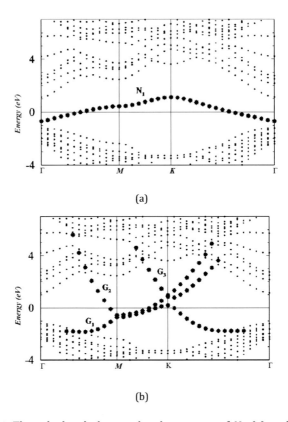

Figure 6.8 The calculated electron band structure of N- (a) and Ga- (b) terminated GaN(0001)(1×1) surface [Gavrilenko and Wu (2002)].

The first-principles full potential linearized augmented plane-wave (FLAPW) method was used by [Gavrilenko and Wu (2002)] for atomic geometries of the GaN(0001) surface optimization and calculations of the electron energy structure. According to the direct measurements the real GaN(0001) exists in two possible atomic configurations: N-terminated (1×1) structure, and Ga-terminated (1×1) surface where additional ad-atoms of Ga are placed directly over the surface nitrogen atoms [Gavrilenko and Wu (2002)]. Significant differences were found between the two cases. The N-terminated GaN(0001)(1×1) surface has one dangling bond per unit cell. The π- interaction between adjacent dangling bonds is responsible for the metallic character of this surface. The electronic

Fundamentals of the Optics of Materials: Tutorial and Problem Solving | **139**

structure of this surface has well pronounced low dispersive metallic surface state in the gap (see bold symbols in Fig. 6.8(a)). The Ga-terminated GaN(0001)(1×1) surface is characterized by three surface states (labeled as G_1, G_2, and G_3 in Fig. 6.8(b)) which disperse strongly in the Brillouin zone, indicating the strong interaction between the neighboring surface atoms.

6.3 Optics of Solid Surfaces

The goal of physical experiments on surface optics is to measure and identify specific features on optical spectra that could be unambiguously related to the surface area, pattern, and local atomic configurations on solid surfaces and interfaces that will result in understanding the mechanisms that govern optical excitation on surfaces and interfaces. Reliable diagnostics and detailed understanding of electronic processes on the surfaces and interfaces is a key point in engineering for a variety of modern electronic and photonic devices.

From previous consideration follows that truncation of a crystal, atomic relaxation, and reconstruction on the surface (and within several near-the-surface mono-layers) results in remarkable modifications of the electron energy structure, i.e. appearing of new electron energy states. Optical excitation of these new electron states contributes to the overall optical response modifying optical functions of solids (i.e. dielectric permittivity, reflectance, absorption spectra).

Incoming electromagnetic radiation penetrates into a large volume of a solid. Within the spectral region of the strong (fundamental) absorption the absorption coefficient reaches the maximum values near or somewhat above $\alpha \sim 10^6 cm^{-1}$. From equation (4.5) (see Section 4.1) follows that under this conditions drop of the light intensity by $1/e$ occur at the thickness of near 100Å, which in different solids involves 60 to 90 atomic mono-layers with further (lower but still substantial) contributions of deeper volume bulk-like layers. On the other hand, the atomic reconstruction involves only few (typically 2 to 4) uppermost mono-layers quickly relaxing to the bulk atomic structure. This was verified both experimentally and theoretically [Bechstedt (2003)] and depicted in Fig. 6.9.

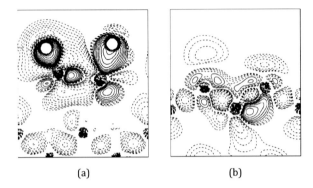

Figure 6.9 Calculated valence charge-density differences $(\rho_{bare} - \rho_{mono})$ between bare and unrelaxed monohydride Si(001)(2 × 1) surfaces. The charge maps are cut through the vertical $[\bar{1}10]$ planes located at $y = 0$ (panel (a)) and at $y = a_y$ (panel (b)) in the unit cell. Contours start from $\pm 1 \times 10^{-3} e/a.u.^3$ and increase successively by a factor of $\sqrt{2}$. Dashed (solid) lines indicate charge accumulation (depletion) in the electron density. Adapted from [Gavrilenko et al. (2001)].

As shown in Chapter 1 the optical functions of materials are determined through the light induced polarization (see Sections 1.2 and 1.3), i.e. by the redistribution of the electron charge located on atomic orbitals induced by incoming electromagnetic radiation. Due to the atomic reconstruction charge located on atomic orbitals is redistributed comparing to bulk. This process is depicted in Fig. 6.9 where the charge maps cut through two vertical planes in surface unit cell for Si(001)(2 × 1) surface are shown. Note that any dashed (accumulation) or solid (depletion) contour line means charge changes. The charge maps in Fig. 6.9 clearly demonstrate that charge redistribution due to the surface reconstruction occur mainly within upper two mono-layers and quickly relaxes to bulk value. These results demonstrate that comparing to the total number of the light excited atomic bonds, direct contribution of the surface related processes is extremely small and in most experiments on linear optics is not detectable.

6.3.1 Probing of Solid Surfaces

There are several experimental techniques applied for solid surfaces studies. In this section few widely used methods are described.

Scanning tunneling microscopy (STM) and atomic force microscopy (AFM) are widely used for direct studies of the surface atomic structures.

6.3.2 Surface-Enhanced Raman Spectroscopy

Raman spectroscopy has been used for decades as a powerful tool to study a variety of properties of materials [Cardona *et al.* (1979); Cardona (1982)]. The Raman process is an inelastic scattering of light by molecular vibrations which results in generation of scattered light at new frequencies which are the combinations of the primary photon frequency and the vibrations [Cardona (1982)]. Today Raman spectroscopy became standard spectroscopic tool for characterization of materials [Yu and Cardona (2010)].

Basic ideas of the Raman spectroscopy could be understood from the following consideration [Cardona (1982)]. Polarization $P(r, t)$ of the material exposed to external optical harmonic field $E(r, t)$ is determined by the susceptibility function $\chi(\omega)$ accordingly to (see also definitions by Eqs. (1.1) and (1.26) in Section 1.1):

$$P(r, t) = P(k, \omega) \cos(kr - \omega t)$$
$$E(r, t) = E_0(k, \omega) \cos(kr - \omega t) \tag{6.8}$$
$$P(k, \omega) = \chi(\omega) E_0(k, \omega)$$

where $|E_0|^2$ is the excitation electric optical-field amplitude. Normal modes of atomic vibrations are quantized to phonons. The atomic displacements associated with a phonon could be presented as plane waves:

$$Q(r, t) = Q(q, \Omega) \cos(qr - \Omega t) \tag{6.9}$$

where q is a wave vector and Ω frequency of phonon. In adiabatic approximation when $\omega \gg \Omega$ the $\chi(\omega)$ function could be expanded in Taylor series in $Q(r, t)$:

$$\chi(k, \omega, \Omega) = \chi_0(k, \omega) + \left(\frac{\partial \chi}{\partial Q}\right)_0 Q(r, t) + \dots \tag{6.10}$$

Substituting Eq. (6.10) into Eq. (6.9) the polarization function reduces to [Yu and Cardona (2010)]:

$$P(r, t, \Omega) = P_0(r, t) + P_{ind}(r, t, \Omega)$$

$$P_0(r, t) = \chi_0(k, \omega)E_0(k, \omega) \cos(kr - \omega t)$$

$$P_{ind}(r, t, \Omega) = \frac{1}{2}\left(\frac{\partial \chi}{\partial Q}\right)_0 Q(q, \omega)E_0(k, \omega)F(k, \omega) \qquad (6.11)$$

where spectral function generally defined by Eq. (1.81) (see Section 8.2.2) is now given by:

$$F(k, \omega) = \cos\left[(k + q)r - (\omega + \Omega)t\right]$$

$$+ \cos\left[(k - q)r - (\omega - \Omega)t\right] \qquad (6.12)$$

The induced part of polarization function P_{ind} consists of two additional waves: a *Stokes* shifted wave with wavevector $k_S = k - q$ and frequency $\omega_S = \omega - \Omega$, and an *anti-Stokes* shifted wave with wavevector $k_{AS} = k + q$ and frequency $\omega_{AS} = \omega + \Omega$. The Raman frequencies ω_S and ω_{AS} are called *Stokes* and *anti-Stokes* shifts, respectively. Note, that Eqs. (6.11) represent one-phonon Raman frequencies. In the same way one can obtain more complex combinations of two phonon combinations (*two-phonon* or second order Raman scattering) and of higher order Raman scattering.

The intensity of the scattered light can be obtained from the time-averaged value of P_{ind} given by the Eq. (6.11). The complex second order Raman tensor ($\mathcal{R}$) could be defined as [Yu and Cardona (2010)]:

$$\mathcal{R} = \left(\frac{\partial \chi}{\partial Q}\right)_0 \frac{Q}{|Q|} \qquad (6.13)$$

Using Eq. (6.13) the scattered light intensity (I_s) will depend on polarization of both incoming and scattered light, which are defined as unit vectors e_i and e_s, respectively. The value of I_s is given by [Yu and Cardona (2010)]:

$$I_s \propto |e_i \mathcal{R} e_s|^2 \qquad (6.14)$$

Practical applications of the Eqs. (6.11), (6.13), and (6.14) for calculations of I_s requires full knowledge of electron energy structure of the system.

Surface-enhanced Raman spectroscopy (or scattering) (SERS) is an optical spectroscopy method extremely sensitive to molecular adsorption on surfaces. Discovery of SERS almost three decades ago made strong impact on modern optics (see [Kneipp *et al.* (1999); Schatz *et al.* (2006); Kiefer (2008)] and references therein). The SERS effect represents strong (several order of magnitude) enhancement of the Raman radiation intensity of molecules bounded to nanostructured metallic surface (like nanoparticles of noble metals). Strongly enhanced Raman signals were verified for many different molecules, which had been attached to the *SERS-active substrates* [Schatz *et al.* (2006); Kiefer (2008)]. These substrates are based on transition metal structures with sizes on the order of tens of nanometers. The most common types of SERS-active substrates exhibiting the largest effects are colloidal silver or gold particles in the 10–150 nm size range, silver or gold electrodes or evaporated films of these metals [Kneipp *et al.* (1999)]. Several mechanisms were proposed to explain the SERS effect: electromagnetic (EM), charge transfer (CT), and chemical mechanism. The electromagnetic theory of SERS can account for all major SERS observations. However, the electromagnetic model does not account for all that is learned through SERS: single molecular resonances, charge-transfer transitions, and other processes at atomic scale [Schatz *et al.* (2006)].

In a regular Raman scattering, the total Stokes Raman signal $(I(\omega_S))$ is proportional to the Raman cross section of free molecules σ_{free}^K, the excitation laser intensity $I(\omega_L)$, and the number of molecules N in the probed volume (see Fig. 6.10(a)). It is given by [Kneipp *et al.* (1999)]:

$$I_{Ram}(\omega_S) = NI(\omega_L)\sigma_{free}^R \tag{6.15}$$

Figure 6.10 presents schematics of the regular Raman effect and the SERS. The Raman scattering has a very low cross section. In the absence of any resonance Raman process, the differential Raman cross sections $(d\sigma/d\Omega)_{RS}$ are less than $10^{-29} cm^2 sr^{-1}$, i.e., generally more than 10 orders of magnitude lower than that of infrared absorption. Because of the extremely small Raman cross sections,

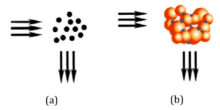

Figure 6.10 Schematics of the regular (a) and surface-enhanced (b) Raman scattering. The regular Stockes Raman scattering is determined through the light interaction with the noninteracting molecules (represented by balls in (a)). The Raman efficiency in SERS process is determined through the light interaction with molecules and by the interaction between molecules and metallic surfaces through molecular adsorption on metallic particles shown by big balls in (b). (Adapted from [Kneipp et al. (1999)].)

at least $\sim 10^8$ molecules are necessary to generate a measurable normal Raman scattering signal [Cardona (1982)].

In a SERS experiment (Fig. 6.10(b)), the molecules are attached to metallic nanostructures which frequently exhibit clusters formed by aggregation of metal colloids (fractals). In this case several processes contribute to the enhancement. Strong contributions are provided by the electromagnetic (EM) field enhancement [Schatz et al. (2006)]. The EM enhancement is associated with large local fields caused by surface plasmon resonance. The chemical mechanism could also provide with substantial contributions [Schatz et al. (2006); Kneipp et al. (1999)]. The chemical enhancement involving a resonance Raman-like process is associated with chemical interactions between the molecule and the metal surface. The latter effects include enhancement of the Raman signal of the adsorbed molecules (scattering cross section, σ_{ads}^R), which are related to the electronic coupling between the molecule and metal. SERS mainly is generated as a combination of the two effects. For the most noble-metal systems, the electromagnetic field enhancement generated from a variety of metal surfaces is dominant. The SERS intensity could be given by [Kneipp et al. (1999)]:

$$I_{SERS}(\omega_S) = NI(\omega_L)|A(\omega_L)|^2|A(\omega_S)|^2\sigma_{ads}^R \qquad (6.16)$$

where $A(\omega_L)$ and $A(\omega_S)$ represent field enhancement factors of the laser and Stokes fields, respectively.

The molecular Raman scattering is a weak process, characterized by cross sections of $\sim 10^{-29} \text{cm}^2$. SERS is commonly used to enhance Raman scattering intensities by up to 6 orders of magnitude. The most significant progress to bring SERS into the forefront was done with a single-molecule Raman spectroscopy [Kneipp *et al.* (1999)]. The high-quality SERS or surface-enhanced resonance Raman scattering (SERRS) spectra from a single molecule adsorbed on the surface of silver and gold particles or the aggregated colloids can be obtained. The estimated signal enhancement amounts up to $10^{14} - 10^{15}$ [Kneipp *et al.* (1999)], which is much higher than the widely accepted value of 10^6 [Kneipp *et al.* (1997)]. The single-molecule Raman spectroscopy is reviewed in [Kneipp *et al.* (1999)].

SERS spectroscopy of the system of R6G and *Ag* colloidal nanoparticles has been used by [Emory *et al.* (1998)] to study the relationship between optical excitation wavelength and particle size. Figure 6.11 shows the SERS spectra of R6G obtained from three different *Ag* nanoparticles.

The enhancement factors on the optically *hot spots* was reported by [Emory *et al.* (1998)] as large as $10^{14} - 10^{15}$. The results presented in Fig. 6.11 are obtained from the system having different averaged nanoparticle size; the 488-nm-excited particles had an average size (diameter) of 70 nm with a standard deviation of 6 nm, and that the 568 nm excited particles had an average size of 140 nm with a standard deviation of 9 nm. For efficient excitation at 647 nm, the nanoparticles were about 190–200 nm with a standard deviation of 20 nm [Emory *et al.* (1998)]. The data presented in Fig. 6.11 are the directly measured the size-tunable optical properties of the R6G–*Ag* system.

Despite remarkable progress in studies of SERS, this technology had not developed so far to be a powerful surface diagnostic tool that can be widely used because of some obstacles. Only three noble metals *Au*, *Ag*, and *Cu* were demonstrated to provide large enhancement, severely limiting the widespread applications involving other metallic materials of both fundamental and practical importance [Schatz *et al.* (2006)]. SERS generated on net transition metals (e.g. *Pt*, *Ru*, *Rh*, *Pd*, *Fe*, *Co*, *Ni*, and their alloys) has been studied by developing various roughening procedures and

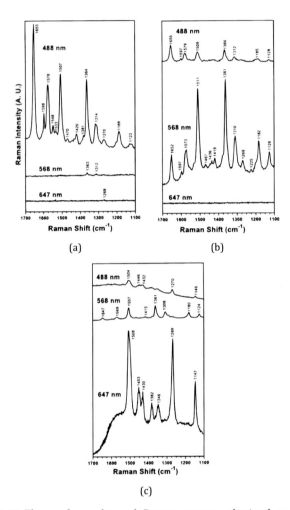

Figure 6.11 The surface-enhanced Raman spectra obtained at different excitation wavelength from the system of Rhodamine 6G molecules coupled with *Ag* nanoparticles. The panels correspond to different particle size *d*, 70 mm (a), 140 mm (b), and 200 mm (c), providing maximum enhancement at 488 nm (a), 567 nm (b), and 647 nm (c) excitation wavelength, respectively. (Adapted from [Emory *et al.* (1998)].)

optimizing the performance of the confocal Raman microscope. A replacement of the randomly roughened surface with ordered nanorod arrays of transition metals was introduced as a promising class of highly SERS-active substrates. The surface enhancement

Fundamentals of the Optics of Materials: Tutorial and Problem Solving | **147**

factor for transition metals ranged from 1 to 4 orders of magnitude has been reported [Kneipp *et al.* (1999)]. The applications of SERS in surface adsorption, electro-catalysis, and corrosion of transition-metal-based systems demonstrated several advantages of *in situ* surface Raman spectroscopy.

6.3.3 Surface Reflectance Differential Spectroscopy

In order to extract surface related electronic processes optical differential methods (e.g. Reflectance Differential Spectroscopy, RDS) are used [Del Sole (1995); Schmidt (1997); Onida *et al.* (2001)].

As a result of atomic reconstruction and/or due to the atomic or molecular adsorption on solid surface additional anisotropy is introduced that is not naturally occur in bulk. Experimentally, the surface sensitive linear optical RDS signal is measured as a normalized difference between two reflectance coefficients when the incoming light is polarized along two mutually perpendicular in-surface vectors, e.g. for the (001) surface, the two vectors are $[110]$ and $[1\bar{1}0]$ [Del Sole (1995)]:

$$\frac{\Delta R}{R} = 2\frac{R_y - R_x}{R_x + R_y}, \tag{6.17}$$

where reflectance coefficients R_y and R_x of linearly polarized light are measured for two mutually polarized directions. For example for the (001) surface $x\|[1\bar{1}0]$ and $y\|[110]$, respectively. The differential technique allows complete elimination of the contributions from bulk thus dramatically enhancing the effect of the surface[1].

Interpretation of experimental RDS spectra is performed through a theoretical calculations and analysis. The theoretical background for RDS is well developed and mechanisms that govern optical response associated with the solid surfaces are well understood [Manghi *et al.* (1990); Del Sole (1995); Schmidt (1997); Onida *et al.* (2001); Schmidt *et al.* (2001)]. It has been demonstrated before that the surface induced changes of the dielectric function

---

[1] The Reflectance Differential Spectroscopy (RDS) is sometimes called in literature Reflectance Anysotropy Spectroscopy (RAS) indicating additional anisotropy on solid surfaces induced by atomic reconstruction and/or adsorption [Schmidt *et al.* (2001)].

of solid is described in terms of the dynamic polarizability function (α^{hs}) associated with a half slab polarization (that eliminates spurious contribution of the back face within the slab model of the solid surface discussed above) that is given by [Onida *et al.* (2001)]:

$$\alpha_{ii}^{hs}(\omega) = \frac{\pi e^2}{m\omega^2 Ad} \sum_{k,c,v} |\langle \phi_v | p_i | \phi_c \rangle|^2 \delta(E_{k,c} - E_{k,v} - \hbar\omega), \quad (6.18)$$

where i denotes the direction of light polarization on the surface, d is the slab thickness, $\mathbf{p}$ stands for the momentum, m is electron mass, and A is the area of the surface unit cell.

The function $\alpha_{ii}^{hs}(\omega)$ defines the RDS spectrum measured experimentally (see equation (6.17)) in the following way [Del Sole (1995)]:

$$\frac{\Delta R}{R} = \frac{4\omega}{c} \mathrm{Im} \frac{4\pi d \alpha_{ii}^{hs}(\omega)}{\varepsilon_b(\omega) - 1}, \quad (6.19)$$

where $\varepsilon_b(\omega)$ is the complex dielectric function spectrum of bulk.

The calculated tridymite atomic configuration of SiO_2 is used as a basic structural model for the initial oxide layer in the Si(001) surface. The fully relaxed atomic geometry of the Si(001)/SiO_2 interface is shown in Fig. 6.12.

The atomic structure presented in Fig. 6.12 corresponds to the reported in literature atomic configuration for the tridymite [Kolasinski (2008)]. In order to understand a contribution of selected orbitals of a specific atom in the unit cell to the optical spectra, one removes the atom of interest and performs electronic structure and optical function calculations. A comparison of energy structure and optics predicted with and without geometry optimization can help separate the effects of structural reconstruction, local stress, and chemical hybridization.

In order to separately study the effects of local strains and chemical bond hybridization, in this book optical spectra are calculated by [Gavrilenko (2008)] for both relaxed and unrelaxed structures after the removal of a specific atom in the unit cell. The contributions of different atoms in the unit cell to optical response functions were also analyzed.

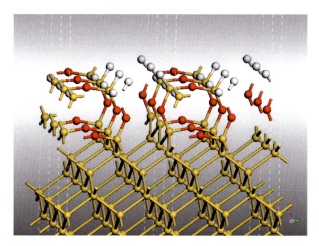

Figure 6.12 Fully relaxed atomic configuration of Si(001)/SiO$_2$ interface. Dashed lines indicate unit cells. Red and white colored balls correspond to oxygen and hydrogen atoms, respectively. (Adapted from [Gavrilenko (2008)].)

The signature of the dimer bridge oxygen local configuration was identified from the analysis of both linear and nonlinear optical spectra [Gavrilenko (2008)]. This structure is created at the initial oxidation step after breaking the Si-Si dimer bond on the Si(001) surface by approaching the oxygen atom and creating the Si-O-Si dimer structure as presented in the tridymite oxide structure (see Fig. 6.12).

The experimental RDS spectra shown in the upper panel of Fig. 6.13 were measured on Si(001) after room temperature adsorption of 160 Langmuir ($1L = 1 \times 10^{-6}$ torr sec) molecular oxygen [Borensztein et al. (2005)]. The spectral positions between predicted (with a QP correction of 0.5 eV [Gavrilenko (2008)]) and measured RDS optical peaks are in good agreement. Structural disorder apparently smears out the optical spectra, which may explain the substantial disagreement in a region around 4.0 eV.

Removal of the bridge oxygen substantially suppressed the RDS response. Calculations on a structurally relaxed system (see upper panel of Fig. 6.13) show that mechanical stress can cause some spectral shifts and a minor redistribution of the oscillator strengths. However, a remarkable decrease in RDS response in the

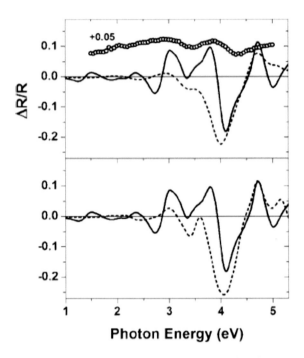

Figure 6.13 Reflectance difference spectra of the Si(001)/SiO₂ interface corresponding to the fully relaxed interface atomic geometry given in Fig. 6.12 (solid lines) are shown in comparison to the RDS spectra calculated after the removal of the bridge oxygen (dashed lines). In the upper (lower) panel, the fully relaxed (unrelaxed) atomic structure of the *Si*(001)/*SiO₂* interface without bridge oxygen is used for optical calculations. The experimental data of [Borensztein et al. (2005)] are shifted upward by 0.05 for convenience. (Adapted from [Gavrilenko (2008)].)

spectral region between 3 and 4 eV presented in both relaxed and unrelaxed model systems (lower panel of Fig. 6.13) are the result of oxygen-related rehybridization of back bonds, which demonstrated that the bridge dimer oxygen configuration provided substantial contribution to the RDS spectra measured by [Borensztein et al. (2005)].

6.3.4 Surface Analysis by the Second Harmonic Generation

Within recent decades the Second Harmonic Generation (SHG) method based on even-order optical excitation processes [Bloembergen (1965); Shen (2003)] became a very important and widely used tool to study surface atomic structure distortion of crystals with inversion symmetry (such as Si, Ge, or C diamond), molecular surface chemistry, physics of quantum confined systems, etc. (see e.g. [Kiselev and Krylov (1985); Downer *et al.* (2001); Wang *et al.* (2005); Murzina *et al.* (2012)] and references therein). In this section we briefly overview and discuss physics and experimental technique of the SHG.

In early 60s the SHG from the GaAs surface was first reported (see [Bloembergen (1965)] and references therein). The electromagnetic wave of SHG is caused by boundary conditions: in nonlinear crystal an initial optical wave induces electric polarization wave at the second harmonic frequency. This causes the SHG process.

$$P_i^{(2)} = \chi_{ijk}^{(2)} E_j E_k + \chi_{ijk}^{(2)S} E_j E_k, \qquad (6.20)$$

where $\chi^{(2)}$ is the second order susceptibility tensor, $\chi^{(2)S}$ is the second order surface susceptibility tensor. The repeated indices stand for summation.

According to the *Neumann's* principle:[2] the symmetry of the SHG components are defined by crystal point group symmetry. The surface nonlinear second order susceptibility can be presented as

$$P_i^{(2)} = \chi_{ijk}^{(2)S} \varepsilon_j(z = 0)\varepsilon_k(z = 0), \qquad (6.21)$$

where $z = 0$ defines a location of the interface, $\varepsilon_{x,y,z} = E_{x,y}$ stand for in-plane electric field vector components at $z = 0$; $\varepsilon_z = D_z$ stands for a dielectric induction vector component normal to the plane $z = 0$.

For an isotropic medium the nonzero $\chi^{(2)S}$ tensor components on interface are given by

$$\chi_{xxz} = \chi_{xzx} = \chi_{yzy} = \chi_{yyz}, \ \chi_{zxx} = \chi_{zyy}, \ \chi_{zzz} \qquad (6.22)$$

---

[2] Neumann's principle, states that, if a crystal is invariant with respect to certain symmetry operations, any of its physical properties must also be invariant with respect to the same symmetry operations. Applying to the SHG that means that the symmetry operations of the second harmonic radiation components of a crystal must include the symmetry operations of the point group of the crystal.

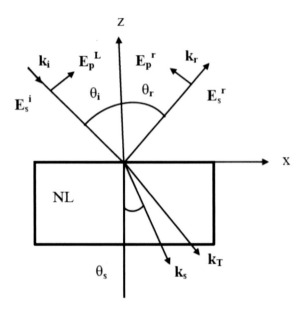

Figure 6.14 Incident, transmitted, and reflected electromagnetic plane waves propagated in a nonlinear medium.

Assume an incident plane electromagnetic wave penetrates at θ angle through the solid surface. The optical wave vector $\mathbf{k}_i$ is located within the $y = 0$ plane (see Fig. 6.14). The wave is given by

$$\mathbf{E} = \frac{1}{2} \mathbf{e} E \exp(ik_z z + ik_x x - i\omega t) + \text{c.c.} \qquad (6.23)$$

where $\mathbf{e} = (\cos\theta_i \cos\varphi, \sin\varphi, -\sin\theta_i \cos\varphi)$ is the direction of vector $\mathbf{E}$; φ is an angle between normal to the $z = 0$ plane and the $\mathbf{e}$ vector; x and z wave vector components are given by $k_z = -\frac{\omega}{c} \cos\theta_i$, and $k_x = \frac{\omega}{c} \sin\theta_i$.

The SHG-wave is characterized by the wave vector $\mathbf{k}_r(2\omega) = 2\mathbf{k}_2(\omega)$; $|\mathbf{k}_s| = 2(\frac{\omega}{c})\sqrt{\varepsilon(\omega)}$. It propagates at the angle of θ_s. Waves with $\mathbf{k}_r(2\omega)$ and $\mathbf{k}_r(\omega)$ propagate in the same direction. However, the situation changes in a medium with dispersion. In this case

$$n_r(2\omega) \sin\theta_r(2\omega) = n_r(\omega) \sin\theta_r(\omega) \qquad (6.24)$$

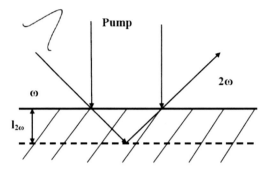

Figure 6.15 Optical SHG-probing of a phase transition on a semiconductor surface.

Electric field vector components of reflected waves are given by

$$E_{2\omega,s} = (E_{2\omega})_y = i\frac{4\pi 2\omega}{c^2(k_{1z} + k_{2z})} P_y^{NL}(2\omega), \quad (6.25)$$

$$E_{2\omega,p} = i\frac{4\pi 2\omega k_1}{c^2(k_z^2 k_{1z} + k_1^2 k_{2z})} (k_{2x} P_z^{NL} + k_{2z} P_x^{NL}) \quad (6.26)$$

In equation (6.25) optical amplitude indices s and p indicate perpendicular and in-plane waves, respectively. Wave vectors of optical field are equal to

$$k_1 = \frac{2\omega}{c}, k_2^2 = (\frac{2\omega}{c})^2 \varepsilon(2\omega), k_s = 2\frac{\omega}{c}\sqrt{\varepsilon(\omega)}. \quad (6.27)$$

As an example consider now a SHG wave generation in non-stationary phase transition experimental study of a crystal surface melting under a powerful laser beam.

The SHG-signal is generated in the near surface layer which has the thickness of $l_{2\omega} \sim \alpha_{2\omega}^{-1}$, where $\alpha_{2\omega}$ is an absorption coefficient for the second harmonic signal. In experiments by [Akhmanov and Khokhlov (1972)] for melting the surface of GaAs crystal a pulse of Ru laser ($\lambda = 694\ nm$) was used. Surface laser emission density was $0.1...1\ J/cm^2$. Nd:YAG pulse laser was user as a probing light source ($\lambda = 1064nm$). The results of experiment are shown in Fig. 6.16. One can see that

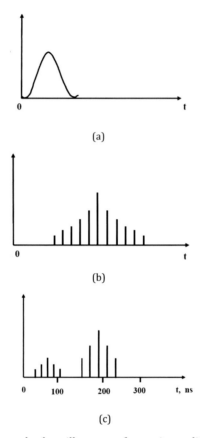

Figure 6.16 Time resolved oscillograms of pumping radiation (a), probing beam (b), and the surface-related SHG signal (c).

SHG-signal disappeared as a result of the liquid phase creation on the GaAs crystal surface as a results of melting.

The fully disordered and amorphous systems show inversion symmetry thus SHG generation in these systems is forbidden. The ideal GaAs crystal does not have an inversion symmetry thus showing a strong SHG generation [Downer *et al.* (2001)]. Melting of the surface creates a thin fully disordered layer with inversion symmetry and no SHG response. This is shown in Fig. 6.16(c). Within first $\sim$ 100 *ns* the intensity of the probing SHG signal is reduced (it is non-zero due to the contribution to the reflected probing beam from the

Fundamentals of the Optics of Materials: Tutorial and Problem Solving | **155**

deeper non-melted layers). After ~ 200 *ns* the probing SHG signal is restored due to recrystallization.

The centrosymmetric crystal (such as Si) can show the SHG signal due to surface atomic reconstruction and breaking the inversion symmetry [Gavrilenko (2008); Murzina *et al.* (2012)].

6.3.5 Si-SiO$_2$ Interface Analysis by the Second Harmonic Generation

The increasing amount of high-quality works on a Si/SiO$_2$ system within the past decades is caused by an extensive development of nanostructured electronics, which requires a good understanding of the processes at the atomic monolayer scale. An understanding of the oxygen-related processes on the Si/SiO$_2$ interface is very important for both fundamental and applied physics as well as for high-technology electronics.

The SHG method can noninvasively probe interfaces buried beneath transparent overlayers. The interior of crystalline Si as well as amorphous SiO$_2$ are centrosymmetric, consequently can contribute only quadrupolar SHG [Downer *et al.* (2001)]. However, centrosymmetry is broken locally at the Si/SiO$_2$ interface. It has been demonstrated earlier [Fuchs *et al.* (2005); Hughes and Sipe (1996); Lim *et al.* (2000); Erley and Daum (1998); Gavrilenko and Wu (2000)] that SHG is a unique method to study centrosymmetric solid interfaces: the SHG response is forbidden in bulk, and the SHG signal is generated within only few interface monolayers. In particular one can expect a stronger contribution of oxygen-related process within the Si/SiO$_2$ interface to the SHG spectra than in linear optics traditionally used to diagnose Si/SiO$_2$ interface.

Consider the effect of the chemical nature of electronic bond contributions on Si(001)-SiO$_2$ interface to the SHG. Figure 6.17 presents the calculated SHG spectra of Si(001)-SiO$_2$ interface using computer simulated model shown in Fig. 6.12 [Gavrilenko (2008)].

As expected the non-zero SHG response is predicted only in the spectral region corresponding to optical two-photon excitations of the near-surface located disturbed Si electron orbitals. Due to the high value of Si-O bond energy corresponding contributions are located in far ultraviolet region. The SHG response from the Si-SiO$_2$

interface located mostly in visible and near-UV regions is attributed to the distorted host atomic bonds [Fuchs *et al.* (2005); Gavrilenko *et al.* (2001); Gavrilenko and Wu (2000)]. From the data presented in Fig. 6.17 one can immediately extract features related to the oxygen. Removal of the bridge oxygen results in dramatic reductions of the SHG features near 2.7, 3.3, 3.8 to 4.0, and 5.1 eV. Comparison between relaxed and unrelaxed structure calculations of the SHG efficiency (upper and lower panel in Fig. 6.17, respectively) shows a more pronounced effect of mechanical stress in SHG than in RDS spectra discussed in Section 6.3.3 (see Fig. 6.17). The SHG features near 3.3 and 5.1 eV are close to the bulk electron transitions E_1 and E'_1.

Note that the theory with the QP correction predicted critical point energies in Si slab at: $E_1 = 3.4$ eV, $E_2 = 4.15$ eV, and $E'_1 = 5.0$ eV. The peaks near 3.3 and 5.1 eV apparently relate to the bulk-like electron excitations. According to the theoretical predictions [Gavrilenko (2008)] the SHG response near 5.1 eV exhibits strongest sensitivity to oxygen. This spectral region however is still unavailable for experimental study.

The SHG peak near 1.8 eV is of the surface nature related to the Si dimer bond. This peak is strongly affected by oxygen and local stress (see Fig. 6.17). One can expect that it will also be sensitive to the size of nanoparticles since the smaller the particle, the higher the curvature of the interface and the more stress that is introduced at the Si/SiO_2 interface [Bechstedt (2003); Kolasinski (2008)]. Contribution of oxygen in this region is clearly shown in RDS spectra in this spectral region, however effect of the stress relaxation is much less important in RDS then in SHG. The SHG peaks located near 2.7 and 3.8 eV are new and they are not predicted on the Si surfaces without oxygen. Accordingly to the analysis of the partial density of states (PDOS) spectra these peaks are directly related to the Si backbonds of the dimer atoms hybridized to the bridge oxygen $2p$-electrons. Peak near 4.0 eV is close to the E_2 transition and it is strongly affected by the new feature near 3.8 eV. Comparison between SHG efficiency predicted for relaxed and unrelaxed tridymite structure after removal of bridge oxygen (upper and lower panels in Fig. 6.17, respectively) demonstrates that effects of local stresses and bond rehybridization are equally important in SHG which is in contrast to the linear optical RDS response [Borensztein *et al.* (2005)].

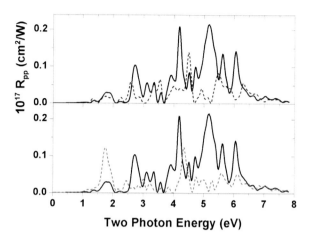

Figure 6.17 Calculated SHG efficiency spectra of the Si(001)-SiO$_2$ interface corresponding to the geometry given in Fig. 6.12 (solid line) are shown in comparison to the SHG spectra calculated after removal of the bridge oxygen (dashed line). In upper (lower) panel fully relaxed (unrelaxed) atomic structure of the Si(001)-SiO$_2$ interface is used for optical calculations. (Adapted from [Gavrilenko (2008)].)

The predicted SHG efficiency spectrum of oxidized Si(001) surface is compared next with available experimental data [Rumpel et al. (2006); Lim et al. (2000)].

In Fig. 6.18 a comparative analysis is presented of the SHG spectra measured on two different systems: the Si(001) surface oxidized [Rumpel et al. (2006)] or with natural oxide [Erley and Daum (1998)], and Si-SiO$_2$ multiple quantum well structure [Avramenko et al. (2006)]. In order to compare shape of the predicted and measured SHG spectra the amplitudes of experimental data were scaled to meet the theoretical values at 4.2 eV (lower panel) and at 2.7 eV (upper panel).

The experimental SHG spectrum shown in lower panel in Fig. 6.18 was measured by [Rumpel et al. (2006)] on Si(001) sample with 10 nm oxide layer grown by thermal oxidation at 1000°C. The dominant SHG peak at 4.3 eV was attributed to the E_2 bulk transitions [Rumpel et al. (2006)]. Note that the theoretical bulk

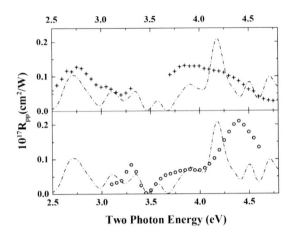

Figure 6.18 Calculated SHG efficiency spectra of the Si(001)-SiO₂ interface in comparison with experimental data measured on oxidized Si(001) surface [Rumpel et al. (2006)] (lower panel) and on Si(001)-SiO₂ multiple quantum well structures [Avramenko et al. (2006)] (upper panel). (Adapted from [Gavrilenko (2008)].)

value of E_2=4.15 eV underestimates experimental one by about 0.1 eV. The results shown in Fig. 6.18 indicate appearance of new SHG response in the region of 3.6 to 4.0 eV and strong enhancement of the E_2 peak. Comparison with SHG data presented in Fig. 6.17 indicates that this is caused by the combined effect of the oxygen related rehybridization of Si backbonds and structural reconfiguration. The strong enhancement of both measured and predicted SHG efficiencies caused by boron doping of the Si(001) surface (which was obtained on the relaxed system) was reported earlier (see [Downer et al. (2001)] and references therein). In both cases by neglect of electric field effect, the physical nature of the predicted strong enhancement of SHG efficiency is impurity related rehybridization and structural reconfiguration of the Si backbonds.

The predicted new SHG response near 3.6 to 4.0 eV is accompanied by dramatic increase of calculated SHG efficiency at 2.7 eV due to the dimer bridge oxygen (see Figs. 6.17, 6.18). This part of the predicted SHG spectra agrees well with another type of experimental results reported by [Avramenko et al. (2006)] who measured strong SHG signals in the region near 2.7 eV on Si(001)-SiO_2 multiple quantum wells. In MQW system the quantum size effect is an additional factor affecting the SHG. However in the presence of multiple Si(001)-SiO_2 boundaries the effect of the interface oxygen should be enhanced. In the absence of microscopic theory the 2.7 eV SHG signal measured in Si(001)-SiO_2 MQW was interpreted as a result of electron transitions from the bound quantum electron states in QW. In addition to the quantum size effect, the chemical nature of the last SHG feature was also suggested by [Avramenko *et al.* (2006)] as an alternative interpretation of their data. The results of the theoretical work suggest that the origin of the feature measured by [Avramenko *et al.* (2006)] and the predicted SHG responses near 2.7 eV is a dominant contribution of the re-hybridized and reconstructed Si back bonds due to creation of the dimer bridge oxygen structure on Si(001)-SiO_2 interface. Additional argument towards oxygen related interpretation of the 2.7 eV signal is that this response should be accompanied by the SHG features near 3.8 eV as discussed above (see Fig. 6.17). Experimental data of [Avramenko *et al.* (2006)] confirm this rule: in addition to the 2.7 eV SHG peak, the authors reported strong SHG response around 3.8 to 4.0 eV which is discussed above.

Analysis of the atoms- and orbitals-resolved projected electron density of states (PDOS) calculated by [Gavrilenko (2008)] clearly indicates strong effect of the bridge oxygen which results in additional contribution (hybridization) of oxygen p-orbitals and silicon backbond orbitals. The $2p$-orbitals of bridge oxygen contribute to the top of the valence band. Modifications of the c-band are less pronounced. The rehybridization of the host valence electron orbitals caused by both $2p$-orbitals of oxygen and by structural distortions seems to be responsible for the measured and predicted optical anisotropy of this system. The dominating effect in optical anisotropy of Si(001) due to the distorted backbonds and additional hybridization to oxygen has been shown by [Fuchs *et al.* (2005)].

6.4 Review Questions and Exercises

6.4.1 Anharmonicity on the Surface

Calculate root-mean-square (RMS) thermal fluctuations of atom moving under elastic forces. Take into account anhamonicity which effect is essential on solid surfaces. Hint: Potential energy of an atom moving in one dimensional lattice is given by: $U(x) = U_0 + 1/2\beta x^2 - 1/3\gamma x^3 + ...$

6.4.2 Carrier Concentration on the Surface

In the absence of external fields (i.e. $E = 0$) calculate spatial distribution of the excess concentration of charge carriers in the direction (x) perpendicular to the surface of the sample. Take into account the surface recombination τ (τ is a recombination constant), the stationary generation rate of charge carriers, G_0. Hints: Apply continuity equation for the diffusion process; consider two limiting values for the recombination current (i.e. at the surface $x = 0$ and at the end of the transition layer, $x = d$) as known, i.e. $j(0) = j_0, j(d) = j_d$.

Chapter 7

Plasma Optics

Collective electronic excitations are also known as plasma excitations in metals, metal surfaces, and metal/dielectric interfaces. They play a key role in the optics of materials. A unified theoretical description of these phenomena is based on the many-body dynamical electronic response of solids, which underlines the existence of various collective electronic excitations at metal surfaces, such as the conventional surface plasmon, multipole plasmons and the acoustic surface plasmon. Detailed description of the surface plasmon polariton (SPP) phenomena and its applications in modern optical spectroscopy is out the scope of present book. Several specialized monographs and reviews can be recommended for advanced reading [Tudos and Schasfoort (2008); Pitarke *et al.* (2007); Liebsch (1997); Ritchie (1973); Venger *et al.* (1999); Raether (1988)]. Here we present the basic descriptions and properties of plasma in solids and SPP which follow from the classical electrodynamics.

7.1 Theoretical Model Based on Classical Electrodynamics

Electrons in solids move in the field of positively charged ions while the entire solid is neutral. The moving electrons and the motionless

Fundamentals of the Optics of Materials: Tutorial and Problem Solving
Vladimir I. Gavrilenko and Volodymyr S. Ovechko
Copyright © 2024 Jenny Stanford Publishing Pte. Ltd.
ISBN 978-981-4877-93-0 (Hardcover), 978-1-003-25694-6 (eBook)
www.jennystanford.com

Plasma Optics

ions (adiabatic approximation) in lattice create the so called *solid-state plasma*. Fluctuations in the homogeneous distribution of electron density cause strong Coulomb restoring forces, which force electrons to return to their equilibrium states. The electron charge density oscillates in time as demonstrated below. The equation of electron motion is given by

$$m^* \frac{d\mathbf{v}}{dt} = -e\mathbf{E}, \tag{7.1}$$

where m^* is effective electron mass, $\mathbf{v}$ is velocity, e is electron charge, and $\mathbf{E}$ is electric field.

The Poisson equation is given by

$$\nabla \mathbf{E} = -4\pi e(n - n_0)/\varepsilon, \tag{7.2}$$

where $n - n_0$ stands for a fluctuation of the electron density, ε is dielectric permittivity. If the number of electrons does not change we have continuity equation

$$\frac{\partial n}{\partial t} = -\nabla(\mathbf{v}n). \tag{7.3}$$

After taking into account the expression for $n\nabla\mathbf{v}$ in Eq. (7.3) from above equations follows:

$$\frac{\partial^2 n}{\partial t^2} = -\frac{4\pi e^2}{\varepsilon m^*}(n - n_0). \tag{7.4}$$

Solution of the equation (7.4) is a harmonic oscillation with the frequency

$$\omega_0^2 = \frac{4\pi e^2 n_0}{\varepsilon m^*}, \tag{7.5}$$

where ω_0 is the *plasma frequency* introduced in Section 1.2 (see Eq. (1.47)).

After taking into account elastic properties of electron gas, the plasma excitation can be presented as a propagating wave described by the following dispersion equation:

$$\omega^2(\mathbf{q}) = \omega_0^2 + v^2(\mathbf{q})\mathbf{q}^2, \tag{7.6}$$

where $\mathbf{v}(\mathbf{q})$ is the velocity of the acoustic waves, $\mathbf{q}$ is wave vector. Additional condition for plasma oscillation is

$$\omega_0 \tau \geq 1, \tag{7.7}$$

where τ stands for the *collision time*. The collision time τ typically ranges from 10^{-13} to 10^{-14} seconds. For the plasma excitation the electron concentration should be high enough, i.e. $n_0 > 10^{18} cm^{-3}$. These concentration is normally observed in metals. In semiconductors it can be easily achieved as a result of a power laser irradiation.

7.2 Plasma in Metals

A response of free electrons to an external driving electric field is described by the approach similar to that given in Chapter 1 (see Section 1.2) with one essential difference: the absence of the restoring force (Drude model). In contrast to Eq. (1.39) the equation of electron motion (7.1) in external electromagnetic field $\mathbf{E}(\omega) = \mathbf{E}_0 exp(i\omega t)$ is given by:

$$m^* \frac{d^2 \mathbf{r}}{dt^2} + \frac{m^*}{\tau} \cdot \frac{d\mathbf{r}}{dt} = -e\mathbf{E}_0 exp(-i\omega t). \tag{7.8}$$

We neglect any elastic interactions of electrons in solids (semiconductors or metals). Solution of the equation (7.8) is given by:

$$\mathbf{r}(t) = \frac{e}{m^*(\omega^2 + i\omega/\tau)} \mathbf{E}_0 exp(-i\omega t). \tag{7.9}$$

The electron displacements ($\mathbf{r}$) results in appearance of the dipole moment $e\mathbf{r}$ that determines polarization given by the vector field $\mathbf{P}$ according to:

$$\mathbf{P} = -n_0 e\mathbf{r} = -\frac{n_0 e^2}{m^*(\omega^2 + i\omega/\tau)} \mathbf{E}_0 exp(-i\omega t), \tag{7.10}$$

Frequency dependence of the dielectric susceptibility function $\chi(\omega)$ is given by:

$$\chi(\omega) = -\frac{n_0 e^2}{m^*(\omega^2 + i\omega/\tau)}. \tag{7.11}$$

After some algebra spectral dependence of the dielectric permittivity function is given by:

$$\varepsilon(\omega) = 1 + 4\pi\chi(\omega) = 1 - \frac{\omega_p^2}{\omega^2 + i\omega/\tau}, \tag{7.12}$$

where $\omega_p = (4\pi n_0 e^2/m^*)^{1/2}$ is the plasma frequency of free electrons.

Dielectric permittivity, $\varepsilon(\omega)$, defined in Eq. (7.12) is a complex function with a real ε' and imaginary ε'' parts given by:

$$\varepsilon' = 1 - \frac{\omega_p^2 \tau^2}{1 + \omega^2 \tau^2}, \qquad \varepsilon'' = \frac{\omega_p^2 \tau}{\omega(1 + \omega^2 \tau^2)}. \tag{7.13}$$

For the optical frequency range, corresponding to the condition $\omega\tau \gg 1$, we have:

$$\varepsilon' \simeq 1 - \frac{\omega_p^2}{\omega^2}, \qquad \varepsilon'' \simeq \frac{\omega_p^2}{\omega^3 \tau}. \tag{7.14}$$

If $\omega\tau \ll 1$ equation (7.13) reduces to the following

$$\varepsilon' \simeq 1 - \omega_p^2 \tau^2, \qquad \varepsilon'' \simeq \omega_p^2 \tau/\omega. \tag{7.15}$$

Other optical functions (e.g. optical reflectance) corresponding to the excitation of the solid-state plasma could be calculated within the Drude model from the functions ε' and ε'' defined by Eq. (7.13).

7.3 Surface Plasma

Plasma excitations (*plasmons*) considered in previous sections are called the volume plasmons. Consider another type of plasmons: the surface plasmons. These excitations can be generated near the

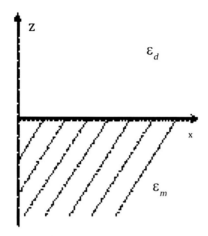

Figure 7.1 Dielectric–metal interface model to study propagation of surface plasmons; ε_d stands for dielectric and ε_m stands for metal permittivity.

surfaces or within the solid interfaces, e.g. dielectric–metal interface shown in Fig. 7.1.

For the interface model shown in Fig. 7.1, a solution of the Maxwell equation for the dielectric region is given by:

$$E_{dx} = E_0 \exp[i(k_{sp}x + k_{zd}z - \omega t)],$$
$$E_{dz} = -\frac{k_{sp}}{k_{zd}} E_0 \exp[i(k_{sp}x + k_{zd}z - \omega t)]. \quad (7.16)$$

For the metal region the solution is given by:

$$E_{mx} = E_0 \exp[i(k_{sp}x - k_{zm}z - \omega t)], \quad (7.17)$$
$$E_{mz} = \frac{k_{sp}}{k_{zm}} E_0 \exp[i(k_{sp}x - k_{zm}z - \omega t)],$$

where

$$k_{zd} = (\varepsilon_d(\omega)(\frac{\omega^2}{c^2}) - k_{sp}^2)^{1/2} \quad (7.18)$$
$$k_{zm} = (\varepsilon_m(\omega)(\frac{\omega^2}{c^2}) - k_{sp}^2)^{1/2}.$$

166 | *Plasma Optics*

From the boundary conditions for fields (i.e. continuity of fields) on the interface follows:

$$\frac{\varepsilon_m(\omega)}{k_{zm}} + \frac{\varepsilon_d(\omega)}{k_{zd}} = 0. \tag{7.19}$$

Equation (7.19) represents a dispersion equation for the wave propagating in the interface region. The wave vectors calculated from Eqs. (7.16) and (7.18) are given by:

$$k_x = k_{sp} = \frac{\omega}{c}\left(\frac{\varepsilon_m(\omega)\varepsilon_d(\omega)}{\varepsilon_m(\omega) + \varepsilon_d(\omega)}\right)^{1/2} \tag{7.20}$$

$$k_{j,z} = \frac{\omega}{c}\frac{\varepsilon_j}{(\varepsilon_m(\omega) + \varepsilon_d(\omega))^{1/2}}.$$

where $j = m, d$. For the localized plasmons the values of k_{sp} must be real and those of $k_{j,z}$ must be imaginary. Therefore

$$\varepsilon_m(\omega) \cdot \varepsilon_d(\omega) < 0, \qquad \varepsilon_m(\omega) + \varepsilon_d(\omega) < 0. \tag{7.21}$$

Within the Drude theory the dispersion relation for metal is given by:

$$\varepsilon_m(\omega) = 1 - \frac{\omega_p^2}{\omega^2}. \tag{7.22}$$

The dispersion function is depicted in Fig. 7.3 and is discussed in Section 7.4.

7.4 Surface Plasmon Polaritons

Surface plasmons have some properties useful for applications. The amplitude of surface plasmons is characterized by rapid decay along z-axis. In quantum confined materials (i.e. nanomaterials, nano-particles, -dots, -rods, etc. that are characterized by many internal interfaces) this feature results in many interesting consequences.

Consider a model (see Fig. 7.1) consisting of two semi-infinite nonmagnetic media with local (frequency-dependent) dielectric functions $\varepsilon_d(\omega)$ (dielectric) and $\varepsilon_m(\omega)$ (metal) separated by a planar interface at $z = 0$ [Pitarke *et al.* (2007)]. The full set of Maxwell's equations in the absence of external sources can be expressed as follows [Jackson (1998)]:

$$\nabla \times \boldsymbol{H}_n = \varepsilon \frac{1}{c} \frac{\partial}{\partial t} \boldsymbol{E}_n, \tag{7.23}$$

$$\nabla \times \boldsymbol{E}_n = -\frac{1}{c} \frac{\partial}{\partial t} \boldsymbol{H}_n, \tag{7.24}$$

$$\nabla \cdot (\varepsilon_n \boldsymbol{E}_n) = 4\pi\rho \tag{7.25}$$

$$\nabla \cdot \boldsymbol{H}_n = 0 \tag{7.26}$$

where the index $n = d$ refers to dielectric and $n = m$ to metal.

Within the classic picture, the metal could be treated as a semi-infinite electron gas with abruptly terminated profile of the electron density function. The surface charge density on the metal-dielectric interface can be presented as [Liebsch (1997); Raether (1988)]:

$$n(\boldsymbol{r}, \omega) \sim e^{j\boldsymbol{q}_\parallel \boldsymbol{r}_\parallel} \delta(z) \tag{7.27}$$

The electric field associated with the density Eq. (7.27) is given by the Gauss law, which follows from Eq. (7.25) in the presence of charge:

$$\nabla \cdot \boldsymbol{E}(\boldsymbol{r}, \omega) = -4\pi e n(\boldsymbol{r}, \omega) \tag{7.28}$$

Neglecting the retardation effects only the longitudinal plasma oscillations are considered here. In terms of the scalar potential ϕ the electric field is given by:

$$\boldsymbol{E}(\boldsymbol{r}, \omega) = \nabla \cdot \phi(\boldsymbol{r}, \omega) \tag{7.29}$$

For the chosen model system the potential ϕ is given by:

$$\phi(\boldsymbol{r}, \omega) = e^{j\boldsymbol{q}_\parallel \boldsymbol{r}_\parallel} \phi(z, \omega) \tag{7.30}$$

The z-components of the field decay evanescently into both media. This follows from the electroneutrality since $\nabla^2 \phi = 0$ must valid everywhere except at $z=0$ [Liebsch (1997)]. Consequently the

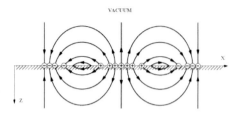

Figure 7.2 Schematics of charge and field distribution for a surface plasmon describing by Eq. (7.31). Adapted from [Hofmann (2008)].

potential in Eq. (7.30) must be taken in form:

$$\phi(\mathbf{r},\omega) = \phi_0 e^{j\mathbf{q}_\parallel \mathbf{r}_\parallel} e^{-q|z|} \tag{7.31}$$

where $q \equiv q_z$. In this case the field determined by Eqs. (7.29) and (7.31) varies continuously within the interface, however normal component is discontinuous. The components above and below the interface are given by:

$$E_z(z+0) = q_\parallel \phi_0 e^{j\mathbf{q}_\parallel \mathbf{r}_\parallel}$$
$$E_z(z-0) = -q_\parallel \phi_0 e^{j\mathbf{q}_\parallel \mathbf{r}_\parallel} \tag{7.32}$$

The field and charge distribution for such a surface mode are represented in Fig. 7.2.

In the long-wavelength limit the boundary condition can be written as:

$$\varepsilon_m(\omega) E_z(0^-) = \varepsilon_d E_z(0^+) \tag{7.33}$$

For the metal–vacuum interface (i.e. by $\varepsilon_d = 1$) the condition for the existence of the surface plasmons followed from Eqs. (7.32) and (7.33) now reads:

$$\varepsilon_m(\omega) = -1 \tag{7.34}$$

The equation (7.34) defines the frequency of the surface plasmon in the **$q_\parallel$**=0 limit [Liebsch (1997)]. Dielectric dispersion in metal could be described by the Drude model (see Eq. (1.46) in Section 1.2) which could be writen in form:

$$\varepsilon_m(\omega) = 1 - \frac{\omega_p^2}{\omega(\omega + j\Gamma)} \qquad (7.35)$$

Pluging Eq. (7.34) into Eq. (7.35) and neglecting losses results in the following:

$$\omega_s = \frac{\omega_p}{\sqrt{2}} \qquad (7.36)$$

The plasma frequency ω_p is given by Eq. (1.47) (see Section 1.2). The existence of the electronic excitations on the interfaces was predicted by [Ritchie (1957)]. If the overlayer has the dielectric constant $\varepsilon_d > 1$ the condition Eq. (7.34) is given by:

$$\varepsilon_m(\omega) = -\varepsilon_d \qquad (7.37)$$

Consequently surface plasmon frequency is red-shifted accordingly to:

$$\omega_s = \frac{\omega_p}{\sqrt{\varepsilon_d + 1}} \qquad (7.38)$$

This red shift has been observed on different metal–dielectric interfaces (see e.g. [Ritchie (1973); Pitarke *et al.* (2007); Raether (1988)] and references therein). It is a key point of Surface Plasmon Resonance (SPR) based sensing spectroscopic tools widely used for different fundamental and applied studies.

Consider now the dispersion of surface plasmons. The solution of the system Eqs. (7.23) to (7.26) will be separated into s- (E vector perpendicular) and p-polarized (E vector parallel to the plane of incidence) EM modes. If there is a wave propagating along the interface, it should contain the electric field E-component perpendicular to the interface (the p-polarized mode) and thus the s-mode is not relevant. Consequently, the problem is now formulated as the search for the conditions of the propagating of p-polarized EM wave along the interface. Choosing the wave propagation direction along the x-axis the solution should be taken in the form

170 | *Plasma Optics*

[Pitarke *et al.* (2007)]:

$$E_n = (E_{n,x}^0, 0, E_{n,z}^0)e^{j(q_n x - \omega t)}e^{-k_n|z|} \tag{7.39}$$

$$H_n = (0, H_{n,y}^0, 0)e^{j(q_n x - \omega t)}e^{-k_n|z|} \tag{7.40}$$

where q_n denotes a two-dimensional wave vector $q_{\parallel}$ of the wave propagating along the interface.

Substituting Eqs. (7.39) and (7.40) into Eqs. (7.23) to (7.26) results in the following set of equations:

$$k_d H_{dy} = \frac{\omega}{c}\varepsilon_d E_{d,x}, \tag{7.41}$$

$$k_m H_{my} = -\frac{\omega}{c}\varepsilon_m E_{m,x} \tag{7.42}$$

and

$$k_n = \sqrt{q_n^2 - \varepsilon_n \left(\frac{\omega}{c}\right)^2}. \tag{7.43}$$

The standard boundary conditions require that the components of both electric and magnetic fields must be continuous [Jackson (1998)]. Consequently the Eqs. (7.41) and (7.42) result in:

$$\frac{k_d}{\varepsilon_d}H_{dy} + \frac{k_m}{\varepsilon_m}H_{my} = 0, \tag{7.44}$$

and

$$H_{dy} = H_{my} \tag{7.45}$$

The system of equations (7.44) and (7.45) has a solution if the determinant is equal to zero (see Eq. (7.19)) that is given by:

$$\frac{\varepsilon_d}{k_d} + \frac{\varepsilon_m}{k_m} = 0 \tag{7.46}$$

Equation (7.46) represents the surface plasmon condition [Pitarke *et al.* (2007)]. The boundary conditions also require continuity of the two-dimensional wave vectror $q_{\parallel}$ in Eq. (7.43), i.e. $q_d = q_m = q$. Based on this condition and combining Eq. (7.46) and Eq. (7.43) one arrives in the following:

$$\varepsilon_d^2 \left(q^2 - \varepsilon_m \frac{\omega^2}{c^2} \right) = \varepsilon_m^2 \left(q^2 - \varepsilon_d \frac{\omega^2}{c^2} \right) \tag{7.47}$$

Equation (7.47) leads to another widely used form of the surface plasmon condition [Ritchie (1957)]:

$$q(\omega) = \frac{\omega}{c} \sqrt{\frac{\varepsilon_d \varepsilon_m}{\varepsilon_d + \varepsilon_m}}. \tag{7.48}$$

For a metal–dielectric interface with the dielectric permittivity ε_d, the solution $\omega(q)$ of equation (7.48) has slope equal to $c/\sqrt{\varepsilon_d}$ at the point $q = 0$ and is a monotonic increasing function of q, which is always smaller than $cq/\sqrt{\varepsilon_d}$ and for large q is asymptotic to the value given by the solution of

$$\varepsilon_d + \varepsilon_m = 0 \tag{7.49}$$

This is the nonretarded surface plasmon condition which follows from Eq. (7.46) at $k_d = k_m = q$. This is valid as long as the phase velocity is much smaller than the speed of light, i.e. $\omega/q \ll c$.

It is instructive now to analyze the dispersion of the SPP propagation on the interface between metal and dielectric. The $q_0 = \omega/c$ represents the magnitude of the light wave vector. Assume for the dielectric $\varepsilon_d = 1$. In this case the Eq. (7.48) yields:

$$q(\omega) = \frac{\omega}{c} \sqrt{\frac{\omega^2 - \omega_p^2}{2\omega^2 - \omega_p^2}}. \tag{7.50}$$

Dispersion relation described by Eq. (7.50) is represented in Fig. 7.3 [Hofmann (2008)].

The upper solid line in Fig. 7.3 represents the dispersion of light in solid. The lower solid line is the surface plasmon polariton dispersion curve which is given by:

$$\omega^2(q) = \omega_s^2 + c^2 q^2 - \sqrt{\omega_s^4 + c^4 q^4} \tag{7.51}$$

where $\omega_s = \omega_p/\sqrt{2}$ represents the classical nondispersive surface-plasmon frequency.

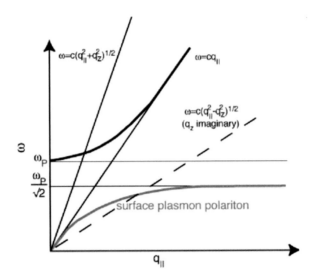

Figure 7.3 Bold solid lines represent the dispersion of light in the retarded (upper line) and the nonretarded surface-plasmon polariton regions (lower line). By thin line the dispersion of light striking the interface at different angles is shown. Thin horizontal lines indicate the values of bulk ω_p and surface plasmon frequencies $\omega_s = \omega_p/\sqrt{2}$. (Adapted from [Hofmann (2008)].)

In the *retarded* region ($q < \omega_s/c$), the surface-plasmon polariton dispersion curve approaches the light line ($\omega = cq_\parallel$, see thin line in Fig. 7.3). At short wavelengths where $q_\parallel \gg \omega_s/c$ the surface-plasmon polariton approaches asymptotically the *nonretarded* surface-plasmon frequency $\omega_s = \omega_p/\sqrt{2}$ (see the horizontal dashed line in Fig. 7.3).

Important conclusions can be made regarding the excitation of the surface plasmon polaritons corresponding to the lower branch in Fig. 7.3. The wave vector of the SPPs has the value of the two-dimensional vector within the interface plane, $\boldsymbol{q}_\parallel$. Depending on the angle of incidence it varies from $\boldsymbol{q}_\parallel = 0$ (normal incidence) to $|\boldsymbol{q}_\parallel| = q_\parallel$ (for grazing incidence, $q_z = 0$). The light dispersion will change from the vertical line to that given by $\omega = cq_\parallel$ (see Fig. 7.3). For any other angle the light dispersion is given by $\omega = c\sqrt{q_\parallel^2 + q_z^2}$. Consequently the light dispersion line and the surface plasmon polariton dispersion curve never cross and hence there can not be any excitation of SPP on ideal interface considered above.

7.5 Application of Plasma Resonance Spectroscopy

There are two basic approaches to generate SPP. The SPP can be generated on a grating. Additional periodic profile on the surface causes modifications of the wave vector selection rules (like additional Bragg reflection in super lattices). According to the super periodicity the dispersion curve will get folded crossing the light dispersion line and thus allowing excitation of the SPPs. This has been observed experimentally by Wood at the beginning of the last century [Wood (1902)], that he described as an "anomalous diffraction gratings" effect [Wood (1935)]. The same effect can be achieved by a rough surface which can be viewed as a superposition of many gratings with different periodicities [Ritchie (1973); Venger *et al.* (1999); Raether (1988)]. The excitation of the SPPs via surface roughness is thought to play a role in surface-enhanced Raman scattering (see [Gavrilenko (2020)] for details).

The other way to achieve the coupling is to use optical system where value of the photon wave vector will increase thus reducing slope of the curve. Widely used are optical systems with a total light reflection inside a prism mounted in a short distance over the surface. In this case, an evanescent electric field penetrates the gap between prism and surface. The field decays exponentially because the wave vector contains an imaginary q-value in the z direction (see dashed line in Fig. 7.3). Complex value of the light wave vector causes slope decrease of the light dispersion curve in Fig. 7.3 that results in the situation when the light dispersion line and the surface plasmon polariton dispersion curve cross thus allowing the excitation of the SPPs. This design is widely used in SPR based optical spectroscopic tools for materials characterization, for example for a characterization of biological materials that is considered in Section 10.4.1.

7.6 Practical Examples

For illustration purposes this section presents several problems with solutions and explanations. Discussion of given examples is helpful for deeper understanding of the subject.

174 | *Plasma Optics*

7.6.1 Absorption Coefficient of Free Electrons

Problem 1. Calculate absorption coefficient α of the free electrons (plasmons) in Cu.

Hint: Use following parameters in Cu for the wavelength value of $\lambda = 2000$ nm, the relaxation time $\tau \simeq 4 \cdot 10^{-14}$ sec, and $n_0 = 8.5 \times 10^{22}$ cm^{-3}.

Solution. Dielectric permittivity function ε for the free electron gas is given by

$$\varepsilon = \varepsilon' - i\varepsilon'' = 1 - \frac{4\pi n_0 e^2}{m^*} \frac{1}{\omega^2 - i\omega/\tau}, \tag{7.52}$$

The complex refraction coefficient $\tilde{n}$ is given by

$$\tilde{n}^2 = (n - i\kappa)^2 = \varepsilon = \varepsilon' - i\varepsilon''. \tag{7.53}$$

After separation into real and imaginary parts follows

$$\varepsilon' = n^2 - \kappa^2 \tag{7.54}$$

$$\varepsilon'' = 2n\kappa. \tag{7.55}$$

In the case of Cu: $\omega^2\tau^2 \simeq 1400 \gg 1$, $\omega_p = 1,34 \cdot 10^{16} sec^{-1}$

$$\varepsilon' \simeq 1 - \frac{\omega_p^2}{\omega^2}, \quad \varepsilon'' \simeq \frac{\omega_p^2}{\tau\omega^3}, \quad \omega_p^2 = \frac{4\pi n_0 e^2}{m^*}. \tag{7.56}$$

From Eqs. (7.52), (7.55), and (7.56) follows

$$\kappa \simeq \frac{\omega_p}{2\omega^2\tau} = 0.19. \tag{7.57}$$

The absorption coefficient α of light is given by

$$\alpha = 2\frac{\omega}{c}\kappa = 1.19 \cdot 10^4 cm^{-1}. \tag{7.58}$$

Table 7.1 Electronic parameters of Ge, Ag, and Au

Constant	n_0	m^*/m_e
Ge	$2.4 \cdot 10^{13}$	1.64
Ag	$5.76 \cdot 10^{22}$	0.96
Au	$5.9 \cdot 10^{22}$	0.99

7.6.2 Plasma Frequency Dependence on Material Parameters

Problem 2. Calculate a functional dependence of the plasma frequency ω_p on the factor of n_0/m^*. On the graph of this function indicate location of points corresponding to Ge, Ag, and Au.

Solution. The plasma frequency is given by

$$\omega_p = (\frac{4\pi n_0 e^2}{m^*})^{1/2}, \tag{7.59}$$

where $e = 4.8 \times 10^{-10}$ in units of CGSE system.

Converting equation (7.59) to the logarithmic scale for $\nu_p = \omega_p/2\pi$ results in the following

$$lg(\nu_p) = lg \left(\frac{n_0 e^2 m_e}{\pi m^* m_e}\right)^{1/2} = lg \left(\frac{e^2}{\pi m_e}\right)^{1/2} + lg \left(\frac{n_0 m_e}{m^*}\right)^{1/2}$$

$$- 3.95 + \frac{1}{2} lg \left(\frac{n_0 m_e}{m^*}\right). \tag{7.60}$$

Plugging numerical values in Table 7.1, one gets data shown in Fig. 7.4.

7.6.3 Reflectance at Plasma Frequency

Problem 3. Calculate reflection coefficient for electromagnetic wave that has plasma frequency of ω_p.

Solution. According to the Fresnel formulae the reflection coefficient the intensity of light at normal incidence on the sample surface is

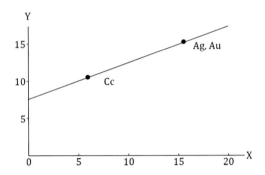

Figure 7.4 Dependence of plasmon frequency $\nu_p = \omega_p/2\pi$ on parameter $n_0 m_e/m^*$ in lg-scale: $x = lg(\frac{n_0 m_e}{m^*}), y = lg\nu_p$.

given by

$$R = \left|\frac{\tilde{n}-1}{\tilde{n}+1}\right|^2, \tag{7.61}$$

where $\tilde{n} = n - i\kappa$. For $n = 1$ the equation (7.61) can be rewritten in the following way

$$R^{min} = \frac{(n-1)^2 + \kappa^2}{(n+1)^2 + \kappa^2} = \frac{\kappa^2}{4+\kappa^2}. \tag{7.62}$$

Next we calculate conditions at which $R^{min} \to 0$. In the case if electromagnetic field has a frequency of ω, plasmonic frequency has value of ω_p, and the high frequency permeability is equal to $\varepsilon_\infty = \varepsilon_L$, we have:

$$\omega_{RES}^2 = \frac{\omega_p^2}{\varepsilon_L - 1}. \tag{7.63}$$

Consequently $R^{min} \to 0$ if $\omega^2 \gg \varepsilon_L - 1$.

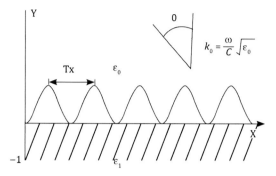

Figure 7.5 Model of a surface profile for generation of surface plasmons.

7.6.4 Geometrical Profile for Generation of Surface Plasmons

Problem 4. From the diagram depicted in Fig. 7.3 follows that the wave vector of surface plasmon exceeds the wave vector of free electromagnetic wave, which has the same frequency. In this case an excitation of a plasmon by a light wave is not possible. In order to achieve equity of the wave vectors modern researchers use metal surfaces with diffraction grate on the surface (see Fig. 7.5).

By taking info account the high frequency approximation $\omega\tau \gg 1$ calculate parameters of the diagram depicted in Fig. 7.5 for Au: $\hbar\omega = 2ev$, $\hbar\omega_p = 9ev$, $T_x = 2\lambda$. Wave vector of the surface plasmon k_{sp} is given by.

$$k_{sp} = k_x = \frac{\omega}{c}\sqrt{\frac{\varepsilon_m(\omega)\varepsilon_d(\omega)}{\varepsilon_m(\omega) + \varepsilon_d(\omega)}}. \qquad (7.64)$$

Solution. Assume $\varepsilon_d = 1$ in vacuum. Within the Drude model a dispersion in metal is described by

$$\varepsilon_m(\omega) = 1 - \frac{\omega_p^2}{\omega^2}, \qquad \omega_p^2 = 4\pi\frac{n_0 e^2}{m^*}. \qquad (7.65)$$

178 | *Plasma Optics*

Diffraction of electromagnetic wave causes change of wave vector according to the following equation:

$$k_x = k_{x0} \pm G, \qquad G = \frac{2\pi}{T_x}.$$ (7.66)

Finally

$$k_0 \sin\theta + \frac{2\pi}{T_x} = k_{sp}.$$ (7.67)

Or

$$\sin\theta = \frac{k_{sp} - (2\pi/T_x)}{k_0}.$$ (7.68)

Angle θ for efficient excitation of the surface plasmons for Au, (i.e. $\hbar\omega = 2\ eV$, $\hbar\omega_{pl} = 9\ eV$, $T_x = 2\lambda$) is given by:

$$\sin\theta = \sqrt{\frac{1 - \omega_p^2/\omega^2}{2 - \omega_p^2/\omega^2}} - \frac{\lambda}{T_x} = \sqrt{\frac{1 - (9/2)^2}{2 - (9/2)^2}} - \frac{1}{2} = 0.527. \quad (7.69)$$

That corresponds to $\theta = 31.81°$.

7.7 Review Questions and Exercises

7.7.1 Free Electron Absorption

Show that free electron can not absorb a photon. Hint: Consider energy and momentum conservation laws in the system.

7.7.2 Free Electron Scattering

Show that a free electron can scatter a photon. Estimate a frequency change of a non-relativistic optical photon, assuming that $\hbar\omega \ll m_e c^2$. Hint: Consider energy and momentum conservation laws in the system.

7.7.3 Dielectric Constant of Plasma

Calculate dielectric constant of plasma if frequency of the electron collision τ is given.

7.7.4 Refractive Index of Plasma

Using solution of Problem 7.7.3, find approximate expression for the real and imaginary parts of the refractive index for the following cases: $\omega > \omega_p$, (a); and $\omega < \omega_p$, (b), where $\omega_p = 4\pi e^2 N_e/m$, and N_e is electron concentration.

7.7.5 Reflection Coefficient of Plasma

Calculate reflection coefficient of optical radiation at λ = 1000 nm from plasma with electron concentration, $N_e = 3 \times 10^{19} cm^{-3}$.
Hint: Calculate plasma frequency (ω_p) and obtain Fresnel equations for a normal incidence.

7.7.6 Dispersion in Plasma

Find dispersion law for transverse electromagnetic waves in plasma. Assume that dielectric function of plasma is described by the Drude's dispersion law. Plot the dispersion curves for both plasma and free space.

7.7.7 Penetration Depth of Electromagnetic Radiation

Calculate penetration depth of electromagnetic radiation (skin layer) in copper at wavelengths $\lambda = 1.9 \times 10^{-4}$cm and $\lambda = 2.9 \times 10^{-4}$cm. Use the following value for the specific resistance of copper, $\rho = 1.6 \times 10^{-6}$ Ohm·cm.

Chapter 8

Optics of Composites, Alloys, and Artificial Materials

A composite material is a material made up of different individual components, known as constituents of the material and may be defined as a homogeneous or non-homogeneous material depending on the structure and functions. Typical example of composite metals is an alloy that is a blend of a metal with one or more metallic or nonmetallic materials. The components of an alloy do not combine chemically but, rather, are very finely mixed. An alloy might be homogeneous or non-homogeneous, i.e. it might contain small particles of components that can be viewed with a microscope. Brass is an example of an alloy, being a homogeneous mixture of copper and zinc. Another example is steel, which is an alloy of iron with carbon and possibly other metals. The purpose of alloying is to produce desired properties in a metal that naturally lacks them. Brass, for example, is harder than copper and has a more gold-like color. Steel is harder than iron and can even be made rust proof (stainless steel).

Unlike metals or ceramics where alloying is mainly targeting mechanical behaviors, creation of semiconductor alloys mostly concerns achieving specific electronic or optical properties. Semiconductor alloys provide an opportunity of tuning the magnitude

---

Fundamentals of the Optics of Materials: Tutorial and Problem Solving
Vladimir I. Gavrilenko and Volodymyr S. Ovechko
Copyright © 2024 Jenny Stanford Publishing Pte. Ltd.
ISBN 978-981-4877-93-0 (Hardcover), 978-1-003-25694-6 (eBook)
www.jennystanford.com

of the band-gap energy and other basic material properties so as to optimize and widen the application of semiconductor devices in modern electronics, optoelectronics, and nanoelectronics.

Compound semiconductor alloys such as $In_xGa_{1-x}As$, $GaAs_xP_{1-x}$, or $CuIn_xGa_{1-x}Se_2$ are increasingly employed in numerous electronic, optoelectronic, and photonic devices due to the possibility of tuning their properties over a wide parameter range simply by adjusting the alloy composition. Interestingly, the material properties are also determined by the atomic-scale structure of the alloys on the nanometer and subnanometer scale. Generally, the semiconductor alloys are disordered solid solutions. However, recent epitaxial techniques, i.e. molecular beam epitaxy (MBE) and metal-organic chemical vapour deposition (MOCVD) have produced alloy films which have a long-range order as confirmed by reflected electron-beam diffraction experiments.

8.1 Effective Medium Approximations in the Optics of Composite Materials

Effective Medium Approximation (EMA) is a powerful method in description and analysis of optical functions of composite materials. It has been proved that the Mie theory is only valid for very low concentrations of the composite particles in a solvent or solid matrix [Kreibig and Vollmer (1995)]. By application of the Mie theory it is assumed that the individual particles are non-interacting and separated from one another. However if the interparticle distances become smaller than the particle dimension or if aggregation occurs, the plasmon resonance red shifts and often a second absorption peak at a longer wavelength is observed [Kreibig and Vollmer (1995)]. This is similar to the optical absorption of dye molecular aggregates at high concentration [Gavrilenko and Noginov (2006)].

The Maxwell-Garnett theory [Maxwell-Garnett (1904)], is based on the Clausius-Mossotti equation. Application of the Clausius-Mossotti equation (1.92) (see Section 1.4) is one of the most widely used methods for calculating the bulk dielectric properties of inhomogeneous materials [Levy and Stroud (1997); Ovechko and

Shur (2005); Ovechko *et al.* (2005)]. It is useful when one of the components can be considered as a host in which inclusions of the other components are embedded. It involves an exact calculation of the field induced in the uniform host by a single spherical or ellipsoidal inclusions and an approximate treatment of their distortions by the electrostatic interactions between the different inclusions. These distortions are caused by the charge dipoles and higher multipoles induced in the other inclusions. The induced dipole moments cause the longest range distortions and their average effect is included in the EMA which results in a uniform field inside all the inclusions. Different EMA calculations are particularly appropriate for composites and polycrystals in which the grains of the various components are randomly and symmetrically distributed, so that none of the components is identifiable as a host in which the others are preferentially embedded.

The EMA assumes that it is justified to describe the composite material containing metal nanoparticles embedded in an inert host medium by an effective complex dielectric constant ε_c such that [Kreibig and Vollmer (1995); Levy and Stroud (1997)]:

$$\frac{\varepsilon_c - \varepsilon_m}{\varepsilon_c + k\varepsilon_m} = f_m \frac{\varepsilon - \varepsilon_m}{\varepsilon + k\varepsilon_m} \tag{8.1}$$

where ε_m and ε are the dielectric constants of the metal nanoparticles and the host medium, respectively. In contrast with the Mie theory, ε_m is a complex function depending on the frequency of light. Constant f_m is the volume fraction of the metal nanoparticles in the composite material. k is a screening parameter which relates to the shape of the nanoparticles. For example, $k = 2$ for spherical nanoparticles, and $k = 1$ for long nanorods oriented with their axes of rotation parallel to the direction of the incident light [Link and El-Sayed (2000)].

For practical application one needs to take functions ε and ε_m from the literature and compute ε_c with the Eq. (8.1). For example, the absorption spectrum of the metal nanoparticles in a transparent non-interacting host medium can then easily be calculated, as the dielectric constant is related to the optical refractive index n_c and the extinction coefficient k_c (see Eq. (1.15)) accordingly to:

$$\varepsilon_c = \tilde{n}_c^2 = \varepsilon_c' + i\varepsilon_c'',$$
$$\varepsilon_c' = n_c^2 - k_c^2, \tag{8.2}$$
$$\varepsilon_c'' = 2n_c k_c$$

Calculated values of n_c and k_c can be used to obtain spectra of functions for direct comparison with experiment. For example, the transmittance of the effective medium film is given by [Foss *et al.* (1994)]:

$$T = \left[1 - \frac{(n_c - 1)^2 + k_c^2}{(n_c + 1)^2 + k_c^2} \right]^2 \exp\left(-4\pi k_c d/\lambda \right) \tag{8.3}$$

where λ is the wavelength of light and d is the film thickness.

The extended Maxwell-Garnett theory has been used by [Foss *et al.* (1994); Hornyak *et al.* (1997)] in order to explain the optical absorption spectra of needle-like and pancake-like gold nanoparticles in a porous alumina membrane. The generalization of the EMA has been performed by [Sarychev *et al.* (2000)] where a theory that takes into account effects of retardation was developed for calculating the effective dielectric constant and magnetic permeability of metal–dielectric composites and photonic crystals containing a metallic component.

8.2 Electronic Band Structure and Optical Functions of Alloys

Optical functions of solid alloys are determined by their electronic band structure [Harrison (1989); Martin (2004)]. In contrast to ideal bulk crystals the atomic structure of solid alloys contains substantial random component as a result of the impurity diffusion by fabrication process. Consequently the wave vector as well as band structure concepts cannot be applied to the random systems. This problem is overcame by well established approach in the solid-state theory—the Virtual Crystal Approximation (VCA) [Podgorny *et al.* (1986); Martin (2004)]. The VCA is successfully used in many applications within the pseudopotential theory of the electron energy band structure of materials.

8.2.1 Method of Pseudopotential for Electronic Band Structure Calculations of Alloys

Pseudopotential approach is a very powerful method in the electronic band structure theory of complex systems. Basics of the pseudopotential method as applied for optics has been outlined in many books and reviews (see e.g. [Martin (2004); Harrison (1989); Gavrilenko (2009)]). The pseudopotentials replace the complicated effects of the motion of the core (i.e. non-valence) electrons of an atom and it's nucleus with an effective potential, or pseudopotential, so that the Schrödinger equation (see Eq. (8.8)) contains a modified effective potential.

This approach is very convenient for calculation of optical functions, and is widely used in the literature. The wave function is given by [Gavrilenko (2009)]

$$\psi_{n,\mathbf{k}}(\mathbf{r}) = \frac{1}{\sqrt{\Omega}} \sum_{\mathbf{G}} d_{n,\mathbf{k}}(\mathbf{G}) e^{i(\mathbf{q}+\mathbf{G})\mathbf{r}}, \tag{8.4}$$

where $d_{n,\mathbf{k}}$ are the expansion coefficients of the wave function characterized by the wave vector $\mathbf{k}$ and related to the n'th electron energy state, $\mathbf{G}$ is the reciprocal lattice vector, and $\mathbf{q}$ is the wave vector of light.

In a spatially periodic system the pseudopotential is given by

$$V(\mathbf{r}, t) = \sum_{\mathbf{qG}} \int_{-\infty}^{\infty} V(\mathbf{q} + \mathbf{G}, \omega) e^{i(\mathbf{q}+\mathbf{G})\mathbf{r}} d\omega, \tag{8.5}$$

where $\mathbf{G}$ is a reciprocal lattice vector.

Method of pseudopotential applied within the VCA allows a realistic modeling of the electronic band structure of solid alloys. As a typical example of the VCA application for the electronic properties of alloys, the SiGe alloy is considered here. According to the basic concepts of the VCA the lattice constant of the alloy is assumed

Optics of Composites, Alloys, and Artificial Materials

linearly changing between the values of the bulk crystal components. Thus for the SiGe alloy it is defined by [Podgorny et al. (1986)]:

$$a_x = xa_{Si} + (1 - x)a_{Ge},\qquad(8.6)$$

where x ($0 \leq x \leq 1$) is the alloy composition and a_{Si}, a_{Ge} are the lattice constants of the pure crystals.

The ionic pseudopotential of the SiGe virtual crystal is given by [Podgorny et al. (1986)]:

$$V_x(\mathbf{r}) = xV_{Si}(\mathbf{r}) + (1 - x)V_{Ge}(\mathbf{r}).\qquad(8.7)$$

As the result of the assumptions (8.6) and (8.7) a new (virtual) crystal is introduced with lattice constant and pseudopotential values depending on the alloy content. Electronic band structure of the alloy with a periodic potential $V_x(\mathbf{r})$ has to be obtained from the Schrödinger equation given by:

$$H|\mathbf{k}, l\rangle = \left[-\frac{1}{2}\nabla^2 + V_x(\mathbf{r}) \right]|\mathbf{k}, l\rangle = E_{\mathbf{k},l}|\mathbf{k}, l\rangle.\qquad(8.8)$$

Note, that ionic pseudopotential given by equation (8.5) contains both local and non-local parts [Podgorny et al. (1986)], i.e.

$$V(\mathbf{r}) = V_{core}^{loc}(\mathbf{r}) + V^{nloc}(\mathbf{r}).\qquad(8.9)$$

The local part of the alloy, $V_{core}^{loc}(\mathbf{r})$, can be obtained within the VCA from equation (8.7). However, in order to achieve a good accuracy, a special treatment of the non-local part, $V^{nloc}(\mathbf{r})$ in real space is needed [Podgorny et al. (1986)].

Comparison with experiment (see [Podgorny et al. (1986); Gavrilenko (2009)] and references therein) showed that in SiGe alloys despite remarkable differences in absolute eigen energy values (due to the LDA) an acceptable accuracy in the dependencies of transition energies on alloy composition can be achieved by using only local pseudopotentials as calculated within the VCA according to Eq. (8.7). The potential energy presented in Eq. (8.7) has to be generated for the SiGe alloy using condition Eq. (8.6) following the general procedure described before (see [Gavrilenko (2009)] and references therein). The electronic band structure of the alloy has to be calculated from Eq. (8.8).

Optical functions (e.g. spectra of the dielectric function) of the SiGe alloys can be calculated from the first principles within the density functional theory (DFT) as outlined in several textbooks (see e.g. [Yu and Cardona (2010); Martin (2004)]). This is a state-of-the-art approach in modern first principles theory of alloys [Martin (2004)]. In order to achieve better agreement of the calculated optical data with experiment, several corrections to the basic DFT should be applied [Gavrilenko and Bechstedt (1997)]. The eigen energies and wave functions obtained from Eq. (8.8) should be used as an input for optical calculations (see [Gavrilenko (2009, 2020)] and references therein).

The VCA theory has been successfully applied for SiGe alloys using different theoretical frameworks. Structural model of SiGe alloy has been proposed by [Mousseau and Thorpe (1993)]. The segregation of Ge during growth on SiGe(001) surfaces was investigated by *ab initio* calculations by [Boguslawski and Bernholc (2002)]. The nonlocal empirical pseudopotentials were used by [Fischetti and Laux (1996)] to calculate electronic band structure of SiGe alloys. Strained Si/Ge superlattices were studied by the tight-binding method by [Tserbak and Theodorou (1994)].

The VCA approach has been also used for more complicated systems like ternary (SiGe:B) and quaternary alloys (SiGeC:B), i.e. the SiGe alloys doped at high concentration level (between 10^{18} to $10^{21} cm^{-3}$) with B and/or C.

8.2.2 Calculation of Optical Functions from the Electronic Band Structure of Alloys

Most suitable for optics is pseudopotential method while allowing extremely simple evaluation of the wave functions in terms of the plane waves (see Eq. (1.53)). The parameterization of the oscillator model of alloys is performed in two steps. First the trends of the band structure are calculated as functions of the alloy content according to the description above. Next the optical functions are calculated using band structure as inputs.

Optics of Composites, Alloys, and Artificial Materials

The dielectric function given by equation (1.84) is defined in Section 1.3 in terms of the polarization function [Gavrilenko (2009)].

The imaginary part of Eq. (1.84) corresponds to the widely used *golden rule* formula [Martin (2004)]. Evaluation of the polarization function in equation (1.84) is described in Section 1.3. The tensor components $(\alpha, \beta = x, y, z)$ of the polarization function $\hat{P}$ are given by equation (1.79) and the spectral function is given by equation (1.81). The constant coefficient in Eq. (1.79) appears after summation over volume Ω of the homogeneous ambient of noninteracting N-dipoles [the random phase approximation (RPA)].

8.3 SiGe Alloys

While the first transistor was fabricated using germanium in 1947 and group III-V semiconductor materials have consistently demonstrated superior high-speed performance, it is silicon which completely dominates the present semiconductor market. Currently the Si-based technologies accounted for more than ninety percent of the integrated circuit (IC) market.

SiGe heterojunction bipolar transistor (HBT) technology is rapidly emerging as a competitor to III-V technologies such as GaAs and InP for many wireless IC applications, because it provides comparable device-level performance while maintaining compatibility with the economy-of-scale, and hence cost benefits, associated with conventional Si IC manufacturing.

This section is focused on basic properties that govern optical functions (dielectric permittivity, refractive index spectra, etc.) of SiGe alloys. Silicon-germanium alloys, $Si_{1-x}Ge_x$, with controllable concentration of Ge are used to built the band-gap engineered devices with higher performance or new functionality that those built on silicon [Gavrilenko (2020)].

8.3.1 Atomic Structure of SiGe Alloys

The lattice constant of Ge is 4.2% larger than that of Si [Paul (1999)]. Therefore if a bulk Ge layer is placed on a bulk Si layer in an attempt to form a single crystal, every 24th Si atom at the interface would

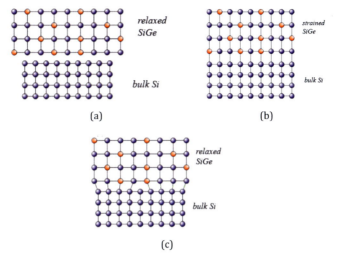

Figure 8.1 Schematic illustration of the fabricated atomic structure of $Si_{1-x}Ge_x$ alloy on top of bulk Si crystal: before the relaxation (a), strained structure (b), and creation of misfit dislocations as a result of the relaxation (c). (Adapted from Ref. [Paul (1999)].)

not be able to form a bond with a Ge atom. Figure 8.1 demonstrates atomic lattice structure of Si and $Si_{1-x}Ge_x$ crystals and its evolution with boron doping.

Provided only a thin layer of $Si_{1-x}Ge_x$ is grown pseudomorphically (lattice matched) to bulk Si then the layer is strained and the symmetry changes from cubic to tetragonal (Fig. 8.1(b)).

For thin layers of $Si_{1-x}Ge_x$ grown on bulk Si there exists a maximum thickness called the critical thickness (Fig. 8.1(c)) above which is costs too much energy to strain additional layers of material into coherence with the substrate. Defects in the system appear to relieve the strain, in this case misfit dislocations (Fig. 8.1(c)).

Numerous models have been proposed to calculate the critical thickness of such strained systems. John Bean and Roosevelt People at Bell Laboratories demonstrated that strained layers well above the equilibrium critical thickness may be grown epitaxially to form metastable strained layers (Fig. 8.2) but these layers may subsequently relax forming defects if the layers are thermally processed.

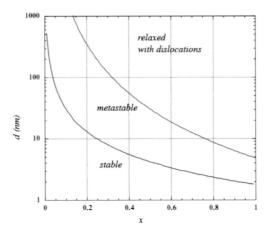

Figure 8.2 Critical thickness (d) of $Si_{1-x}Ge_x$ alloy as a function of Ge concentration (x). (Adapted from [Paul (1999)].)

The SiGe devices are used in many applications and will also allow many fruitful research projects to be performed in the future but low defect density will remain an issue for the high-quality electronic devices.

8.3.2 Optical Functions of SiGe Alloys

Optical functions of Si, Ge, and SiGe alloys (i.e. dielectric functions, reflection coefficient, etc.) are well understood and described based on their electron band structure [Yu and Cardona (2010); Gavrilenko (2020); Adachi (1999)]. Spectro-ellipsometry (SE) studies of $Si_{1-x}Ge_x$ alloys showed continuous evolution of dielectric function spectra from Si to Ge with changing Ge content from 0 to 1 [Humliček et al. (1989); Pickering and Carline (1994); Rowell et al. (2002); Chen et al. (1997); Sieg et al. (1993)]. This is depicted in Fig. 8.3.

The state-of-the-art quantum theory allows unambiguous identification of the electron transitions in the band structure and detailed interpretation of their nature [Yu and Cardona (2010)]. For numerous practical applications in modern optical engineering simple semiempirical models are developed describing experimental

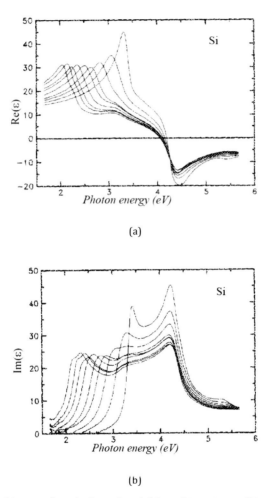

Figure 8.3 Measured evolution of real (a) and imaginary (b) parts of the dielectric permittivity function with the Ge concentration of the $Si_{1-x}Ge_x$ alloys. (Adapted from [Humlíček et al. (1989)].)

spectra with high accuracy. These models are based on the Lorentz theory of the optical response that is described in Section 1.2. The classical Lorentz theory describes spectra of optical susceptibility function ($\chi(\omega)$) and that of optical dielectric function ($\varepsilon(\omega)$) in the form of the oscillator function given by equation (1.43). It has been demonstrated before (see e.g. [Adachi (1999); Yu and Cardona

Optics of Composites, Alloys, and Artificial Materials

(2010)] and references therein) that optical spectra of $\varepsilon(\omega)$ of most semiconductors in the region between near infrared (IR) to the near ultraviolet (UV) are well by a set of only few non-interacting oscillators of the type of equation (1.43).

Optical functions of SiGe alloys are available in literature (see e.g. [Palik (1985); Humliček *et al.* (1989); Adachi (1999)]). Semiempirical modeling of the dielectric function spectra of SiGe alloys is based on application of the multiple oscillator model that incorporates actual electronic band dispersion, damping, and phase mixing of electron transitions in the alloy and evolution with composition [Adachi (1999); Chen *et al.* (2007); Yu and Cardona (2010)]. Accordingly to this model, a spectrum of the dielectric function, $\varepsilon(\omega) = \varepsilon_1(\omega) + j\varepsilon_2(\omega)$ is given by:

$$\varepsilon(\omega) = C + \sum_{i=1}^{N} A_i e^{j\Theta_i} \left(\hbar\omega - E_{gi} - j\Gamma_i\right)^{-1} \tag{8.10}$$

where C is a static (macroscopic) dielectric constant, A_i, Θ_i, Γ_i stand for amplitude, phase, and damping parameters of the i-th oscillator [Adachi (1999)]; a value of E_{gi} corresponds to the direct energy gap in the band structure related to the i-th oscillator contributing to the total optical response within the spectral area.

After some algebra the real ($\varepsilon_1(\omega)$) and imaginary parts ($\varepsilon_2(\omega)$) of complex dielectric function are given by:

$$\varepsilon_1(\omega) = C + \sum_{i=1}^{N} A_i \frac{\cos\Theta_i \left(\hbar\omega - E_{gi}\right) - \Gamma_i \sin\Theta_i}{\left(\hbar\omega - E_{gi}\right)^2 + \Gamma_i^2} \tag{8.11}$$

$$\varepsilon_2(\omega) = \sum_{i=1}^{N} A_i \frac{\sin\Theta_i \left(\hbar\omega - E_{gi}\right) + \Gamma_i \cos\Theta_i}{\left(\hbar\omega - E_{gi}\right)^2 + \Gamma_i^2} \tag{8.12}$$

Dielectric function of SiGe alloys in visible and near ultraviolet (UV) regions are described with a very good accuracy by using five oscillators (N=5 in Eqs. (8.11), (8.12)) corresponding to the fife critical points in the joint density of states of the face-centered lattice Brillouin zone (BZ) [Yu and Cardona (2010)]. Energies (in eV) of

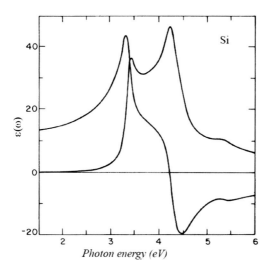

Figure 8.4 Real and imaginary parts of bulk Si dielectric function measured by spectro-ellipsometry. (Adapted from [Aspnes and Studna (1983)].)

these energy gaps at room temperature are given by:

$$E'_0 = 3.33 \quad eV,$$
$$E_1 = 3.40 \quad eV,$$
$$E_1 + \Delta_1 = 3.42 \quad eV, \quad (8.13)$$
$$E_2 = 4.38 \quad eV,$$
$$E'_1 = 5.32 \quad eV.$$

where Δ_1 represents spin-orbit splitting of the E_1 band [Martin (2004)].

The theoretical spectra of both real ($\varepsilon_1(\omega)$) and imaginary ($\varepsilon_2(\omega)$) parts of the dielectric function calculated by Eqs (8.11) and (8.12) using gap values listed by Eq. (8.14) fit experimental spectra of bulk Si (depicted in Fig. 8.4) with a very good accuracy (see e.g. [Adachi (1999); Chen et al. (2007)]).

It should be noted, that other oscillator models were used in literature applying less number of single oscillators (see e.g. [Fursenko et al. (2008)]) where only three oscillators were used for

Table 8.1 Amplitude, broadening, and phase factors providing the best fit of fife-oscillator model to the experiment of SiGe shown in Fig. 8.4

Osc. No.	A_i	Γ_i	Θ_i
1	-23.885	1.0	0.4
2	-3.68	0.15	-5.8
3	11.67	0.5	-0.5
4	-6.28	0.2	-7.2
5	4.880	1.0	-10.0

the analysis. However, despite a simplicity the energy gaps provided by these models do not have physical interpretations. Consequently these models are not compatible with the electronic band structure theory and they can not be used for the detailed analysis of alloys.

In contrast, the energy gaps listed by Eq. (8.14) have direct interpretation by the band structure theory. In this case theoretical predictions for the alloys can be applied directly to this model.

Increase of Ge-content (x) results in changes of the energy gap energies given by Eq. (8.14). Concentration dependencies of SiGe alloy energy gaps are somewhat different for strained and relaxed alloys. For the relaxed SiGe alloys energy gaps are given by:

$$E_0'(x) = (3.33 - 0.19x) \quad eV,$$
$$E_1(x) = [3.40 - 1.3x - 0.15x(1-x)] \quad eV,$$
$$E_1 + \Delta_1(x) = [3.42 - 1.13x - 0.06x(1-x)] \quad eV, \qquad (8.14)$$
$$E_2(x) = [4.38 - 0.0007(1-x)] \quad eV,$$
$$E_1'(x) = (5.32 - 0.15x) \quad eV.$$

The set of the fife oscillators provides with most realistic model spectra of different SiGe alloys in the wide range from near IR up to vacuum UV. The model is developed based on real electron band structure calculations of the SiGe alloys [Yu and Cardona (2010); Gavrilenko (2020)]. Optical electron transitions contributing to the optical functions of Si and Ge occur between filled and empty levels located in two valence bands and three conduction bands. Within this model and according to the selection rules the optical functions in SiGe alloys are formed by direct electron transitions between five direct energy gaps listed by Eq. (8.14). Therefore the five oscillator model allows physically justified interpretation of optical spectra evolution with the alloy content based on electron band structure.

According to the electron band structure analysis (see e.g. [Yu and Cardona (2010)]) in bulk Si the direct optical transitions start at the photon energies near 3.0 eV and completely expire the energies above 6.0 eV. With increase of the Ge concentration (x) in SiGe alloys the whole picture shifts to red direction (up to 2 eV in bulk Ge).

8.4 $A^{III}B^{V}$, and $A^{III}B^{V}$-Based Alloys

The III-V semiconductor materials have consistently demonstrated superior high-speed performance that caused extensive application of these materials in modern high-tech industry. This chapter addresses optical functions dispersions and their modelling with applications for transparent Focused Beam Ellipsometry and Ultra Violet Reflectometry, i.e. FBE+UVR, measurement technology.

Group III–nitrides semiconductors are widely used in modern electronics. AlGaN/GaN heterostructures have been attracting much attention for nitride high-electron-mobility transistor HEMT applications [Cörekci *et al.* (2007)]. Therefore, research on HEMTs has mainly focused on the growth of AlGaN/GaN heterostructures with high crystal quality. It is rather well known that III-V nitrides are generally grown on Al_2O_3 substrates due to their low cost, stability at high temperatures, and mature growth technology. However, their lattice parameters and thermal expansion coefficients are not well matched. The large lattice mismatch leads to high dislocation densities of $10^7 - 10^{11} cm^{-2}$ in the epilayer. Therefore, AlGaN/GaN heterostructures with high crystal quality are typically grown for various buffer layers on the substrate. Recently, it has been reported that the crystal quality of GaN epitaxial films grown on an AlN buffer/Al_2O_3 template is considerably improved.

8.4.1 Gallium Arsenide

Optical functions data of bulk GaAs obtained by spectro-ellipsometry and other optical metrology methods have been reported in literature [Adachi (1999); Aspnes and Studna (1983); Palik (1985)].

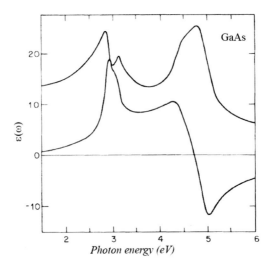

Figure 8.5 Real and imaginary parts of the GaAs dielectric function measured by spectro-ellipsometry. (Adapted from [Aspnes and Studna (1983)].)

Spectra of real and imaginary part of the GaAs dielectric function measured by spectro-ellipsometry [Aspnes and Studna (1983)] are shown in Fig. 8.5.

In terms of the oscillator model these spectra can be fitted using up to seven independent oscillators [Kim et al. (1992)]. Band structure parameters of these oscillators (critical point energies and related broadening parameters) extracted from the fit are listed in Table 8.2.

In Fig. 8.6 the best fit of the oscillator model [Kim et al. (1992)] containing six oscillators is shown representing individual contributions of every spectral region.

Fitting procedure presented presented in [Kim et al. (1992)] is one of the most accurate models described in literature. Previous

Table 8.2 Energy gap values (E_j) and related broadening (Γ_j) of dominating critical points in the band structure of bulk GaAs

Critical point	$E_0(\Gamma)$	$E_0 + \Delta$	$E_1(L)$	$E_1 + \Delta_1$	$E_0'(\Delta)$	$E_1(X)$	$E_1(K)$
E_j (eV)	1.411	1.748	2.930	3.171	4.478	4.811	5.003
Γ_j (meV)	5.0	8.8	43.9	77.5	76.2	111.3	125.8

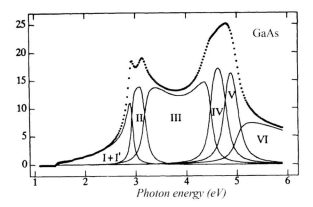

Figure 8.6 Contributions of each of the six regions to the $\varepsilon_2(\omega)$ spectrum of bulk GaAs. The dotted line depicts the simulated spectrum, which is indistinguishable from the measured $\varepsilon_2(\omega)$ spectrum by [Aspnes and Studna (1983)] shown in Fig. 8.5. (Adapted from [Kim et al. (1992)].)

model [Terry (1991)] provided with somewhat better accuracy but requires more sophisticated description of electronic structure.

8.4.2 Aluminium Gallium Arsenide

The $Al_xGa_{1-x}As$ alloy system is of great technological importance for high-speed electronic and optoelectronic devices due to the nearly perfect lattice match of GaAs to AlAs. Optical functions data of bulk AlGaAs obtained by spectro-ellipsometry and other optical metrology methods have been reported in literature [Adachi (1999); Logothetidis et al. (1990); Palik (1985)].

The electronic transitions of $Al_xGa_{1-x}As$ alloys are studied and compared with theoretical band structure calculations in Ref. [Logothetidis et al. (1990)]. In Fig. 8.7 evolution of dielectric function spectra in $Al_xGa_{1-x}As$ alloys with variation of the content are shown.

8.4.3 GaP, InAlP, and Other Phosphorous-Based Alloys

The quaternary semiconductor alloys $In_{1-x}Ga_xAs_yP_{1-y}$ lattice-matched to InP are important in the communications industry

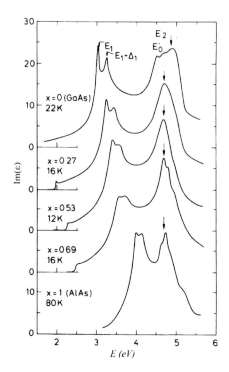

Figure 8.7 Imaginary part of the dielectric function of the GaAs, $Al_xGa_{1-x}As$ alloys, and AlAs measured at low temperatures. (Adapted from [Logothetidis et al. (1990)].)

because the band gap can be tuned with composition over the spectral range that gives the lowest loss and dispersion for optical fibers. Dielectric function spectra of $In_{1-x}Ga_xAs_yP_{1-y}$ alloy are measured by spectro-ellipsometry [Kelso et al. (1982)].

Figures 8.9 and 8.10 present real and imaginary parts of the dielectric function of $In_{1-x}Ga_xAs_yP_{1-y}$ alloy measured by spectro-ellipsometry [Kelso et al. (1982)]. Evolution of the band structure parameters is obtained from the spectral shape analysis.

8.4.4 Group III Nitride-Based Alloys

In the past recent decades, there has been rapid development of wide-band-gap nitride-based blue laser diodes that requires

Fundamentals of the Optics of Materials: Tutorial and Problem Solving | 199

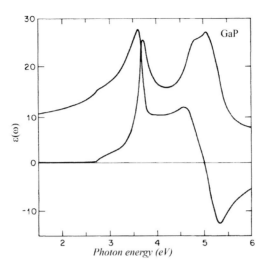

Figure 8.8 Dielectric function spectra of bulk GaP measured by spectro-ellipsometry. (Adapted from [Aspnes and Studna (1983)].)

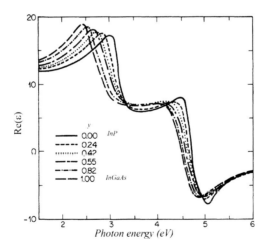

Figure 8.9 Real part of the dielectric function spectra of $In_{1-x}Ga_xAs_yP_{1-y}$ alloy measured by spectro-ellipsometry for different y-values. (Adapted from [Kelso et al. (1982)].)

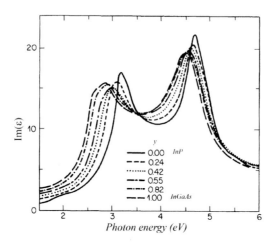

Figure 8.10 Imaginary part of the dielectric function spectra of $In_{1-x}Ga_xAs_yP_{1-y}$ alloy measured by spectro-ellipsometry for different y-values. (Adapted from [Kelso et al. (1982)].)

accurate optical metrology of different nitride based materials like $Al_xGa_{1-x}N$ and/or $In_xGa_{1-x}N$ alloys, etc.

Optical constants of both cubic and hexagonal GaN crystals have been studied and reported in literature [Logothetidis et al. (1994); Kawashima et al. (1997); Schmidtling et al. (2005)]. In Fig. 8.11 the dielectric-function spectra of cubic and hexagonal GaN thin films grown on silicon and sapphire substrates, respectively, are shown in the energy region 3–10 eV and at room temperature [Logothetidis et al. (1994)].

By arrows in Fig. 8.11 the locations of direct electron transitions in electronic band structure of GaN are shown. At the energies below 3 eV, the GaN films are transparent. The spectra in this region are dominated by interference fringes originating from the multiple reflection of the incident light beam at the film/substrate interface. Figure 8.11 clearly demonstrates the differences of the optical response between the two polytypes of GaN. Interpretation of the measured spectra in terms of the band structure parameters based on first-principles modeling and simulation is given in literature [Gavrilenko and Wu (2000)].

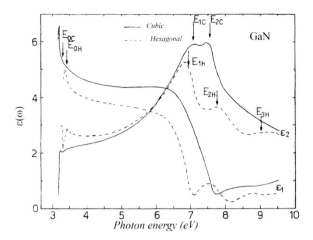

Figure 8.11 Real (ε_1) and imaginary (ε_2) parts of the dielectric function spectra of cubic and hexagonal GaN (solid and dashed lines, respectively). Arrows indicate locations of direct interband transitions in the band structure of GaN. (Adapted from [Logothetidis et al. (1994)].)

Variations of optical refractive index in these materials for different alloy concentrations (x) is presented in Fig. 8.12 [Bergmann and Casey (1998)].

Spectro-ellipsometry has been applied in [Cobet et al. (2001)] to measure optical functions of $Al_xGa_{1-x}N$ samples with small (x < 0.1) content of Al. The measured ordinary and extraordinary dielectric functions data are shown in Fig. 8.13.

Both independent components of the dielectric tensor of wurtzite $Al_{0.1}Ga_{0.9}N$ have been determined by spectroscopic ellipsometry between 1.9 and 9 eV [Cobet et al. (2001)]. The measurements were done on a M-plane film grown on $\gamma-LiAlO_2$ substrate. This sample contained a small fraction of 10 percent aluminium, responsible for a slight shift of the peak positions with respect to those of pure GaN.

8.5 Inorganic–Organic Composites

Inorganic–organic composites have gained growing interest for clinical diagnosis, drug screening, and multiplexed bioassays as a

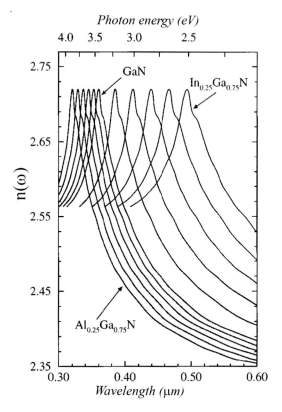

Figure 8.12 Dispersion of the refractive index for $Al_xGa_{1-x}N$ and $In_xGa_{1-x}N$ obtained for different x content. The alloy content x changes by 0.05 between adjacent curves [Bergmann and Casey (1998)].

basis for sensors, because they provide a highly flexible platform on the micrometer scale that can be simultaneously functionalized by many types of chemistry in a cost-effective manner. Incorporating quantum dots into the microsphere moiety can yield tremendous advantages of miniaturized sensors with highly stable fluorescence labels. Quantum dots that are capped with organic ligands such as trioctylphosphine (TOP) have a hydrophobic nature and are therefore compatible with lipophilic polymers [Yin and Alivisatos (2005); Alivisatos *et al.* (2005)].

Generally, the optical properties of nanostructured systems are governed by both the intrinsic properties of the nanostructures and

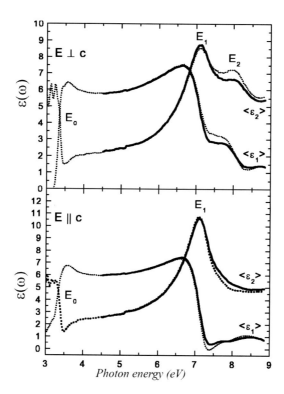

Figure 8.13 Dielectric function spectra of wurtzite $Al_{0.1}Ga_{0.9}N$ measured by specro-ellipsometry for two different light polarization. The dotted (solid) lines show the dielectric function as measured for the c axis oriented parallel (perpendicular) to the plane of incidence. (Adapted from [Cobet et al. (2001)].)

their interactions with the matrix. Therefore optics of the composite systems can be controlled by the size, shape, chemical composition, and structure of the nanoparticles and/or by the nature of the matrix in which they are embedded [Dmytruk et al. (2000)].

Depending upon the nature of association between the inorganic and polymeric components, nanocomposites can be classified into two categories: one in which the inorganic material (nanocrystals, colloids, polyoxometalates) is embedded in a polymeric matrix, and the other where the organic polymer is confined into an inorganic template [Li et al. (2008); Holder et al. (2008)]. In the first case (inorganic embedded in organic composites), the materials are characterized by the good processability, low density of the polymer component, and well-defined optoelectronic properties of the inorganic component. Optical properties of this class of lumophores are determined by the quantum confinement effect, so that their emission color and the electron affinity can be finely controlled, not only by the material choice, but also by size within a single synthetic route. State-of-the-art syntheses, which can be carried out either in organic solvents or in water, provide a broad palette of II-VI, III-V, and IV-VI NCs with variable size and a narrow size distribution leading to narrow emission spectra 25–35 nm full width at half maximum in solution tunable from the UV to the near-infrared spectral region [Holder *et al.* (2008)]. Proper surface passivation leads to improved chemical stability and high photoluminescence (PL) quantum yields (of 50% or higher) for so-called core-shell NCs like *CdSe/ZnS* or *CdSe/CdS*, where the large bandgap semiconductors (ZnS or CdS) epitaxially overgrow the core material (CdSe) and the band edges of the core material lie inside the bandgap of the outer material (type I structures). Variable surface chemistry of NCs allows for the ease of their processability from different solvents and for their incorporation into different organic matrices.

The conjugated polymers (that are organic semiconductors) represent large polymer chains that are soft and flexible. The inorganic semiconductors are well known to change their emission color over a very wide spectral range, while in organic semiconductors this effect is less pronounced and often goes in hand with changed electrical properties. Although organic molecules span the entire visible spectrum in terms of emission wavelength, a change of material required to tune the color of emission can result in a dramatic modification of the charge transport properties. This will affect the device characteristics. Therefore a key goal in research and applications of organic–inorganic nanocomposites is the optimization of emission and charge transfer properties.

The organic semiconductors are typically hole transporting materials, but the nanocomposite solids generally display strongly n-type behaviour. Therefore combinations of these two classes of materials are very promising for organic optoelectronics [Yin and Alivisatos (2005)].

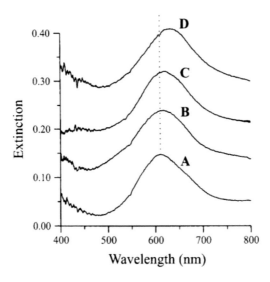

Figure 8.14 Optical absorption spectra of Ag nanoparticles on various substrates: (A) fused silica ($\lambda_{max} = 611$ nm). (B) Borosilicate optical glass ($\lambda_{max} = 616$ nm). (C) Mica ($\lambda_{max} = 622$ nm). (D) SF-10 ($\lambda_{max} = 635$ nm) in a N_2 environment. The spectra have been vertically spaced for clarity. (Adapted from [Malinsky et al. (2001)].)

The combinations of polymers with inorganic nanoparticles has been demonstrated to be an efficient way for the charge transfer optimization through the separation of the charge in organic–inorganic substances [Li et al. (2008); Holder et al. (2008)]. The charge transfer could be made faster through the chemical bonding between organic and inorganic materials. The quantum confinement enhances optical absorption in the nanoparticles with respect to the bulk materials thus contributing to the overall optimization of the optoelectronic properties of the organic–inorganic composites.

It has been understood that optical properties of metallic nanoparticles in visible and near UV spectral range are determined by the plasma resonance (see Chapter 7). The plasmon related optical spectra are significantly affected by the ambient that has been demonstrated by optical absorption studies of metallic nanoparticles embedded or adsorbed on different substrates [Sun et al. (2005); Li et al. (2008); Malinsky et al. (2001)]. Absorption spectra for solvent-treated Ag nanoparticles fabricated with aspect ratio 4:1

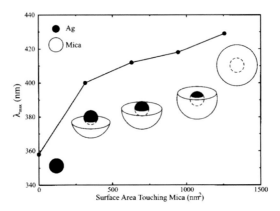

Figure 8.15 Spectral position of the plasma resonance wavelength as a function of particle area in contact with substrate. The calculations were performed for a Ag core of radius 10 nm and a partial mica shell of radius 30 nm. The embedding was changed by truncating the mica shell along vertical axis at different level, as depicted. (Adapted from [Malinsky et al. (2001)].)

(width:height) on the following substrates: fused silica ($n = 1.46$), borosilicate glass ($n = 1.52$), mica ($n = 1.6$), and SF-10 ($n = 1.73$) are presented in Fig. 8.14 [Malinsky et al. (2001)].

The observed red shift of the absorption maximum shown in Fig. 8.14 has been interpreted within the classical electromagnetic theory as an effect of the ambient with varying dielectric function [Malinsky et al. (2001)]. Simulation of optical spectra using ordered array of classical dynamic dipoles clearly demonstrates the red shift of the plasma resonance with the increased effect of the ambient. This is demonstrated in Fig. 8.15.

The results of the simulations presented in Fig. 8.15 clearly demonstrate that the plasma resonance wavelength shifts dramatically to the red as the sphere goes from free to being partially embedded (25% of the sphere area is exposed to mica). As the sphere becomes increasingly more embedded (area exposed to mica varies from 25% to 100%), the shift to the red is much slower, and is approximately linear with exposed area.

Microscopy pictures demonstrate that the particle shape evolves from a polyhedron to a rounded and finally to a nearly ideal sphere [Sun et al. (2005)]. Transmission electron microscopy (TEM) study

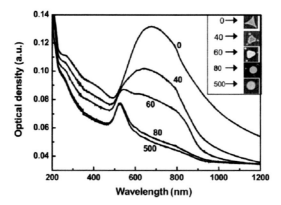

Figure 8.16 Optical absorption spectra of gold particle arrays on quartz versus laser irradiation with the indicated number of laser pulses. Inset: the corresponding morphology of individual dots at each irradiation dose. (Adapted from [Sun et al. (2005)].)

has shown that the atomic structure of individual gold particles in the array were transformed from polycrystalline (prior to irradiation) to a single crystal structure.

Changes of the Au nanoparticles morphologies were accompanied by dramatic variations of the optical absorption spectra [Sun et al. (2005)]. This is demonstrated in Fig. 8.16. The as-prepared samples show a very broad absorbance peak centred around 680 nm together with a shoulder extending well into the near infrared region. As the laser irradiation increased the peak at 680 nm decreased and almost completely disappeared. After irradiation by about 60 pulses, another peak emerges around 550 nm, and after about 100 pulses, the peak shifted to 530 nm. This peak showed very little dependence on the number of laser pulses beyond 100 laser pulses [Sun et al. (2005)].

The results presented in Fig. 8.16 were qualitatively explained in terms of the plasma resonance changes in Au nanoparticles due to the controlled morphology changes [Sun et al. (2005)]. These results demonstrate that laser irradiation by nanostructured lithography is an efficient way to control the morphology of nanostructured materials and their optical properties.

Results by [Sun *et al.* (2005)] demonstrate possibilities to control the morphology of composite materials and hence their optical properties. Laser irradiation induced a morphology evolution from triangularly to nearly spherically shaped particles and a structural evolution from poly to single crystal. The optical absorption spectra changed significantly with particle shape. This can be considered as a promising tool for the fabrication of specific future nanodevices by area-selective treatment.

8.6 Metamaterials

A *metamaterial* (the name is from Greek word *meta*, meaning "beyond") is a special class of materials that are artificially engineered to have properties not found in nature. They can be made for example from assemblies of multiple elements built from composite materials such as metals or plastics. These materials are frequently arranged in a form of repeating patterns, at scales smaller than the wavelength of light. Basic properties of metamaterials arise mainly not from the properties of the base materials and composites but from their structures specially designed at precise shape, geometry, size, and orientation.

In metamaterials an interaction with both components of electromagnetic radiation is important [Sarychev and Shalaev (2007)]. Metamaterials can allow both field components of light to be coupled to meta-atoms, enabling entirely new optical properties and exciting applications [Cai *et al.* (2007); Shalaev (2007)]. Appropriately designed metamaterials can affect waves of electromagnetic radiation or sound in a manner not observed in bulk materials. Among the fascinating properties is a negative refractive index [Agranovich *et al.* (2004)]. Potential applications of metamaterials are diverse and include optical filters, medical devices, remote aerospace applications, sensor detection, infrastructure monitoring, etc.

As shown before in this book the refractive index is one of the most fundamental characteristics of light propagation in materials. It characterizes the electromagnetic field profile and penetration of the energy in media. In materials having simultaneously negative values of dielectric permittivity (ε) and magnetic permeability (μ) theory

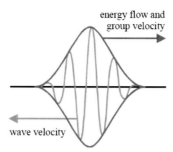

Figure 8.17 Materials with negative refraction are sometimes called left handed materials because the Poynting vector has the opposite sign to the wave vector. (Adapted from [Pendry (2004)].)

[Veselago (1968)] predicted that the energy flow as dictated by the Poynting vector would be in the opposite direction to the wave vector, which means that rays travel in the opposite direction to waves. This situation is shown in Fig. 8.17 [Pendry (2004)].

The reversal of energy flow follows from the Maxwell's equations: flipping the sign of both ε and μ is equivalent to flipping the sign of the magnetic field but having the same wave vector. The Poynting vector (given by $\boldsymbol{E} \times \boldsymbol{H}$) describing energy flow changes the direction (see Fig. 8.17). This type of material was called *left handed* [Veselago (1968)]. At an interface between doubly positive and doubly negative materials light would be bent the wrong way relative to the normal as predicted by the theory [Veselago (1968)] and is illustrated in Fig. 8.18.

The *negative index* concepts generated considerable interest in both fundamental and applied science. A material in which the sign of the electrical permittivity, ε, and magnetic permeability, μ, were simultaneously negative was designed and fabricated by [Smith *et al.* (2000)]. Experimental verification of the predicted material with negative refraction angle was presented by [Shelby *et al.* (2001)].

Refraction in a material is described by the refractive index n. The causality principle causes negative n when both ε and μ are negative [Pendry (2004)]. The causality requires that both ε and μ have small positive imaginary parts representing the fact that real systems have losses. This gives a positive value to $\sqrt{\varepsilon}$ ($\sqrt{\mu}$) when $\varepsilon < 0$ ($\mu < 0$).

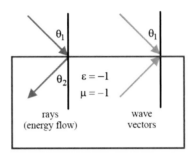

Figure 8.18 In a double negative material as described by [Veselago (1968)], light makes a negative angle with the normal, but the energy flow is opposite to the wave vector. (Adapted from [Pendry (2004)].)

Therefore:

$$n = -\sqrt{\varepsilon\mu}, \quad Re(\varepsilon) < 0, \quad Re(\mu) < 0 \tag{8.15}$$

The term *negative refractive* materials is therefore applied also to the double negative materials. There are many theoretical simulations as well as experimental verifications of the wave propagation and field configurations in negative index materials. Experimental study of Snell's law on a negative index of refraction material in free space in microwave regime (in frequency range 12.6 to 13.2 GHz) has been reported by [Parazzoli et al. (2003)]. The authors prepared two prisms: one constructed from the negative index material, NIM, (the 901 high wire density (HWD) structure, i.e. 901 HWD NID), and the other made from a positive index material, a Teflon wedge. In Fig. 8.19 the results of the numerical simulations are shown [Parazzoli et al. (2003)].

Experimental results and numerical simulations obtained by [Parazzoli et al. (2003)] are shown in Fig. 8.20. The deviation of a microwave beam freely propagating in air in NIM and Teflon wedge agreed with the theory very well thus experimentally verifying the concept of the *negative index* materials. Metamaterials with negative refraction may lead to the development of a superlens capable of imaging objects and fine structures that are much smaller than the wavelength of light [Pendry et al. (2006)].

Other exciting applications of metamaterials include antennae with superior properties, optical nanolithography and nanocircuits,

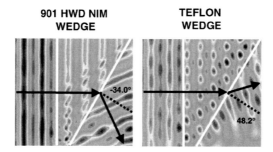

Figure 8.19 Contour plots of the E_z component of the electric field as computed at 12.6 GHz. The radiation propagates from left to right. The wedge angle was 32.19°. Right (left) panel corresponds to the Teflon wedge $n = 1.4$ (NIM wedge). (Adapted from [Parazzoli *et al.* (2003)].)

and special coatings that can make objects invisible [Shalaev (2007); Huang *et al.* (2007)]. The word *meta* means *beyond* in Greek, and in this sense the name *metamaterials* refers to *beyond conventional materials*. Metamaterials are typically man-made and have properties that are not found in nature. What is so magical about this simple merging of *meta* and *materials* that has attracted so much attention from researchers and has resulted in exponential growth in the number of publications in this area?

The notion of metamaterials, which includes a wide range of engineered materials with pre-designed properties, has been used, for example, in the microwave community for some time. The idea of metamaterials has been quickly adopted in optics research, thanks to rapidly developing nanofabrication and subwavelength imaging techniques.

Metamaterials are expected to open a new gateway to unprecedented electromagnetic properties and functionality unattainable from naturally occurring materials. The structural units of metamaterials can be tailored in shape and size. One of the most exciting opportunities for metamaterials is the development of negative-index materials.

The refractive index is a complex number $n = n' + in''$, where the imaginary part n'' characterizes light extinction (losses). The real part of the refractive index n' gives the factor by which the phase velocity of light is decreased in a material as compared with

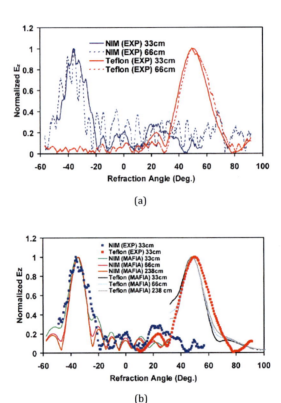

Figure 8.20 Measured angular profile of the normalized electric-field amplitude $E_z(\theta)$, at a constant frequency $f = 12.6$ GHz for detector distances of 33 and 66 cm from the Teflon and 901 HWD NIM wedges (a). Measured 33 cm data compared to simulated results at 33, 66, and 238 cm (100λ) from the Teflon and 901 HWD NIM wedges (b). (Adapted from [Parazzoli et al. (2003)].)

vacuum. NIMs have a negative refractive index, so the phase velocity is directed against the flow of energy in a NIM. This is highly unusual from the standpoint of *conventional* optics.

At an interface between a positive- and a negative-index material, the refracted light is bent in the *wrong* way with respect to the normal. Furthermore, the vectors **E**, **H**, and **k** form a left-handed system (hence NIMs are also called *left-handed* materials).

Negative phase velocity and its consequences were discussed in works by Sir Arthur Schuster [Schuster (1904)] and H. Lamb

[Lamb (1904)] as early as 1904. Later the optical properties of negative-index materials were studied by L. I. Mandelstam [Mandelstam (1945)], D. V. Sivukhin [Sivukhin (1957)], and V. G. Veselago [Veselago (1968)]. Sir John Pendry made a number of critical contributions to the field including his prediction of the superlens based on negative-index materials with resolution beyond the diffraction limit [Pendry (2000)].

The optical properties of materials are governed by the permittivity ε and the permeability μ, describing the coupling of a material to the electric- and magnetic-field components of light, respectively. A negative refractive index in a passive medium could be obtained in a material where the (isotropic) permittivity, $\varepsilon = \varepsilon' + i\varepsilon''$, and the (isotropic) permeability, $\mu = \mu' + i\mu''$, obey the equation $\varepsilon'|\mu| + \mu'|\varepsilon| < 0$ (see [Depine and Lakhtakia (2004); McCall et al. (2002)]). This results in a negative real part of the complex refractive index $n = n' + in'' = -\sqrt{\varepsilon\mu}$.

The inequality above is always satisfied if both ε' and μ' are negative. The negative real part of the complex refractive index could be realized in magnetically active media (that is, $\mu \neq 1$) with a positive real part μ' for which the inequality is fulfilled and which therefore show a negative real part of the refractive index n'. However, the figure of merit $F = |n'|/n''$ in the latter case is typically small [Shalaev (2007)].

At optical frequencies the light field interacts (couples) with bound and free electrons of materials. Consequently the dielectric permittivity ε is different from that in vacuum. In contrast, the magnetic permeability μ for naturally occurring materials is close to its free space value in the optical range. This is because the weak coupling of the magnetic-field component of light to atoms. The magnetic coupling to an atom is proportional to the Bohr magneton $\mu_B = e\hbar/2m_e c = \alpha e a_0/2$ (where e is the electron charge, m_e is the electron mass, a_0 is the Bohr radius) and the electric coupling is proportional to $e a_0$. The induced magnetic dipole also contains the fine-structure constant $\alpha \approx 1/137$ so that the effect of light on the magnetic permeability is α^2-times weaker than that on the electric permittivity. This also explains why all naturally occurring magnetic resonances are limited to relatively low frequencies. As a magnetic response is a precursor for negative refraction, it is of

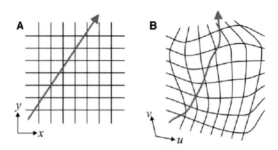

Figure 8.21 A field line within the undistorted (A) coordinate system and after material related controllable distortions (B). The field line can represent the electric displacement field **D**, the magnetic field intensity **B**, and/or the Poynting vector **S**. (Adapted from [Pendry et al. (2006)].)

critical importance to address the fundamental problem above by engineering optical magnetism.

The problem of low coupling to the magnetic-field component of light can be overcome by using metamaterials that mimic magnetism at high frequencies. For the gigahertz range, two concentric split-ring resonators (SRRs) of subwavelength dimensions, facing in opposite directions were predicted [Pendry (2000)] to give rise to $\mu' < 0$. This can be understood as an electronic circuit consisting of inductive and capacitive elements. The rings form the inductances and the two slits as well as the gap between the two rings can be considered as capacitors. A magnetic field oriented perpendicular to the plane of the rings induces an opposing magnetic field in the loop owing to Lenz's law. This leads to a diamagnetic response and hence to a negative real part of the permeability.

A similar structure was predicted by [Kildishev and Shalaev (2006)] and produced by [Shvets and Urzhumov (2006)]. A negative magnetic response with $\mu' = -1.7$ at a wavelength $\lambda = 725$ nm, has been obtained in arrays of pairs of parallel silver strips [Shalaev (2007)]. The magnetic response in the pairs of metal strips results from asymmetric currents in the metal structures induced by the perpendicular magnetic-field component of light [Kildishev and Shalaev (2006)].

Negative index materials have enabled unprecedented flexibility in manipulating electromagnetic waves and producing new

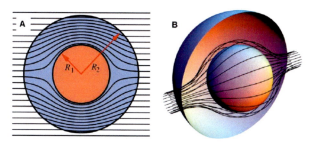

Figure 8.22 A field line within the undistorted (A) coordinate system and after material related controllable distortions (B). The field line can represent the electric displacement field D, the magnetic field intensity B, and/or the Poynting vector S. (Adapted from [Pendry et al. (2006)].)

functionalities [Cai et al. (2007); Pendry et al. (2006); Silveirinha et al. (2008); Alu and Engheta (2005)]. By proper design one can take control over the material properties and form inhomogeneous composites. The conserved quantities of electromagnetic field, the electric displacement field D, the magnetic field intensity B, and the Poynting vector S, can all be directed at will. As an illustration of this idea, the controlled distortions of EM field in metamaterial is shown in Fig. 8.21 presenting field lines in undistorted and distorted material related coordinate systems [Pendry et al. (2006)].

As an example the cloak of invisibility could be prepared based on special design of metamaterials [Cai et al. (2007); Pendry et al. (2006); Silveirinha et al. (2008); Alu and Engheta (2005)].

Unlike other cloaking approaches, which are typically limited to subwavelength objects, the coordinate transformation method [Pendry et al. (2006)] allows the design of cloaking devices that render a macroscopic object invisible.

In the example presented by [Pendry et al. (2006)] for spherical object the assumptions implied that no radiation can get into the concealed volume, nor can any radiation get out. By proper design of the negative index metamaterials [Silveirinha et al. (2008); Pendry et al. (2006); Alu and Engheta (2005)] any radiation attempting to penetrate the secure volume is smoothly guided around the object as demonstrated in Fig. 8.22. Rays in Fig. 8.22 are generated by numerical integration of a set of equations obtained by taking the geometric limit of Maxwell's equations with anisotropic,

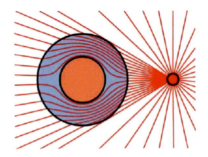

Figure 8.23 A ray-tracing picture for the point light source located near the cloaked sphere. Optical field is excluded from the cloaked region, but emerges from the cloaking sphere undisturbed. (Adapted from [Pendry *et al.* (2006)].)

inhomogeneous media [Pendry *et al.* (2006)]. Similar picture can be predicted for the point light source that is shown in Fig. 8.23.

For the observer far away (as in Fig. 8.22) or close to the light source (as in Fig. 8.23) the light field will remain undisturbed by the object (sphere) that will be invisible.

However, the problem of the metamaterials production for the cloak with ε and μ varying in a quite wide region is not solved yet by modern nanotechnology. Another issue is a frequency dependence of the cloaking. In addition, the design is not sensitive to the object that is being cloaked. The first experimental demonstration of such a cloak at microwave frequencies was reported (see [Pendry *et al.* (2006); Silveirinha *et al.* (2008)] and references therein). One should note, however, that the design still cannot be implemented for an optical cloak, which is certainly of particular interest because optical frequencies are where the word *invisibility* is conventionally defined. This area is extensively studied both experimentally and theoretically.

Chapter 9

Anisotropic Material Optics

Anisotropy is the property of a material being directionally dependent. This implies different properties in different directions, as opposed to *isotropy*. The anisotropy can be defined as a difference, when measured along different axes, in a material's physical or mechanical properties [Dmytruk *et al.* (2001)]. For example the light absorbance, refractive index, electric and/or thermal conductivity, tensile strength, etc.

9.1 Basics of Anisotropic Materials

Until now it was assumed that our sample is isotropic, i.e. that $\varepsilon(\omega)$ is a scalar function and that vectors D and E are parallel. The situation is different in an anisotropic material where $\varepsilon(\omega)$ is a tensor (see Eqs. (1.29) and (1.84)). In general the ε-tensor has nine components. However, for transparent crystals the conservation law for the electromagnetic field energy requires that ε-tensor being symmetric, i.e. $\varepsilon_{ij} = \varepsilon_{ji}$ with only six independent components. From linear algebra follows that every symmetric 3×3 tensor can be brought into diagonal form by a suitable rotation of the Cartesian coordinate system. By proper choice of the coordinate system the

---

Fundamentals of the Optics of Materials: Tutorial and Problem Solving
Vladimir I. Gavrilenko and Volodymyr S. Ovechko
Copyright © 2024 Jenny Stanford Publishing Pte. Ltd.
ISBN 978-981-4877-93-0 (Hardcover), 978-1-003-25694-6 (eBook)
www.jennystanford.com

Table 9.1 Relationship between Cartesian and crystallographic (fractional) coordinates

Singony	z	y	x
Triclinic	[001]	[010]	plane orthogonal to [001]
Monoclinic	[001]	[010]	In plane (100)
Rhombic	[001]	[010]	[100]
Tetragonal	[001]	[010]	[100]
Hexagonal	[0001]	[01$\bar{1}$0]	[2$\bar{1}\bar{1}$0]
Cubic	[001]	[010]	[100]

$\varepsilon_{ij} = 0$ if $i \neq j$ and the ε-tensor will have only the three elements on the main diagonal (ε_{xx}, ε_{yy}, ε_{zz}), which are all different in the general case, i.e. for so-called biaxial crystals (see following).

9.1.1 Crystallography and Crystallophysical Coordinate Systems

Crystal optics is a branch of optics that is focused on light propagation in anisotropic media, such as crystals in which light behaves differently depending on the propagation direction. The index of refraction depends on both composition and crystal structure.

Ideal crystal is characterized by a lattice, which is defined by a unit cell (see Section 5.2). Arrangement of atoms in crystalline solids is concerned by a branch of physics named the *crystallography*.

Any lattice plane and directions in a crystal are defined by *Miller indices* that are introduced and discussed in Section 5.2. Two main coordinate systems are used in crystallography: *fractional coordinates* and Cartesian coordinates.

Relationship between Cartesian and crystallographic (fractional) coordinates is given in Table 9.1.

Fractional coordinates are expressed as fractions of the unit cell in each of the three directions $\mathbf{e}_1$, $\mathbf{e}_2$, $\mathbf{e}_3$ separated by the angles α, β, γ. Fractional coordinates are parallel to the crystallographic axes and thus are not necessarily at right angles to each other. Cartesian coordinates (x, y, z) are converted into fractional crystallographic coordinates (X, Y, Z) in order to perform crystallographic operations, and inversely, geometric computations are more easily performed in Cartesian space. In orthonormal systems (cubic, tetragonal, and

Fundamentals of the Optics of Materials: Tutorial and Problem Solving | **219**

orthorhombic) the coordinate transformation reduces to a simple division of the coordinate values by the corresponding cell constants, e.g. $x = Xa, y = Yb, z = Zc$, and $X = x/a, Y = y/b, Z = z/c$.

For example, a center of the unit cell in the orthorhombic Ru_2Ge_3 crystal (with $a = 11.436$ Å, $b = 9.238$ Å, and $c = 5.716$ Å) is defined as $(5.718, 4.619, 2.858)$ in Cartesian and as $(0.5, 0.5, 0.5)$ in fractional coordinates.

Here are few examples for a calculation of some crystallographic parameters. Consider a bond-length calculation between two neighboring atoms in a crystal. The bond-length (L) between two atoms occupying positions (x_1, y_1, z_1) and (x_2, y_2, z_2) is given by

$$L = \sqrt{[(x_2 - x_1)\,\mathbf{a}_1 + (y_2 - y_1)\,\mathbf{a}_2 + (z_2 - z_1)\,\mathbf{a}_3]}$$
$$\cdot\sqrt{[(x_2 - x_1)\,\mathbf{a}_1 + (y_2 - y_1)\,\mathbf{a}_2 + (z_2 - z_1)\,\mathbf{a}_3]} \qquad (9.1)$$

The result depends on the relationship between vectors $\mathbf{a}_1$, $\mathbf{a}_2$, and $\mathbf{a}_3$. For an orthorhombic lattice, i.e. if all angles between all translational vectors are $90°$ equation (9.1) transforms into:

$$L = \sqrt{\left[(x_2 - x_1)^2\, a^2 + (y_2 - y_1)^2\, b^2 + (z_2 - z_1)^2\, c^2\right]} \qquad (9.2)$$

where $|\,\mathbf{a}_1\,| = a, |\,\mathbf{a}_2\,| = b$, and $|\,\mathbf{a}_3\,| = c$.

As an example, let us find the bond-length between neighboring C and O atoms in calcite $(CaCO_3)$ structure. Calcite is a carbonate mineral. It occurs in a great variety of shapes and colors. Consider a very stable hexagonal structure of $CaCO_3$. A shortest $C–O$ bond is within the planar carbonate group (CO_3) having the following atomic (fractional) coordinates: C $(0, 0, 0)$, O $(x, 0, 0)$, $(0, x, 0)$, $(x, x, 0)$ with $x = 0.257$. Relevant translations have equal length, i.e. $|\,\mathbf{a}_1\,| = |\,\mathbf{a}_2\,| = a = 4.990$ Å, and the angle between them $\gamma = 120°$.

For the calculation one should use equation (9.1) that after plugging numerical values results in the following: $L_{C-O} = x \cdot a = 1.282$ Å.

It is instructive to calculate a shortest bond-length between two carbon atoms in two stable carbon based crystals: graphite and diamond. Atomic structure of these crystals is shown in Fig. 9.1

Diamond has a cubic structure with carbon atoms occupying the cube vortices and the centers of cube faces (see Fig. 9.1(a)).

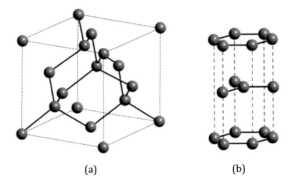

Figure 9.1 Atomic structure of diamond (a) and graphite (b) crystals.

Fractional coordinates of the unit cell atoms (that create the shortest bond-length) are (0, 0, 0) and (0.25, 0.25, 0.25). Lattice constant of the cubic diamond is a = 3.567 Å. For cubic crystal one can use equation (9.2) with $a = b = c$ = 3.567 Å that gives L_{C-C} = 1.545 Å.

Graphite is characterized by the hexagonal structure (see Fig. 9.1(b)) with lattice parameters $|\mathbf{a}_1| = |\mathbf{a}_2| = a$ = 2.456 Å and $|\mathbf{a}_3| = c$ = 6.694 Å. Angle between translation vectors $\mathbf{a}_1$ and $\mathbf{a}_2$ is 120° vector $\mathbf{a}_3$ is perpendicular to the plane of the $\mathbf{a}_1$, $\mathbf{a}_2$ vectors. Atoms within the unit cell occupy the following positions: (0, 0, 0), (0, 0, 1/2), (1/3, 2/3, 0), (2/3, 1/3, 1/2). Remarkable feature of graphite is that it has layered structure along the $\mathbf{a}_3$-axis seen in Fig. 9.1(b). The shortest interatomic distance between carbon atoms within a layer follows from equation (9.1) after plugging coordinates of the neighboring atoms, i.e. (0, 0, 0) and (1/3, 2/3, 0) that gives $L_{C-C} = a/\sqrt{3}$ = 1.418 Å.

The interlayer spacing is L^l_{C-C} = $c/2$ = 3.347 Å, which is much larger than L_{C-C} and causes a relative weak bonding between layers. Due to the weak interlayer bonding the graphite is a soft material. It is used in pencils and as a lubricant. For more detailed analysis see Section 5.3.2.

9.1.2 Linear Optical Anisotropy of Non-cubic Crystals

Important characteristic feature of a non-cubic crystal medium is anisotropy of it's optical properties. The dielectric permittivity function ε is described by a symmetrical tensor of the 2nd rank that in principal coordinate system is given by:

$$\varepsilon = \begin{bmatrix} \varepsilon_{11} & 0 & 0 \\ 0 & \varepsilon_{22} & 0 \\ 0 & 0 & \varepsilon_{33} \end{bmatrix} \tag{9.3}$$

Matrix (9.3) directly relates to a 2nd order characteristic surface:

$$\frac{x^2}{\varepsilon_{11}} + \frac{y^2}{\varepsilon_{22}} + \frac{z^2}{\varepsilon_{33}} = 1, \tag{9.4}$$

Refractive index of a non-cubic crystal depends upon direction of optical wave propagation and its polarization. Refractive index n, dielectric permittivity ε_{ij}, and polarization constant α_{ij} relate to each other according to

$$\varepsilon_{ij} = n^2 = \frac{1}{\alpha_{ij}}. \tag{9.5}$$

Thus Eq. (9.4) can be rewritten as

$$\alpha_{11}^0 x^2 + \alpha_{22}^0 y^2 + \alpha_{33}^0 z^2 = 1 \tag{9.6}$$

or

$$\left(\frac{x}{n_{xx}}\right)^2 + \left(\frac{y}{n_{yy}}\right)^2 + \left(\frac{z}{n_{zz}}\right)^2 = 1 \tag{9.7}$$

Index 0 in α_{ij}^0 indicates an absence of any external excitation. Principal refractive index components are given by

$$\begin{cases} n_{xx} = n_1 = n_{100}, \\ n_{yy} = n_2 = n_{010}, \\ n_{zz} = n_3 = n_{001}. \end{cases} \tag{9.8}$$

The surface defined by equation (9.7) is called *optical indicatrix*. By definition optical indicatrix is a geometric figure that shows the index of refraction and vibration direction for light passing in any direction

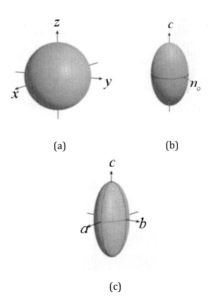

Figure 9.2 Optical indicatrix of isotropic (a), uniaxial (b) and biaxial (c) crystals.

through a material. The surfaces defined by Eq. (9.7) for isotropic, uniaxial, and biaxial crystals are shown in Fig. 9.2

An optical ray of unpolarized light propagating through an anisotropic crystal is split into two rays: an ordinary (*o*) and an extraordinary (*e*) rays. Vector of an electric field amplitude of the *o*-ray is oriented orthogonal to the principal plane and that of the *e*-ray is located within the principal plane (see Fig. 9.2(b)). The principal plane intersects optical axis and wave vector. Crystals that separate ordinary and extraordinary rays are called double-refracting crystals. The double-refraction is also called *birefringence*.

The biaxial anisotropic crystal is characterized by the following relationship between refractive index tensor components: $n_{xx} \neq n_{yy} \neq n_{zz}$. The uniaxial crystals are characterized by $n_{xx} = n_{yy} = n_0$, $n_{zz} = n_e$ and equation (9.7) in this case reads:

$$\frac{x^2}{n_0^2} + \frac{y^2}{n_0^2} + \frac{z^2}{n_e^2} = 1, \qquad (9.9)$$

Equation (9.9) represent an *index ellipsoid* that depicts the orientation and relative magnitude of refractive indices in a crystal [Born

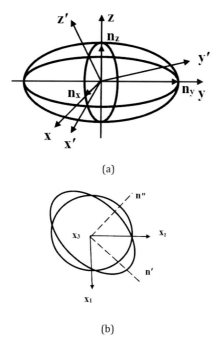

Figure 9.3 Optical indicatrix (a) and optical axes plane view (b) in an anisotropic crystal: x_1, x_2 denote optical axes, ellipsoid semi-axes n_x, n_y correspond to the radii of the H_x circle in a biaxial crystal, i.e. at $n_z < n_x < n_y$.

and Wolf (1999)]. Depending on the relationship between n_o and n_e an anisotropic crystal is called positive (if $n_o < n_e$) or negative (if $n_o > n_e$). In general case, when optical indicatrix axes (x, y, z) differ from the coordinat axes (x', y', z') the equation (9.6) reads:

$$a_{11}^0 x^2 + a_{22}^0 y^2 + a_{33}^0 z^2 + 2a_{23}^0 yx + 2a_{31}^0 zx + 2a_{12}^0 xy = 1 \quad (9.10)$$

where $a_{ij}^0 = C_{ik}C_{jk}a_{kk}^0$. The coefficients C_{ik}, C_{jk} are direction cosines (i.e. $C_{11} = \widehat{\cos(x, x')}$, $C_{21} = \widehat{\cos(y', x)}$, etc.) and are given in Table 9.2.

Table 9.2 Off-direction coefficients

	x	y	z
x'	C_{11}	C_{12}	C_{13}
y'	C_{21}	C_{22}	C_{23}
z'	C_{31}	C_{32}	C_{33}

224 | *Anisotropic Material Optics*

9.2 Jones Vectors and Jones Matrices

Jones vectors and Jones matrix represent a simple and effective method of the linear algebra that is widely used for analysis of the polarized light propagation in materials. The Jones vector is defined as following:

$$E = \begin{pmatrix} E_x(t) \\ E_y(t) \end{pmatrix}. \tag{9.11}$$

Polarized wave is assumed to travel along z-direction. Using the complex representations of the field amplitudes, the equation (9.11) can be written in the following form:

$$E = \begin{pmatrix} a_x e^{i\delta_x} \\ a_y e^{i\delta_y} \end{pmatrix}. \tag{9.12}$$

Light linearly polarized in the x-direction is characterized by the following Jones vector

$$E_{L,x} = \begin{pmatrix} a_x e^{i\delta_x} \\ 0 \end{pmatrix}. \tag{9.13}$$

The Jones vector of the right-handed circularly polarized light is given by

$$E_{c,r} = a \begin{pmatrix} 1 \\ i \end{pmatrix}. \tag{9.14}$$

The Jones vector of the left-handed circularly polarized light is given by

$$E_{c,l} = a \begin{pmatrix} 1 \\ -i \end{pmatrix}. \tag{9.15}$$

While describing the state of polarization, an optical sample (or system) can be considered by transforming income Jones vector $\bar{E}_i$ into outcome Jones vector E_o. This transformation is represented by the (2×2) Jones matrix $\hat{J}$, and can be written as:

$$E_o = \hat{J}E_i \tag{9.16}$$

that in more details is given by

$$\begin{pmatrix} E_{ox} \\ E_{oy} \end{pmatrix} = \begin{pmatrix} J_{11} & J_{12} \\ J_{21} & J_{22} \end{pmatrix} \cdot \begin{pmatrix} E_{ix} \\ E_{iy} \end{pmatrix}. \tag{9.17}$$

Isotropic material does not change the state of the polarization. In this case the $\hat{J}$ matrix is the identity matrix: $(J_{11} = J_{22} = 1, J_{12} = J_{21} = 0)$.

The Jones matrices for some basic optical polarization elements are given as follows:

Ideal linear polarizer with pass-plane oriented along the x-axis

$$J_{x,p} = \begin{pmatrix} 1 & 0 \\ 0 & 0 \end{pmatrix}. \tag{9.18}$$

Quarter wave plate with the fast axis oriented along the x-axis

$$J_{x,\frac{\lambda}{4}} = \begin{pmatrix} 1 & 0 \\ 0 & -i \end{pmatrix}. \tag{9.19}$$

Quarter wave plate with fast axis along the y-axis

$$J_{y,\frac{\lambda}{4}} = \begin{pmatrix} 1 & 0 \\ 0 & i \end{pmatrix}. \tag{9.20}$$

Homogeneous right circular polarizer

$$J_{rcp} = \frac{1}{2} \begin{pmatrix} 1 & -i \\ i & 1 \end{pmatrix}. \tag{9.21}$$

Homogeneous left circular polarizer

$$J_{ecp} = \frac{1}{2} \begin{pmatrix} 1 & i \\ -i & 1 \end{pmatrix}. \tag{9.22}$$

Design of any anisotropic optical element involves selection of a proper material. Polarization of light depends on amplitudes and

226 | *Anisotropic Material Optics*

phases of the electric vector **E** components. It is instructive to consider two classes of anisotropic material:

(1) birefringent (possessing phase anisotropy);

(2) dichroic (possessing amplitude anisotropy).

Within these two classes four types of anisotropic properties were defined: linear and circular phase, as well as linear and circular amplitude anisotropies. Below the Jones matrices for these four types of anisotropic materials are given.

1. The linear phase anisotropy matrix $J^{LP}(\Delta, \alpha)$ is given by:

$$J^{LP}(\Delta, \alpha) = \begin{pmatrix} \cos^2 \alpha + \exp(-i\Delta) \sin^2 \alpha & [1 - \exp(-i\Delta)] \cos \alpha \sin \alpha \\ [1 - \exp(-i\Delta)] \cos \alpha \sin \alpha & \sin^2 \alpha + \exp(-i\Delta) \cos^2 \alpha \end{pmatrix},$$

$$(9.23)$$

where Δ is the value of the linear phase anisotropy (i.e. phase shift between two orthogonal linear components of the light electric vector **E**), α is an azimuth of the fast axis of the phase plate. The ranges of Δ and α values are defined as:

$$0 \leq \Delta \leq 2\pi; \quad -\pi/2 \leq \alpha \leq \pi/2. \tag{9.24}$$

2. The linear amplitude anisotropy matrix $J^{LA}(P, \theta)$ is given by:

$$J^{LA}(P, \theta) = \begin{pmatrix} \cos^2 \theta + P \sin^2 \theta & (1 - P) \cos \theta \sin \theta \\ (1 - P) \cos \theta \sin \theta & \sin^2 \theta + P \cos^2 \theta \end{pmatrix}, \tag{9.25}$$

where $P = \rho_\perp / \rho_\parallel$ is the value of the linear amplitude anisotropy (i.e. relative transparency of the sample), $\rho_\parallel \geq \rho_\perp$, θ is an azimuth of the axes of maximum transperency $\rho_\parallel$. The ranges of P and θ values are defined as

$$0 \leq P \leq 1; \quad -\pi/2 \leq \theta \leq \pi/2. \tag{9.26}$$

Fundamentals of the Optics of Materials: Tutorial and Problem Solving | **227**

3. The circular phase anisotropy matrix $J^{CP}(\varphi)$ is given by

$$J^{CP}(\phi) = \begin{pmatrix} \cos\phi & \sin\phi \\ -\sin\phi & \cos\phi \end{pmatrix}, \tag{9.27}$$

where ϕ is a value of the circular phase anisotropy (i.e. phase shift between two orthogonal circular components of the electric vector **E**). The range of ϕ values is defined as

$$0 \leq \phi \leq 2\pi. \tag{9.28}$$

4. The circular amplitude anisotropy matrix $J^{CA}(R)$ is given by

$$J^{CA}(R) = \begin{pmatrix} 1 & -iR \\ iR & 1 \end{pmatrix} \tag{9.29}$$

where R is the value of the circular amplitude anisotropy ($R = (r_r - r_e)/(r_r + r_e)$, $r_{r,e}$ is the transparencies of the dichroic plate for right and left circular polarisation of the light electric vector **E**).
The range of R values is defined as

$$-1 \leq R \leq 1. \tag{9.30}$$

For isotropic material the values of anisotropic parameters are defines as: $\Delta = R = \phi = 0$, $P = 1$.

Anisotropic materials are materials whose properties (e.g. absorption and refractive indices) depend on the propagation direction of optical waves and light polarization. In the microscopic model we can assume that the oscillator strength depends on light polarization with respect to the crystallographic axis(es). In the principal coordinate system (where the dielectric tensor $\hat{\varepsilon}(\omega)$ is diagonal) two diagonal elements in an uniaxial crystal are equal, i.e.

$$\varepsilon_{xx}(\omega) = \varepsilon_{yy}(\omega) \neq \varepsilon_{zz}(\omega), \tag{9.31}$$

where $\varepsilon_{ij}(\omega)$ stands for dielectric function. Crystals related to trigonal, tetragonal, and hexagonal crystal systems have one principal axis, for example, quartz, $CaCO_3$, KH_2PO_4, KD_2PO_4, $LiJO_3$, $LiNbO_3$, GaSe, CdSe, HgS, Se, Te, etc.

Anisotropic Material Optics

In the case of the biaxial crystals, which have lower symmetry all diagonal elements of the dielectric tensor are different, i.e.

$$\varepsilon_{xx}(\omega) \neq \varepsilon_{yy}(\omega) \neq \varepsilon_{zz}(\omega). \tag{9.32}$$

Crystals that belong to triclinic, monoclinic, and orthorhombic crystal systems have two axes (e.g. mica, gypsum, LiNbO$_3$ at specific temperatures, etc.) Keeping in mind the Kramers-Kronig relations, one can state that every birefringent optical material has dichroic properties. These properties can be observed in spectral regions with high light absorption. Consider now closely anisotropic (birefringent) properties of transparent materials.

9.3 Birefringence

Consider an optical wave propagating through an anisotropic phase media (without absorption). In anisotropic material, direction of the electric induction vector **D** differs from the direction of the **E** vector:

$$\mathbf{D} = \hat{\varepsilon}\mathbf{E}, \tag{9.33}$$

where

$$\hat{\varepsilon} = \begin{pmatrix} \varepsilon_{11} & \varepsilon_{12} & \varepsilon_{13} \\ \varepsilon_{21} & \varepsilon_{22} & \varepsilon_{23} \\ \varepsilon_{31} & \varepsilon_{32} & \varepsilon_{33} \end{pmatrix} \tag{9.34}$$

is an electric permeability tensor $(1 \equiv x, 2 \equiv y, 3 \equiv z)$. In the shorthand representation the Eq. (9.33) reads

$$D_i = \varepsilon_{ij}E_j. \tag{9.35}$$

In Eq. (9.35) the Einstein summation is assumed, i.e. a summation over repeated indices (j). In lossless material the dielectric tensor is symmetric:

$$\varepsilon_{ij} = \varepsilon_{ji}. \tag{9.36}$$

Using a coordinate transformation a real symmetric matrix can be diagonalized, and in Cartesian coordinates in can be given as:

$$\hat{\varepsilon} = \begin{pmatrix} \varepsilon_{xx} & 0 & 0 \\ 0 & \varepsilon_{yy} & 0 \\ 0 & 0 & \varepsilon_{zz} \end{pmatrix} \tag{9.37}$$

In terms of the three axes all crystals could be classified in the following way:

1. Cubic ($\varepsilon_{xx} = \varepsilon_{yy} = \varepsilon_{zz}$) e.g. sodium chloride, diamond ;
2. Uniaxial ($\varepsilon_{xx} = \varepsilon_{yy} \neq \varepsilon_{zz}$) e.g. quartz (positive , $\varepsilon_{xx} < \varepsilon_{zz}$);
3. Uniaxial ($\varepsilon_{xx} = \varepsilon_{yy} \neq \varepsilon_{zz}$) e.g. calcite (negative, $\varepsilon_{xx} > \varepsilon_{zz}$);

Consider now a propagation of a plane-wave in an uniaxial crystal. By applying the following relationships:

$$\partial/\partial t \rightarrow j\omega_0, \quad \nabla \rightarrow -j\mathbf{k}_0 \tag{9.38}$$

the Maxwell equations for a monochromatic plane wave, $\exp[j(\omega_0 t - \mathbf{k}_0 \mathbf{r})]$, can be written in the following way:

$$\mathbf{H} = \frac{\mathbf{k}_0 \times \mathbf{E}}{\omega_0 \mu_0}, \tag{9.39}$$

$$\mathbf{D} = \frac{\mathbf{k}_0 \times \mathbf{H}}{\omega_0}. \tag{9.40}$$

From Eqs. (9.39), (9.40) one can derive the expression for $\mathbf{D}(\mathbf{E})$:

$$\mathbf{D} = \frac{k_0^2}{\omega_0^2 \mu_0} \left[\mathbf{E} - \left(\frac{\mathbf{k}_0}{|\mathbf{k}_0|} \mathbf{E} \right) \frac{\mathbf{k}_0}{|\mathbf{k}_0|} \right], \tag{9.41}$$

where the following identity for the vector triple product was used (see Appendix C):

$$\mathbf{a} \times [\mathbf{b} \times \mathbf{c}] = \mathbf{b} (\mathbf{a} \cdot \mathbf{c}) - \mathbf{c} (\mathbf{a} \cdot \mathbf{b}). \tag{9.42}$$

In Fig. 9.4 a vector diagram for field components $\mathbf{D}, \mathbf{E}, \mathbf{B}, \mathbf{H}, \mathbf{S}$, and wave vector $\mathbf{k}_0$ is shown, where $\mathbf{S} = [\mathbf{E} \times \mathbf{H}]$ is the *Poynting vector*.

230 | Anisotropic Material Optics

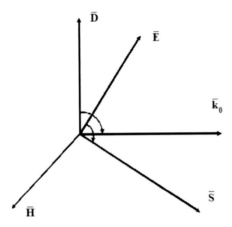

Figure 9.4 Vector diagram of electromagnetic field vectors in anisotropic medium.

Two cases can be considered differing on mutual orientation of $\mathbf{E}$ and $\mathbf{k}_0$ vectors:

(1) Vector $\mathbf{E}_a$ is orthogonal to both the $\mathbf{k}_0$ vector and to the optical axis. In this case from equation (9.41) follows

$$\varepsilon_{yy} = \frac{k_0^2}{\omega_0^2 \mu_0}. \tag{9.43}$$

where $v_0 = c/\sqrt{\varepsilon_{yy}}$, is the phase velocity and $c = (\varepsilon_0 \mu_0)^{-1/2}$.

The velocity of this wave does not depend on the propagation direction. Therefore it is called the *ordinary* wave.

(2) The $\mathbf{E}_b$ vector is located within the plane of both vector $\mathbf{k}_0$ and the optical axis (xz-plane). In this case $\mathbf{E}_b = \mathbf{E}_{bx} + \mathbf{E}_{bz}$. Consequently from Eq. (9.41) follows

$$\varepsilon_{xx} E_{bx} \omega_0^2 \mu_0 = k_0^2 [E_{bx} - (E_{bx} \sin\theta + E_{bz} \cos\theta) \sin\theta], \tag{9.44}$$

$$\varepsilon_{zz} E_{bz} \omega_0^2 \mu_0 = k_0^2 [E_{bz} - (E_{bx} \sin\theta + E_{bz} \cos\theta) \cos\theta]. \tag{9.45}$$

The nontrivial solution can be obtained if determinant of the system of equations (9.44) and (9.45) is zero, i.e.

$$D = \begin{vmatrix} \varepsilon_{xx}\omega_0^2\mu_0 - k_0^2\cos^2\theta & k_0^2\cos\theta\sin\theta \\ k_0^2\cos\theta\sin\theta & \varepsilon_{zz}\omega_0^2\mu_0 - k_0^2\sin^2\theta \end{vmatrix} = 0. \qquad (9.46)$$

which results in the following

$$\frac{1}{n^2(\theta)} = \frac{\sin^2\theta}{n_e^2} + \frac{\cos^2\theta}{n_0^2}, \qquad (9.47)$$

where $n_o = \sqrt{\varepsilon_{xx}} = \sqrt{\varepsilon_{yy}}$ is an *ordinary* refractive index marked by index o and $n_e = \sqrt{\varepsilon_{zz}}$ is an *extraordinary* refractive index marked by index e. Phase velocity $v(\theta)$ for the e-wave is given by

$$v(\theta) = \sqrt{v_e^2\sin^2\theta + v_0^2\cos^2\theta}, \qquad (9.48)$$

where $v_e = c/n_e$ and $v_o = c/n_o$.

Phenomenon in which the phase velocity of an optical wave depends on the polarization and direction of propagation is called *birefringence*.

Transforming polar coordinates to Cartesian, the Eq. (9.46) results in the following

$$\frac{x^2}{n_e^2} + \frac{z^2}{n_0^2} = 1, \qquad (9.49)$$

where $\sin\theta = x/n(\theta)$, $\cos\theta = z/n(\theta)$. Equation (9.48) defines the *index ellipsoid* (see Fig. 9.5). The index ellipsoids characterize positive (Fig. 9.5(a)) and negative (Fig. 9.5(b)) anisotropic crystals.

Dielectric tensor of a crystal corresponds to the crystal symmetry. The crystallographic symmetry classes are divided into three groups according to the number of symmetry axes in the crystal. The groups and the crystal classes that are parts of the groups are listed in Table 9.3.

Table 9.3 Classification of anisotropic materials

Group	Orientation of quadric	Crystal class	Dielectric tensor	Number of independent components	Type
Three equivalent orthogonal directions	Unimportant	Cubic	$\begin{bmatrix} \varepsilon & 0 & 0 \\ 0 & \varepsilon & 0 \\ 0 & 0 & \varepsilon \end{bmatrix}$	1	Isotropic
Two orthogonal directions in a plane normal to crystal symmetry axis	One principal axis directed parallel to the crystal symmetry axis	Trigonal, Tetragonal, Hexagonal	$\begin{bmatrix} \varepsilon_{11} & 0 & 0 \\ 0 & \varepsilon_{11} & 0 \\ 0 & 0 & \varepsilon_{33} \end{bmatrix}$	2	Uniaxial
No equivalent directions	Principal axis aligned with symmetry axis	Orthorombic	$\begin{bmatrix} \varepsilon_{11} & 0 & 0 \\ 0 & \varepsilon_{22} & 0 \\ 0 & 0 & \varepsilon_{33} \end{bmatrix}$	3	Biaxial
"	Principal axis directed parallel to symmetry axis	Monoclinic	$\begin{bmatrix} \varepsilon_{11} & 0 & \varepsilon_{13} \\ 0 & \varepsilon_{22} & 0 \\ \varepsilon_{13} & 0 & \varepsilon_{33} \end{bmatrix}$	4	"
"	Not specified	Triclinic, no symmetry	$\begin{bmatrix} \varepsilon_{11} & \varepsilon_{12} & \varepsilon_{13} \\ \varepsilon_{12} & \varepsilon_{22} & \varepsilon_{23} \\ \varepsilon_{13} & \varepsilon_{23} & \varepsilon_{33} \end{bmatrix}$	6	"

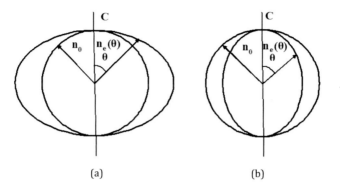

Figure 9.5 Index ellipsoid for positive (a) and negative (b) anisotropic crystals. Refractive indexes $n_o, n_e(\theta)$ are shown for positive $n_o < n_e$ (a) and negative $n_o > n_e$ (b) crystals.

9.4 Practical Examples

In this section several typical problems with solutions in the field of crystal anisotropy are considered with detailed discussion for tutorial purposes.

9.4.1 Polarization of Light by Calcite Dichroic Polarizer

Problem 1. Calculate quality of polarization P of a dichroic polarizer made from calcite crystal ($CaCO_3$). The thickness of crystal is $d = 1$ cm. Absorption coefficients for ordinary wave (α_o) and extraordinary wave (α_e) are shown in Table 9.4.
Solution. The quality of polarization P is defined by

$$P = \frac{I_{max} - I_{min}}{I_{max} + I_{min}} \tag{9.50}$$

Table 9.4 Absorption coefficient of calcite crystal

λ (μm)	α_o, cm^{-1}	α_e, cm^{-1}
2.6	2.5	0.07
3.0	4.0	0.28
3.3	15.0	0.92

234 | Anisotropic Material Optics

For nonpolarized light

$$I_{max} = \frac{1}{2}I_0 e^{-\alpha_e d},$$ (9.51)

$$I_{min} = \frac{1}{2}I_0 e^{-\alpha_o d},$$ (9.52)

where I_0 is the intensity of a nonpolarized light striking the dichroic crystal.

After some algebra from equations (9.50), (9.51), and (9.52) one can derive the following

$$P = \frac{1 - e^{(\alpha_e - \alpha_0)d}}{1 + e^{(\alpha_e - \alpha_0)d}},$$ (9.53)

Under condition of $(\alpha_o - \alpha_e)d > 1$, Eq. (9.53) reduces to the following

$$P \simeq 1 - 2e^{-(\alpha_0 - \alpha_e)d},$$ (9.54)

Plugging numerical values of absorption coefficient into equation (9.54) results in the following

$$P(\lambda = 2.6 \; \mu m) = 0.824$$

$$P(\lambda = 3.0 \; \mu m) = 0.952$$

$$P(\lambda = 3.3 \; \mu m, d = 0.5 \; cm) = 0.998.$$

9.4.2 Snell's Law of Refraction in Anisotropic Crystal

Problem 2. Derive Snell's law of refraction for anisotropic crystal under following conditions: a plane wave is traveling from the outer space into uniaxial crystal at an angle of incidence θ_i; the phase of the plane wave does not change by passing the interface.

Solution. Phase conservation for a plane wave passing an interface means $\omega t - \mathbf{k}_i \mathbf{r} = \omega t - \mathbf{k}_t \mathbf{r}$. Therefore

$$\mathbf{r}\left(\mathbf{k}_t - \mathbf{k}_i\right) = 0$$ (9.55)

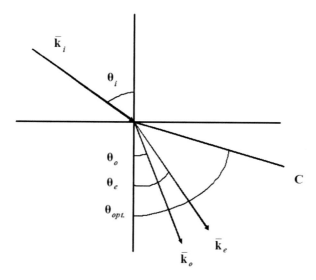

Figure 9.6 Schematics of the Snell's law for ordinary (index *o*) and extraordinary (index *e*) optical waves: *C* denotes the optical axis.

Equation (9.55) means mutual orthogonality of vectors ($\mathbf{k}_t - \mathbf{k}_i$) and $\mathbf{r}$. It can be rewritten as the following

$$\mathbf{r}\mathbf{k}_i = \mathbf{r}\mathbf{k}_o = \mathbf{r}\mathbf{k}_e, \tag{9.56}$$

where $\mathbf{k}_o$ and $\mathbf{k}_e$ are wave vectors of ordinary and extraordinary waves, respectively.

Graphically orientation of the wave vectors in an anisotropic crystal with respect to the crystal axis (*C*) is depicted in Fig. 9.6.

In two dimensional case the equation (9.56) reads

$$xk_{ix} + yk_{iy} = xk_{ox} + yk_{oy} = xk_{ex} + yk_{ey}, \tag{9.57}$$

that should be independent on x, y thus

$$\frac{k_{ox}}{k_{ix}} = \frac{k_{oy}}{k_{iy}} = \frac{k_{ex}}{k_{ix}} = \frac{k_{ey}}{k_{iy}}. \tag{9.58}$$

236 | *Anisotropic Material Optics*

Using boundary conditions on the interface after some algebra one obtains the Snell's law in the conventional form, i.e.

$$\sin \theta_i = n_o \sin \theta_o, \tag{9.59}$$

$$\sin \theta_i = n_e(\theta_{opt} - \theta_e) \sin \theta_e, \tag{9.60}$$

where refraction angles θ_o and θ_e relate to ordinary and extraordinary waves, respectively; angle θ_{opt} denotes orientation of the optical axis.

Equations (9.59) and (9.60) for ordinary and extraordinary waves, respectively, differ by the term $n_e(\theta_{opt} - \theta_e)$ describing propagation direction of the extraordinary wave in an anisotropic crystal.

9.4.3 Light Beam Propagation in Calcite Crystal

Problem 3. A light beam at a wavelength of $\lambda = 633$ nm propagates through a 1 cm thick calcite crystal at an angle of $45°$ with respect to the optical axis (see Fig. 9.7). Calculate displacement (Δd) of the outgoing extraordinary beam with respect to the ordinary one on the back face of the crystal.

Solution. Assume that optical axis of the crystal is oriented along the z-axis. First an angle θ_x between $\mathbf{k}$ and $\mathbf{S}$ as well as between $\mathbf{D}$ and $\mathbf{E}$ vectors will be calculated.

Scalar product of $\mathbf{DE}$ is given by

$$\mathbf{D} \cdot \mathbf{E} = |\mathbf{D}| |\mathbf{E}| \cos \theta_x = D_z E_z + D_y E_y, \tag{9.61}$$

where

$$E_z = E \cos(90 - \theta - \theta_x), \tag{9.62}$$

$$D_z = \varepsilon_{zz} E_z, \tag{9.63}$$

$$E_y = E \cos(\theta + \theta_x), \tag{9.64}$$

$$D_y = \varepsilon_{yy} E_y, \tag{9.65}$$

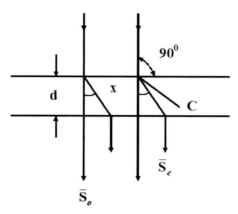

Figure 9.7 Ordinary (o) and extraordinary (e) beams propagation through an anisotropic crystal; C marks an optical axis.

Using Eqs. (9.62) to (9.65), from equation (9.61) one obtains the following

$$n^2(\theta) \cos^2 \theta_x = \varepsilon_{zz} \sin^2(\theta + \theta_x) + \varepsilon_{yy} \cos^2(\theta + \theta_x), \qquad (9.66)$$

$$n^{-2}(\theta) = n_e^2 \sin^{-2}\theta + n_o^{-2} \cos^2 \theta \qquad (9.67)$$

$$\varepsilon_{zz} = n_e^2, \qquad (9.68)$$

$$\varepsilon_{yy} = n_o^2. \qquad (9.69)$$

Angle θ_x will be obtained from the expression for $\tan \theta_x$

$$n_e^2 (\sin \theta + \cos \theta \tan \theta_x)^2 + n_o^2 (\cos \theta - \sin \theta \tan \theta_x)^2 - n^2(\theta) = 0, \qquad (9.70)$$

Solution of equation (9.70) is given by

$$\tan \theta_x = \frac{(n_o^2 - n_e^2) \sin \theta \cos \theta}{n_o^2 \sin^2 \theta + n_e^2 \cos^2 \theta}. \qquad (9.71)$$

Using refractive index values for calcite, $n_o = 1.65$, $n_e = 1.48$ and plugging $\theta = 45°$ one obtains $\tan \theta_x = 0.108$ and $\theta_x = 6.17°$ that

238 | *Anisotropic Material Optics*

results in the following value of the displacement

$$\Delta d = 1\ cm \cdot \tan(6.17°) = 0.11\ cm \tag{9.72}$$

The biggest value of the angle θ_x^{max} follows from the condition for θ^{max} i.e. $\tan\theta^{max} = n_e/n_o = 0.897$ thus $\theta^{max} = 41.89°$ and $\theta_x^{max} = 6.20°$.

Problem 4. Using Jones matrix expression for a horizontal linear polarizer (i.e. when the principal axis is directed along x-axis) given by

$$\hat{T}_x = \begin{pmatrix} 1 & 0 \\ 0 & 0 \end{pmatrix} \tag{9.73}$$

obtain general expression for the Jones matrix $(\hat{T}_p)$ of a linear polarizer with a principal axis oriented at an arbitrary angle ϕ with respect to x.

Solution. The two matrices $\hat{T}_x$ and $\hat{T}_p$ are defined in the coordinate systems rotated to each other by an angle ϕ. According to the general rule from the matrix algebra a transformation of matrix $\hat{T}_x$ from the reference coordinate system to a new (rotated) one is done as follows

$$\hat{T}_p = \hat{R}^{-1}\hat{T}_x\hat{R}, \tag{9.74}$$

where transformation matrix for a rotation is given by

$$\hat{R} = \begin{pmatrix} \cos\phi & \sin\phi \\ -\sin\phi & \cos\phi \end{pmatrix} \tag{9.75}$$

and the inverse matrix $\hat{R}^{-1}$ is given by

$$\hat{R}^{-1} = \begin{pmatrix} \cos\phi & -\sin\phi \\ \sin\phi & \cos\phi \end{pmatrix} \tag{9.76}$$

Consequently matrix $\hat{T}_p$ is given by

$$\hat{T}_p(\phi) = \begin{pmatrix} \cos\phi & -\sin\phi \\ \sin\phi & \cos\phi \end{pmatrix} \begin{pmatrix} 1 & 0 \\ 0 & 0 \end{pmatrix} \begin{pmatrix} \cos\phi & \sin\phi \\ -\sin\phi & \cos\phi \end{pmatrix}$$

$$= \begin{pmatrix} \cos^2\phi & \sin\phi\cos\phi \\ \sin\phi\cos\phi & \sin^2\phi \end{pmatrix} \tag{9.77}$$

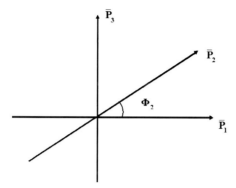

Figure 9.8 Principal axis orientations of the three polarizers optical system: $\hat{P}_1, \hat{P}_2, \hat{P}_3$.

9.4.4 Light Transmition through the System of Three Polarizers

Problem. Obtain transmittance of a system of three polarizers, if the first ($\hat{P}_1$) and the third ($\hat{P}_3$) polarizers are oriented mutually orthogonal but the second polarizer ($\hat{P}_2$) is oriented at an arbitrary angle Φ_2 with respect to ($\hat{P}_1$) as depicted in Fig. 9.8
Applying Eq. (9.74) and taking into account that $\Phi_3 = 90°$ one obtains

$$T = \begin{pmatrix} 0 & 0 \\ 0 & 1 \end{pmatrix} \begin{pmatrix} \cos^2\phi_2 & \sin\phi_2\cos\phi_2 \\ \sin\phi_2\cos\phi_2 & \sin^2\phi_2 \end{pmatrix} = \begin{pmatrix} 0 & 0 \\ \sin\phi_2\cos\phi_2 & \sin^2\phi_2 \end{pmatrix}. \tag{9.78}$$

The field vector $\mathbf{E}_3$ is given by

$$\begin{pmatrix} E_{3x} \\ E_{3y} \end{pmatrix} = \begin{pmatrix} 0 & 0 \\ \sin\phi_2\cos\phi_2 & \sin^2\phi_2 \end{pmatrix} \begin{pmatrix} E_{1x} \\ 0 \end{pmatrix} = \begin{pmatrix} 0 \\ E_{1x}\sin\phi_2\cos\phi_2 \end{pmatrix}. \tag{9.79}$$

Finally the light intensity at the exit of the third polarizer is given by

$$I_{3y} = |E_{3y}|^2 = I_{1x}\sin^2\phi_2\cos^2\phi_2 = 1/4 I_{1x}\sin^2 2\phi_2 \tag{9.80}$$

The relative intensity (transmittance) (9.80) is shown in Fig. 9.9.

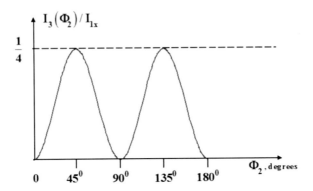

Figure 9.9 Dependence of the relative intensity $I_3(\Phi_2)$ on orientation Φ_2 of the $\hat{P}_2$ polarizer.

9.5 Stokes Vector and Mueller Matrices

In addition to the Jones matrix method the widely used in optics is the method of Mueller matrices. This method is very convenient to be used for description of partially coherent optical radiation.

The stokes vector (**S**) has been introduced in 1852 by J. Stokes. It is defined as follows

$$\mathbf{S} = \begin{pmatrix} S_0 \\ S_1 \\ S_2 \\ S_3 \end{pmatrix} = \begin{pmatrix} \langle E_x E_x^* \rangle + \langle E_y E_y^* \rangle \\ \langle E_x E_x^* \rangle - \langle E_y E_y^* \rangle \\ \langle E_x E_y^* \rangle + \langle E_y E_y^* \rangle \\ \langle E_x E_y^* \rangle - \langle E_y E_y^* \rangle \end{pmatrix} \quad (9.81)$$

where symbol $\langle ... \rangle$ stands for averaging in time. In order to treat fully, partially, or unpolarized light, one defined *Stokes parameters*. The Stokes parameters S_0, S_1, S_2, S_3 can be given in terms of the polarization ellipse angles, i.e. azimuth (or orientation) α and ellipticity β angles.

$$\begin{pmatrix} S_0 = I_0 \\ S_1 = I_0 \cos(2\alpha) \cos(2\beta), \\ S_2 = I_0 \sin(2\alpha) \cos(2\beta), \\ S_3 = I_0 \sin(2\beta) \end{pmatrix} \quad (9.82)$$

From equation (9.82) follow expressions for polarization angels α and β, i.e.

$$\alpha = \frac{1}{2}\arctan(\frac{S_2}{S_1}), \tag{9.83}$$

$$\beta = \frac{1}{2}\arcsin(\frac{S_3}{S_0}) = \frac{1}{2}\arcsin(\frac{S_3}{\sqrt{S_1^2 + S_2^2 + S_3^2}}) \tag{9.84}$$

Partially polarized light is described by a *degree of polarization* Δ defined as

$$\Delta = \frac{\sqrt{S_1^2 + S_2^2 + S_3^2}}{S_0} = \frac{I_p}{I_p + I_{np}}, \tag{9.85}$$

with $I_0 = I_p + I_{np}$, where I_p and I_{np} denote intensities of fully polarized and nonpolarizated light components, respectively. In accordance to the Stokes theorem a partially polarized light can be represented as a sum of two vectors referred to the polarized ($\mathbf{S}_p$) and non-polarized ($\mathbf{S}_{np}$) light as follows

$$\mathbf{S} = \mathbf{S}_p + \mathbf{S}_{np} = \begin{pmatrix} \sqrt{S_1^2 + S_2^2 + S_3^2} \\ S_1 \\ S_2 \\ S_3 \end{pmatrix} + \begin{pmatrix} S_0 - \sqrt{S_1^2 + S_2^2 + S_3^2} \\ 0 \\ 0 \\ 0 \end{pmatrix} \tag{9.86}$$

Stokes parameters can be interpreted in terms of the light intensity. Suppose there are four light detectors and three of them are provided in front with polarizers. These devices detect different intensities, i.e.

1. First detector measures a total irradiance (I_0);
2. Second detector measures only horizontally polarized irradiance (I_1);
3. Third detector measures $+45°$ polarized irradiance (I_2);
4. Fourth detector measures right circularly polarized irradiance (I_3).

242 | *Anisotropic Material Optics*

Note that all these quantities are time-averaged, therefore even randomly polarized light will have a well-defined value. In terms of the intensities I_0, I_1, I_2, and I_3 the Stokes parameters are defined in the following way:

$$S_0 = I_0$$
$$S_1 = 2I_1 - I_0$$
$$S_2 = 2I_2 - I_0 \qquad (9.87)$$
$$S_3 = 2I_3 - I_0$$

In order to describe the Stokes vectors in anisotropic media, H. Mueller introduced in 1943 the 4×4 $\hat{M}$ matrices according to the following:

$$\mathbf{S}^{out} = \hat{M} \cdot \mathbf{S}^{in} = \begin{pmatrix} m_{00} & m_{01} & m_{02} & m_{03} \\ m_{10} & m_{11} & m_{12} & m_{13} \\ m_{20} & m_{21} & m_{22} & m_{23} \\ m_{30} & m_{31} & m_{32} & m_{33} \end{pmatrix} \begin{pmatrix} S_0^{in} \\ S_1^{in} \\ S_2^{in} \\ S_3^{in} \end{pmatrix} = \begin{pmatrix} S_0^{out} \\ S_1^{out} \\ S_2^{out} \\ S_3^{out} \end{pmatrix} \qquad (9.88)$$

where $\mathbf{S}^{in}$ and $\mathbf{S}^{out}$ stand for incoming and outgoing Stokes vectors, respectively.

All matrix elements are real and describe both polarization and coherency of light in homogeneous or scattering medium. Anisotropic properties of an object can be characterized by four types of anisotropy: circular phase $\hat{M}_{CP}$, linear phase $\hat{M}_{LP}$, linear amplitude $\hat{M}_{LA}$, and circular amplitude $\hat{M}_{CA}$ anisotropies. The transformation matrices are given by

$$\hat{M}_{CP} = \begin{pmatrix} 1 & 0 & 0 & 0 \\ 0 & \cos(2\varphi) & \sin(2\varphi) & 0 \\ 0 & -\sin(2\varphi) & \cos(2\varphi) & 0 \\ 0 & 0 & 0 & 1 \end{pmatrix} \qquad (9.89)$$

where φ is an azimuth rotation angle of the polarization ellipse

$$\hat{M}_{LP} = \begin{pmatrix} 1 & 0 & 0 & 0 \\ 0 & C_1^2 + S_1^2\beta & C_1 S_1(1-\beta) & -S_1\mu \\ 0 & C_1 S_1(1-\beta) & S_1^2 + C_1^2\beta & C_1\mu \\ 0 & S_1\mu & -C_1\mu & \beta \end{pmatrix} \tag{9.90}$$

where $\beta = \cos(\delta), \mu = \sin(\delta), C_1 = \cos(2\alpha), S_1 = \sin(2\alpha), \delta$ is a phase displacement of the phase plate, α defines orientation of the fast axis of the phase plate ($\delta = \pi/2$ for the $\lambda/4$ plate).

$$\hat{M}_{LA} = \begin{pmatrix} 1+P & (1-P)C & (1-P)S & 0 \\ (1-P)C & C^2(1+P) + 2S^2\sqrt{P} & CS(1-\sqrt{P})^2 & 0 \\ (1-P)S & CS(1-\sqrt{P})^2 & S^2(1+P) + 2C^2\sqrt{P} & 0 \\ 0 & 0 & 0 & 2\sqrt{P} \end{pmatrix} \tag{9.91}$$

where $P = P_\perp/P_\parallel$ defines linear amplitude anisotropy ($P_\parallel \geq P_\perp$), $C = \cos(2\theta), S = \sin(2\theta), \theta$ is an azimuth of the axes of maximum transparency, ($P = 0$) stands for an ideal linear polarizer.

$$\hat{M}_{CA} = \begin{pmatrix} 1+R^2 & 0 & 0 & \mp 2R \\ 0 & 1-R^2 & 0 & 0 \\ 0 & 0 & 1-R^2 & 0 \\ \pm 2R & 0 & 0 & 1+R^2 \end{pmatrix} \tag{9.92}$$

where

$$R = \frac{r_r - r_l}{r_r + r_l}, \tag{9.93}$$

and coefficients r_r, r_l stand for transmission of right- and left-circularly polarized light components, respectively. Note that for an ideal circular polarizer $R = \pm 1$.

In order to model the effects of more than one medium on the polarization state (e.g. media 1, 2, and 3), one needs just to multiply the input polarization Stokes vector by all of the Mueller matrices:

$$S_{out} = \hat{M}_3 \hat{M}_2 \hat{M}_1 S_{in} \tag{9.94}$$

244 | *Anisotropic Material Optics*

Below application of Mueller matrices for characterization of polarimeters is given.

Thus Mueller matrix method is more universal but more complicated one. Besides it has excess information that does not use for polarization analysis.

Problem 5. In terms of Stokes vector $\mathbf{S}$ and Mueller matrices $\hat{M}$ develop a theory for laser polarimeter.

Solution. A complete function of a polarization analyzer represent matrices of both analyzer (A) and a phase plate (PP), i.e.

$$\mathbf{S}^{out} = \hat{M}_A \hat{M}_{PP} \mathbf{S}^{in}, \tag{9.95}$$

where
$\mathbf{S}^{in}$ is a Stokes vector of a light flux entering the PP,
$\mathbf{S}^{out}$ is a Stokes vector of a light flux exiting the A,
$\hat{M}_A$ is a Mueller matrix of A,
$\hat{M}_{PP}$ is a Mueller matrix of PP.
Required Mueller matrices (see appendix D) are given by

$$\hat{M}_A = \begin{pmatrix} 1 & 1 & 0 & 0 \\ 1 & 1 & 0 & 0 \\ 0 & 0 & 0 & 0 \\ 0 & 0 & 0 & 0 \end{pmatrix} \tag{9.96}$$

$$\hat{M}_{PP} = \begin{pmatrix} 1 & 0 & 0 & 0 \\ 0 & \cos^2 2\alpha + \sin^2 2\alpha \cos \delta & \cos 2\alpha \sin 2\alpha (1 - \cos \delta) & -\sin 2\alpha \sin \delta \\ 0 & \cos 2\alpha \sin 2\alpha (1 - \cos \delta) & \sin^2 2\alpha + \cos^2 2\alpha \cos \delta & \cos 2\alpha \sin \delta \\ 0 & \sin 2\alpha \sin \delta & -\cos 2\alpha \sin \delta & \cos \delta \end{pmatrix} \tag{9.97}$$

Plugging equations (9.96) and (9.97) into Eq. (9.95) results in an expression for the first Stokes element S_0^{out} given by

$$S_0^{out} = S_0^{in} + S_1^{in}(\cos^2 2\alpha + \cos \delta \sin^2 2\alpha)$$

$$+ S_2^{in} \sin^2 \frac{\delta}{2} \sin 4\alpha - S_3^{in} \sin \delta \sin 2\alpha, \tag{9.98}$$

where angles α and δ stand for the fast axis orientation and the phase displacement of the phase plate PP, respectively.

Fundamentals of the Optics of Materials: Tutorial and Problem Solving | **245**

Table 9.5 Outgoing Stokes parameters for selected angular values of the system, azimuth (α) and phase displacement (δ)

δ	α	S_{0i}^{out}
$\frac{\pi}{2}$	0	$S_{01}^{out} = S_0^{in} + S_1^{in}$
$\frac{\pi}{2}$	$\frac{\pi}{4}$	$S_{02}^{out} = S_0^{in} - S_3^{in}$
π	$\frac{\pi}{8}$	$S_{03}^{out} = S_0^{in} + S_2^{in}$
π	$\frac{3\pi}{8}$	$S_{04}^{out} = S_0^{in} - S_2^{in}$

Problem 6. Using solution of the Problem 5 (see equation (9.98)) develop a method for measurements of Stokes parameters: S_0^{in}, S_1^{in}, S_2^{in}, and S_3^{in}.

Solution. The requested Stokes parameters can be obtained after simplifying the equation (9.98) by taking $\delta = \pi/2$:

$$S_0^{out} = S_0^{in} + S_1^{in} \cos^2 2\alpha + \frac{1}{2} S_2^{in} \sin 4\alpha - S_3^{in} \sin 2\alpha, \qquad (9.99)$$

Next one has to chose four values of α_i in order to define S_0^{in}, S_1^{in}, S_2^{in}, and S_3^{in}, e.g. from Table 9.5 follows

$$S_0^{in} = \frac{1}{2}(S_{03}^{out} + S_{04}^{out}),$$

$$S_?^{in} = \frac{1}{2}(S_{03}^{out} - S_{04}^{out}),$$

$$S_1^{in} = S_{01}^{out} - \frac{1}{2}(S_{03}^{out} + S_{04}^{out}), \qquad (9.100)$$

$$S_3^{in} = -S_{02}^{out} + \frac{1}{2}(S_{03}^{out} + S_{04}^{out})$$

Parameters (δ and α) given in Table 9.5 result in a simple solution of Eqs. (9.100).

Comparison of Jones and Mueller matrix methods is summarized in Table 9.6.

246 | *Anisotropic Material Optics*

Table 9.6 Comparison between the Jones and Mueller matrix methods

	Jones matrix method	Mueller matrix method
1	Suitable for an object that does not change light polarization after an interaction	Suitable for an object that changes light polarization after an interaction
2	Accounts phase correlation between beam components	Only seven matrix elements are independent for an object that does not depolarize the light
3	Any complex 2×2 matrix is a Jones matrix for a real object	Not any real 4×4 matrix is a Mueller matrix for a real object
4	Any Jones matrix can be transformed into a Mueller matrix	Any Mueller matrix can be transformed into a Jones matrix, however, only for an optically homogeneous object

9.6 Review Questions and Exercises

9.6.1 Permittivity Tensor

Show that the dielectric permittivity tensor is symmetric for a transparent anisotropic medium, i.e. $\varepsilon_{ij} = \varepsilon_{ji}$. To prove, use the continuity equation for an anisotropic medium.

9.6.2 Electric Field Vectors Relationship

In a real space obtain a relationship between electric field vectors ($\mathbf{E}$), displacement ($\mathbf{D}$), Poynting vector ($\mathbf{S}$), and the wave vector ($\mathbf{k}$).

9.6.3 Poynting Vector in the Iceland Spar Crystal

Find a maximum value of the angle between the Poynting vector ($\mathbf{S}$) and wave vector ($\mathbf{k}$) in the Iceland spar crystal, i.e. $n_o = 1.658$, $n_e = 1.486$.

9.6.4 Circularly Polarized Light on a Quartz Plate

A white circularly polarized light beam strikes a quartz plate, cut parallel to the optical axis and normal to the plate. Above the plate a polarizer is placed whose transmission direction makes an angle of 45°with the optical axis of the quartz plate. The transmitted light passes through the spectrometer. How many dark strips are observed in the spectral range of 400 nm to 500 nm? The thickness of the quartz plate is $d = 2 \times 10^{-3}$ m, and ordinary and extraordinary refractive index values of the plate are $n_o = 1.54$, $n_e = 1.55$, respectively.

9.6.5 Huygens' Construction

Make Huygens' construction and determine directions of the propagating wave fronts and rays on the boundary of a uniaxial anisotropic crystal for the following cases:

1. Optical axis is perpendicular to the plane of incidence and parallel to the surface of the crystal;
2. Optical axis is located in the plane of incidence and oriented parallel to the surface of the crystal;
3. Optical axis is located in the plane of incidence and makes an angle of 45° with the surface of the crystal.

9.6.6 Birefringence of a Quartz Plate

Obtain a birefringence value of an arbitrarily oriented quartz plate. Answer a question: how it should be cut in order to achieve (a) maximum birefringence (i.e. obtain angle α with respect to the crystallographic axis X_3); (b) zero birefringence.

9.6.7 Refraction in a Calcite Crystal

Calculate an angle of refraction (β) of an extraordinary light beam in a calcite crystal for the light incidence angle of 30°. The plate is cut

248 | *Anisotropic Material Optics*

at the following conditions: the optical axis makes an angle of $15°$ with the normal to the plate and it is located in the plane of the light incidence. Use calcite crystal refractive index values of $n_o = 1.658$ and $n_e = 1.486$.

9.6.8 Stokes Vector

Polarization of light can be described by the Stokes vectors: $S_0 = a^2 + b^2$, $S_1 = a^2 - b^2$, $S_2 = 2ab \cos \phi$, and $S_3 = 2ab \sin \phi$, where a, b stand for optical wave amplitudes in two orthogonal directions (x, y), and ϕ is a phase difference between them.

1. Find relationships between Stokes vector squares;
2. Prove the following relationships: $S_1 = S_0 \cos 2\beta \cos 2\theta$, $S_2 = S_0 \cos 2\beta \sin 2\theta$, $S_3 = S_0 \sin 2\beta$, where β is an angle, the tangent of which is equal to the axes ratio of the elliptically polarized vibrations, θ is an angle between the principal axis of the ellipse and the x-axis.

Chapter 10

Optics of Organic and Biological Materials

Organic molecules, polymers, molecular crystals, etc. are interesting materials because of their structures and their electronic, electrical, magnetic, optical, biological, and chemical properties. Researchers continue to face great challenges in the design and construction of well-defined organic compounds that aggregate into larger molecular materials such as tubes, rods, particles, walls, films, and other structural arrays. Such materials fabricated at nanoscale could serve as direct device components that is a central point of modern nanotechnology [Gavrilenko (2020)].

Optical response from organic molecules is caused by an interplay between electronic and vibronic excitations that is currently well understood and addressed in different text books and monographs (see e.g. [Stroscio and Dutta (2001); Kasai *et al.* (2000); Kolasinski (2008); Saleh and Teich (2007)]). Within recent decades organic molecular crystals have been demonstrated to be very promising for variety of applications in modern materials engineering, electronics, and optics. Their electronic and optical properties are fundamentally different from those of inorganic solids, metals, dielectrics, and semiconductors. A unique aspect of organic

---

Fundamentals of the Optics of Materials: Tutorial and Problem Solving
Vladimir I. Gavrilenko and Volodymyr S. Ovechko
Copyright © 2024 Jenny Stanford Publishing Pte. Ltd.
ISBN 978-981-4877-93-0 (Hardcover), 978-1-003-25694-6 (eBook)
www.jennystanford.com

molecular crystals is that the electronic properties of these materials bear the marks of both molecular and condensed matter physics. In a crystal, the organic molecules preserve their identity, since molecular crystals are held together by van der Waals interactions without a formation of intermolecular covalent bonds [van den Brink *et al.* (2005)]. Due to weak intermolecular interaction forces of the van der Waals type, the molecular crystals offer a rich vein of physical phenomena considerably different from the optical and electronic properties exhibited by conventional solids such as covalent or ionic crystals.

A molecular crystal could be defined as a solid that is formed by electrically neutral molecules interacting via weak non-bonding interactions, primarily by van der Waals forces [Silinsh and Capek (1994)]. If the constituent molecules possess specific functional groups then the possibility also exists for the formation of hydrogen bonds and dipolar interactions that will also serve to stabilize the crystal. In general, there is little electronic charge overlap between molecules, and therefore the constituent molecules retain their identity to a large extent. This is in contrast to covalent or ionic solids, where the individual properties of constituent particles in the crystal are completely lost. This chapter overviews optical properties of molecular aggregates, molecular crystals, and molecular-solid systems.

10.1 Optical Properties of Molecular Systems

Optical properties of organic molecules are modified due to their aggregation in molecular complexes that has been observed in a number of molecular systems (see e.g. [Kasai *et al.* (2000); Pope and Swenberg (1982); Schiek *et al.* (2008)] and references therein). Analysis of different molecular aggregates is out of scope of this chapter. Here we are focusing on organic molecular dyes as an example.

The use of dyes in photonic devices, in dye lasers, and various photoinduced reactions caused extensive studies of their electronic and optical properties [Pope and Swenberg (1982)]. More recently, dyes have become the subject of renewal interest, which has been focused on the phenomenon of molecular aggregation (see e.g.

Fundamentals of the Optics of Materials: Tutorial and Problem Solving | **251**

[Martinez *et al.* (2004); Gavrilenko and Noginov (2006); Gavrilenko *et al.* (2008)] and references therein). This section is focused on optical properties of Rhodamine 6G (*R6G*) dye molecules and molecular complexes that are widely used in research and numerous applications for decades. The *R6G* dye is one of the most efficient laser dyes characterized by a high-efficiency luminescence band around 560 nm (see, e.g., [Kikteva *et al.* (1999)] and references therein). However, optical properties of high-concentrated *R6G* dyes differ from those of isolated dye molecules. The *R6G* dye has a high tendency to form *H*-type nonluminescent dimer sandwich-like aggregates with parallel-aligned molecular axes. Another configuration of *R6G* molecules with head-to-tail aligned dipole moments (*J*-type dimers) was normally observed in the presence of an additional external potential (by intercalation, on solid surfaces, etc.) [Martinez *et al.* (2004)]. *H*-type *R6G* dimers are associated with a remarkable blueshift in the optical absorption band with respect to *R6G* monomers. On the other hand, the *J*-type *R6G* dimers are characterized by a redshifted luminescence band [Kikteva *et al.* (1999); Martinez *et al.* (2004)].

Optical properties of high concentrated dye molecules are characterized by a formation of dimers and higher aggregates [Gavrilenko and Noginov (2006); Kikteva *et al.* (1999)]. The earlier interpretation of the absorbance, fluorescence, and dynamics of dye molecules physically bound dimers was based upon the theory of exciton splitting [Kasha (1959); Pope and Swenberg (1982)]. The optical properties of physically bound dimers or higher aggregates can be expressed in terms of monomeric wave functions which have been slightly perturbed through their mutual interaction. In the case of a dimer, the total wave function for the ground state may be written as the product of the two monomeric wave functions Ψ_1, Ψ_2. The total Hamiltonian of the system is given by $H = H_1 + H_2 + V_{12}$, where the H_i are the Hamiltonians of the individual isolated molecules and V_{12} their interaction potential. If the excited-state wave functions of the individual unperturbed molecules are given by Ψ_1^* and Ψ_2^*, then in the absence of any interaction, the two excited states defined by Ψ_1^*, Ψ_2 and Ψ_1, Ψ_2^* are degenerate. For $V_{12} \neq 0$, this degeneracy is lifted and the two dimer excited electronic states are formed with the wave function given by:

$$\Psi_{E\pm} = \frac{1}{\sqrt{2}} \left(\Psi_1^* \Psi_2 \pm \Psi_1 \Psi_2^* \right) \tag{10.1}$$

This predicted exciton splitting for dimers of various geometries is illustrated in Fig. 10.1.

The expected spectral shift due to dimer formation can either be to the blue or the red, depending upon the dimer geometry and therefore on the orientation of the monomer transition moments. The expected spectral shift due to dimer formation can either be to the blue or the red, depending upon the dimer geometry and therefore on the orientation of the monomer transition moments. For parallel molecules, the transition moments are either parallel or antiparallel and only one excited state of the dimer will be optically dipole allowed. For parallel, sandwich-type dimers where the center-to-center angle of inclination lies between $54.7°$ and $90°$, the lower energy excited state is dipole forbidden resulting in dimers which show a blue shift in their absorption spectra but are nonfluorescent. Parallel dimers for which the inclination angle lies between $0°$ and $54.7°$ are characterized by an absorption red shift and fluorescence which is also red shifted. For nonparallel dimers, both excited states have nonzero transition moments and will be optically allowed [Kasha (1959); Pope and Swenberg (1982); Kikteva et al. (1999)]. The result will be an observable splitting in the form of both red- and blue-shifted features in the optical absorption spectrum and red-shifted fluorescence.

Experimental studies confirm this model. Fig. 10.2 shows the optical absorbance spectra of the fused silica surface measured with various Rhodamine surface coverage [Kikteva et al. (1999)].

The optical absorbance spectrum obtained at a surface coverage of 0.5 monolayer (Fig. 10.2a) was very similar to that obtained in solution and shows an absorbance maximum at approximately 530 nm, only slightly shifted from the solution-phase spectrum. Increasing the surface coverage to 1.0 monolayers (Fig. 10.2b) leads to a spectral red shift of approximately 5 nm, but little or no change in any other spectral characteristics. This behavior is typical for all submonolayer surface coverage examined, where any

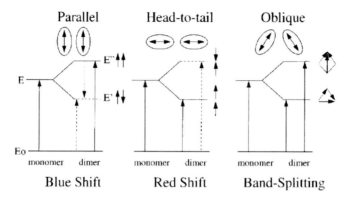

Figure 10.1 Exciton splitting in dimers of various geometries. Orientations of the monomer transition dipoles are represented by short arrows. Dipole-forbidden transitions are indicated by dashed lines, dipole-allowed transitions by solid lines. The dotted line present for parallel geometries represents nonradiative deactivation to the nonfluorescent, lower energy state. (Adapted from [Kikteva et al. (1999)].)

increase in surface concentration leads to a systematic red shift in the absorbance spectrum but no discernible changes otherwise.

Fig. 10.2c shows the effects of increasing the surface coverage beyond that of one monolayer. At a coverage of 1.3 monolayers, the spectrum has undergone an additional red shift of approximately 10 nm, and has an absorption maximum at 546 nm. However, in addition to this red shift, the absorbance spectrum shows the onset of a weak absorbance feature which appears as a shoulder on the blue side of the absorption maximum at approximately 505 nm. Increasing the surface coverage to 1.6 monolayers (Fig. 10.2d) leads once again to a further red shift in the absorption feature but also shows the growth of the blue shoulder, now appearing at approximately 510 nm.

At a surface coverage of 2.2 monolayers (Fig. 10.2e), the blue shoulder, now appearing at a wavelength of approximately 518 nm, has grown in intensity and is now quite pronounced and easily resolved from the absorbance maximum appearing at 555 nm. Fig. 10.2f shows the effect of increasing the surface coverage to ~10 monolayers. At these relatively large surface concentrations, the primary spectral changes are a further red shift in absorbance and the continued growth of the blue absorbance feature to the point

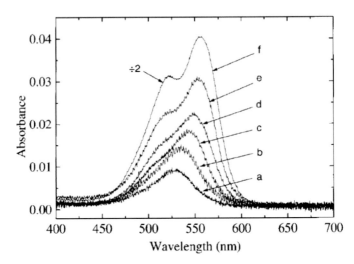

Figure 10.2 Optical absorbance spectra of *R6G* films at surface coverage corresponding to (*a*) 0.5 monolayers, (*b*) 1.0 monolayers, (*c*) 1.3 monolayers, (*d*) 1.7 monolayers, (*e*) 2.2 monolayers, and (*f*) 10 monolayers. Note that the absorbance spectrum labelled *f* has been scaled by a factor of 2 for ease of comparison with the lower surface coverage spectra. (Adapted from [Kikteva *et al.* (1999)].)

where this blue feature now appears as a separate absorption band. As the surface concentration is increased further, even to surface coverage of greater than 30 monolayers, the absorption spectrum of the Rhodamine-covered silica surface undergoes very modest changes in appearance, primarily in the form of a broadening of the spectrum to the red.

Nonradiative and luminescent characters of *H*- and *J*-dimers, respectively, have been semi-qualitatively interpreted in the literature using excitonic theory [Kasha (1959); Pope and Swenberg (1982)]. The modifications of the optical absorption spectrum with concentration were interpreted as the formation of molecular aggregates [Kikteva *et al.* (1999)]. It has been argued that observed blue (*H*-type) or red (*J*-type) shifts in optical absorption spectra are due to the interplay of optical selection rules modified through intermolecular interactions as demonstrated above.

Molecular dimers have been studied by the first principle theories (see e.g. [Xu and Goddard (2004)] and references therein).

The combined quantum chemistry and physics computational approach based on *ab initio* pseudopotentials has been used by [Gavrilenko and Noginov (2006)] to study optical properties of Rhodamine 6G molecular dimers. It has been found that *H*-type configuration is energetically favourable against *J*-type that indicates the difference in the predicted total energies of these systems given by:

$$\Delta E_{tot} = E_{tot}^H - E_{tot}^J = -0.064 \, eV \tag{10.2}$$

The relatively small total energy difference ($\Delta E_{tot} < 3kT$) indicates that both configurations of free standing *R6G* dimers could coexist at room temperature with a concentration ratio of *H*- to *J*-type dimers of at least one order of magnitude. In Fig. 10.3 fully relaxed atomic configurations of *H*- and *J*-type *R6G* molecular dimers calculated by [Gavrilenko and Noginov (2006)] are shown.

Molecular axes in *H*-type are anti-parallel with twisted angle of $28°$ (see Fig. 10.3(a)). The predicted angular value perfectly matches experimental one (of $29°$) reported by [Sasai *et al.* (2002)] for *R6G* molecules intercalated in fluor-taeniolite (TN) film. For *J*-type dimers, the analysis by [Gavrilenko and Noginov (2006)] predicts only one equilibrium configuration with the angle between molecular axes of of $93°$ (*T*-shape configuration shown in Fig. 10.3(b)). This value is in reasonable agreement with a reported torsion angle values of $105°$ and $109°$ obtained by [Martinez *et al.* (2004)] using excitonic theory [Kasha (1959)] to interpret the optical absorption spectra of *R6G* dimers intercalated in Laponite Clay. Predicted distance between geometrical centres of the molecules in *J*-dimer, 0.803 nm, is lower then the value of 0.95 nm obtained from optical absorption by [Martinez *et al.* (2004)].

However predicted intermolecular distance of free standing *H*-type dimer, is much smaller than reported experimental value of 0.71 nm of intercalated dimers [Kikteva *et al.* (1999)]. These discrepancies require some comments. On one hand experimental data by [Kikteva *et al.* (1999); Martinez *et al.* (2004)] were obtained from optical absorption using semi-quantitative analysis based on the excitonic theory [Kasha (1959); Pope and Swenberg (1982)]. Within this model intermolecular interaction is assumed of pure van der Waals type, and neglect of the short-range interaction may

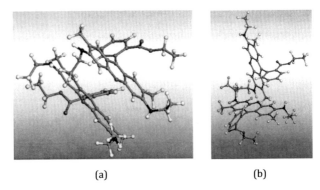

(a)   (b)

Figure 10.3 Predicted equilibrium geometry of *H*- (a) and *J*-type (b) dimers of *R6G* molecules. Note the predicted twisted angle of 28° and opposite orientation of dipole moments of neighbouring molecules in *H*-dimer. Atom type is indicated on a gray scale as follows: hydrogen (white), carbon (gray), oxygen (dark gray), and nitrogen (black). (Adapted from [Gavrilenko and Noginov (2006)]).

cause errors. The intermolecular distance may also increase by intercalation in the TN system [Kikteva et al. (1999)] or in Laponite Clay [Martinez et al. (2004)].

On the other hand, the predicted value of intermolecular distances should also be taken cautionary. The model of [Gavrilenko and Noginov (2006)] was based on the first principles DFT approach and incorporating both short- and long-range (dipole–dipole) interactions. The short-range interaction was determined straightforward from DFT-GGA electronic structure through the calculations of corresponding overlap integrals. However dipole–dipole interaction was incorporated through self-consistent calculations of the charge distribution over molecules in dimers. For better understanding of the numerical results, it is instructive to refer to the electrostatic theory. The calculated static dipole moment (μ) is found to be parallel to the molecular axis oriented along benzene rings of *R6G* molecule. Assuming all molecules been identical, the attractive potential energy describing dipole–dipole interaction is given by [Davydov (1980)]:

$$V(r) = \frac{|\mu|^2}{|r|^3} \left(1 - 3\cos^3\theta\right), \tag{10.3}$$

where θ is an angle between r (the location in direct space) and μ. The energy of dipole–dipole interaction is determined through calculation of overlap integrals with the potential Eq. (10.3) after the integration over entire space. It is proportional to r^{-6}, which represents the van der Waals type of interaction. Equation (10.3) demonstrates that strong attractive (negative) value is expected for dipoles aligned anti-parallel if aggregated in H-type dimers which should be the most energetically favorable configuration resulting in quenching the radiation. This is in agreement with direct computations of this work: total energy of H-dimer is lower than that of J-dimer (see Eq. (10.2)), and predicted value of μ for H-type dimer is 34 times smaller than that of J-dimer.

The dominant calculated absorption peak of $R6G$ monomer is located at 511 nm, while the same peak is seen in the experimental absorption spectrum at about 528 nm [Kikteva *et al.* (1999)]. The difference of 17 nm in the optical absorption peak position should be kept in mind for comparative analysis between theory and experiment.

In Fig. 10.4 calculated optical absorption of the $R6G$ dimers are shown [Gavrilenko and Noginov (2006)]. The predicted optical absorption spectrum corresponding to the singlet transitions of H-type dimers shows the dominant peak located at 483 nm (2.567 eV), Fig. 10.4(a). There is also another small peak located at 526 nm (2.357 eV). The splitting (ΔE_H=0.210 eV) is caused by the strong intermolecular interaction between π-type electrons of the neighbouring molecules in H-dimer. In terms of excitonic theory [Kasha (1959)] these peaks correspond to anti-bonding (dipole-allowed) and bonding (dipole-forbidden) electronic states, respectively. Due to twisting of molecular axes (see Fig. 10.3(a)), optical transitions to dipole forbidden unoccupied electronic state in H-type dimers are weakly allowed. The predicted absorption maximum of $R6G$ monomers is located at 511 nm. Therefore, the theory predicted a strong blue shift ($\Delta\lambda$=28 nm) in optical absorption of $R6G$ due to the creation of H-type dimers. Very close to this value are reported experimental data of absorption maxima at 501 nm [Kikteva *et al.* (1999)] and 500 nm [Martinez *et al.* (2004)] (as compared as to measured 528 nm maximum in initial spectrum) of $R6G$ molecules intercalated in a fluor-taeniolite film [Kikteva *et al.*

258 | *Optics of Organic and Biological Materials*

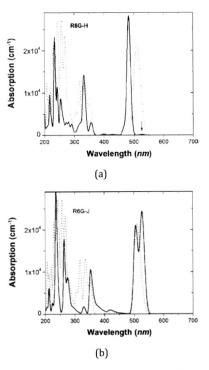

Figure 10.4 Calculated optical absorption spectra of *H*- (a) and *J*-type (b) dimers of *R6G* molecules (bold lines) in comparison with that of single *R6G* molecule (dotted line). Weakly allowed transitions to the anti-symmetrical states in *H*-dimer (due to the twisting) result in small red-shifted peak indicated by the arrow. (Adapted from [Gavrilenko and Noginov (2006)].)

(1999)] or in a laponite clay [Martinez *et al.* (2004)], which were interpreted as an indication of *H*-type dimers.

For *T*-shaped *J*-type *R6G* dimers, the theory by [Gavrilenko and Noginov (2006)] predicts splitting of the main absorption peak into two maxima located at 527 nm (2.353 eV) and at 506 nm (2.451 eV), Fig. 10.4(b). The splitting ((ΔE_J = 0.098 eV) is caused by the interaction of boundary atoms from one molecule with delocalized π-electrons from the other molecule in *J*-dimer. In terms of the excitonic theory [Kasha (1959)], this situation corresponds to the oblique mutual orientation of molecular dipoles when electronic transitions to the both electronic unoccupied levels are allowed. Obtained values of the total energy minima as well as values of ΔE_H

and ΔE_J clearly indicate that intermolecular interaction in H-type is substantially stronger than that in J-type $R6G$ dimers.

Therefore the reviewed results of optical absorption of the Rhodamine 6G dye molecules clearly indicate substantial modifications of their spectra due to the aggregations. These modifications are caused by the strong contributions of the delocalized π electrons that was observed in different molecular systems [Schiek *et al.* (2008)].

10.2 Optics of Molecular Crystals

Molecular crystals is an important class of materials for a variety of applications in modern optics. A direct approach to molecular technology is to start with organic molecules as building blocks. The molecular wires can be grown by self-assembly in liquids or by template wetting of mesoporous materials (see e.g. [Lebedenko *et al.* (2006); Schiek *et al.* (2008)] and references therein). However, the controlled growth of crystallites of predefined shapes and predefined mutual orientations and their transfer onto more complicated target substrates are serious challenges in the fabrication of these materials. The huge advantages, however, are large design flexibility, excellent device integrability, and potentially much improved performance of resulting devices—similar to the evolution from liquid-crystal-based flat screens to organic light-emitting-device-based screens.

The bottom-up fabrication approach of organic nanofibers was described by [Schiek *et al.* (2008)]. The approach is based on directed self-assembled surface growth of nanofibers from functionalized molecules and on the integration of the ordered arrays into a device. The rod-shaped, polarizable organic molecules such as para-hexa-phenylenes ($p6P$) formed needlelike, crystalline, and mutually well-aligned organic crystals upon molecular-beam epitaxial growth or by hot-wall epitaxy on muscovite mica or TiO_2 substrates. The molecular crystals made from $p6P$ were semiconducting with 3.1 eV bandgap which makes them interesting for the photonics applications (e.g. as LEDs) [Schiek *et al.* (2008)]. Chemical structure is shown in Fig. 10.5.

The shorter chain molecular building block (the para-quaterphenylene) has been substituted in the 4,4$''$ positions by a variety of functional groups: methoxy, chloride, cyano, amino,

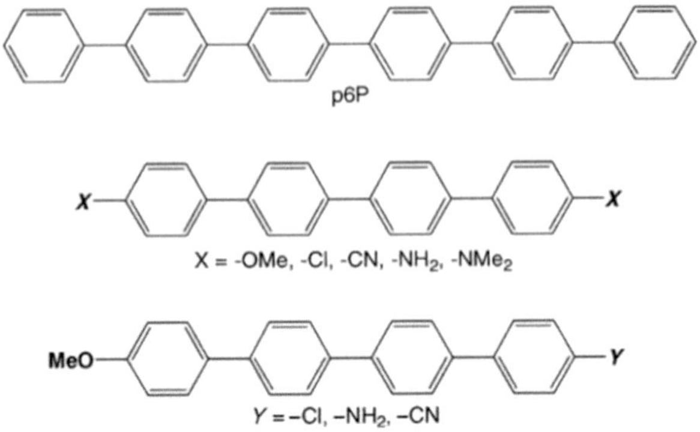

260 | *Optics of Organic and Biological Materials*

Figure 10.5 Chemical formulas for para-hexaphenylene (top), symmetrically functionalized para-quaterphenylenes (middle), and nonsymmetrically functionalized para-quaterphenylenes (bottom). (Adapted from [Schiek *et al.* (2008)].)

and dimethylamino (see Fig. 10.5). Symmetrically substituted compounds like dimethoxy-p-quaterphenylene (MOP4) and dichloro-p-quaterphenylene (CLP4) as well as nonsymmetrically functionalized oligomers such as methoxycyano-p-quaterphenylene (MOCNP4), methoxyamino-p-quaterphenylene (MONHP4), and methoxychloro-p-quaterphenylene (MOCLP4) have been synthesized [Schiek *et al.* (2008)]. The fabricated organic nanofibers showed high degree of crystallinity. The controllable crystallinity and waveguiding resulted in a preferred polarized light emission from the tip of the nanofibers. The oriented array of the nanofibers is shown in Fig. 10.6.

Under the appropriate growth conditions of surface temperature and molecular flux, the growth of mutually oriented nanofibers has been observed (see Fig. 10.6).

The linear and nonlinear optical properties of organic nanofibres can be controlled by appropriate choice of the functional groups. This is demonstrated in Fig. 10.7. The nonlinear optical emission of the nanofibers fabricated from *p6P*, MOCNP4, and MONHP4 was studied by [Schiek *et al.* (2008)]. In the case of *p6P* only two-photon luminescence spectra were observed, whereas the introduction of push and pull groups resulted in an increased molecular hyperpolarizability and thus in the appearance of optical

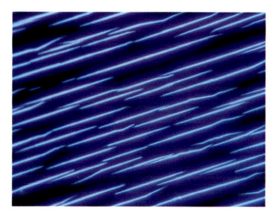

Figure 10.6 Fluorescence microscopy image obtained at the excitation wavelength of 365 nm of mutually parallel, blue-light-emitting methoxy-cyano-p-quaterphenylene (MOCNP4) fibers on muscovite mica. Image area was 135 × 100 µm^2. (Adapted from [Schiek *et al.* (2008)].)

second-harmonic generation (SHG) from the bulk of the crystals [Schiek *et al.* (2008)].

The intensity of SHG depended on the functional groups introduced. The linear luminescence peak wavelength for the two functionalized molecular nanofibers was nearly the same—around 430 nm. However, by changing the functional end groups the ratio between second-harmonic generation at 395 nm and two-photon luminescence for wavelengths above 400 nm can be varied to a large extent, see Fig. 10.7. This can be used for the upconversion of light from submicrometer-scale infrared light sources in integrated photonic circuits.

These results demonstrate that the organic molecular technology allows for a generation of mutually aligned, morphologically well-defined light-emitting organic nanofibers from functionalized molecules, which bridge the gap between the nanoscopic and microscopic worlds. They can be transferred easily and destruction free as individual entities or in a massive parallel fashion onto prestructured target substrates. Due to their crystalline perfection and due to the morphological control organic nanofibers are perfectly suited for fundamental studies of optics, mechanics, and electronics on the mesoscale. Applications as passive and active

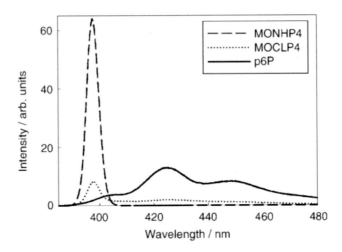

Figure 10.7 Nonlinear optical emission spectra of MOCLP4, MONHP4, and p6P nanofibers following 100 fs infrared laser excitation at 790 nm (4.5 mW). (Adapted from [Schiek et al. (2008)].)

elements in printed all-optical chips are within reach. Core/shell wires and segmented nanowires from different organic materials or even from combinations of organic and inorganic compounds will be challenged for advanced applications.

A variety of organic nanostructures show size-dependent resonance of electronic transitions. The poly-1,6-di(N-carbazolyl)-2,4-hexadiyne (poly-DCHD) molecular crystals demonstrate dependence of the electronic resonance in optical extinction when the size of the material is reduced to the submicron range [Kasai et al. (2000); Patra et al. (2007); Volkov et al. (2004)].

Poly-DCHD has received much attention as a typical polydiacetylene (PDA). The π-electron conjugation along the polymer backbone guaranteed a large third-order optical nonlinearity, and thus promised this group of materials for applications to fast optoelectronics. Furthermore, the delocalization of π-electrons makes PDA a good model system for experimental and theoretical studies on electronic and structural properties in one-dimensional photoconductors. As a result, the optically allowed electronic transitions, the vibrational relaxation and the carrier recombination pathways are well-documented for a variety of PDA.

Fundamentals of the Optics of Materials: Tutorial and Problem Solving | **263**

The high-resolution TEM observation on PDA structural properties in epitaxially grown thin films of DCHD monomers clearly showed a variety of coexisting monomer and polymer microdomains formed at elevated temperature.

Both the steady-state and the time-resolved optical studies in aqueous suspensions of poly-DCHD nanocrystals revealed a number of interesting dependencies of extinction resonance energy and recovery times on the size of the prepared nanocrystals [Kasai *et al.* (2000)]. The size-dependent optical properties of poly-DCHD nanocrystals in ensemble and in a single particle level were studied by [Volkov *et al.* (2004)]. Single poly-DCHD nanocrystals showed resonant Rayleigh scattering spectra with a lower resonance energy peak for nanocrystals having larger cross section against its long axis. The extinction and the resonant Rayleigh scattering spectra in ensemble and the resonant Rayleigh and Raman scattering spectra and their anisotropy of single nanocrystals were studied.

The lower-energy part of the optical extinction consists of two spectral features: the resonance with the $\pi - -\pi^*$ electronic transition to the first allowed excited state in the range of 630 to 660 nm (1.58×10^4 to 1.53×10^4 cm^{-1}) and the blue side series of weaker transitions due to the vibronic progression of C=C stretching as discussed in the literature [Kasai *et al.* (2000)].

10.3 Polymers

Conjugated polymers attract considerable attention for their use in polymer light-emitting diodes, as an active component of optically pumped lasers, in photovoltaic devices, and in all-plastic electronics [Barford (2005); Günes *et al.* (2006)]. These materials combine some of the attractive functional properties of traditional semiconductors with solubility in common organic solvents and, hence, easy processability. The π-conjugated polymers usually have a strongly dipole-allowed electronic transition in the visible region of the spectrum and certain types, e.g. poly (*p*-phenylene-vinylenes) or PPV's show efficient luminescence in solution and thin solid films [Peeters *et al.* (2000)].

As demonstrated in Chapters 5 and 6, the inorganic semiconductors are well suited and are widely used for the design of different optoelectronic devices because of their simple processability combined with their optical properties.

A new generation of photonic devices based on the hybrid materials, consisting of both conjugated polymers and inorganic semiconductor nanocrystals, combine the unique properties of inorganic semiconductors with the film-forming properties of conjugated polymers. There are two general concepts for making use of nanocrystals to improve polymer photovoltaic solar cells: The first is the bulk-heterojunction concept, in which charge-transfer junctions with high interfacial area are formed by blending semiconductor nanocrystals into conjugated polymers [Alivisatos (1996); Arici *et al.* (2003); Huynh *et al.* (2002)]. Operation of such solar cells, caused by photocurrent generation at the interface of nanocrystal/polymer composite materials, has been demonstrated in various blends containing CdS, CdSe [Huynh *et al.* (2002)], CuInS$_2$ [Arici *et al.* (2003)], and PbS [Lü *et al.* (2005); McDonald *et al.* (2005)] nanocrystals. In the second concept, which is based on the Grätzel type of solar cells [Nazzeruddin *et al.* (1993)], dyes are used for sensitizing a nanoporous TiO$_2$ film, which acts as an electrode in these devices [Günes *et al.* (2006)]. While in liquid-electrolyte-containing classical Graetzel cells, dyes like *Ru*-bipyridyl are used for light absorption and electron injection into the nanoporous TiO$_2$ conduction band, in solid state dye-sensitized solar cells the liquid electrolyte is replaced by a layer of semiconductor like CuI [Tennakone *et al.* (1999)], CuSCN [Arici *et al.* (2003); Nazzeruddin *et al.* (1993)], or by conjugated polymers [Gebeyehua *et al.* (2001)]. The use of inorganic nanocrystals instead of organic dyes results in several advantages: the bandgap and, thereby, the absorption range are easily adjustable by the size of the crystals, as band edge type of absorption behavior is the most favourable for effective light harvesting, and the surface properties of the particles can be modified in order to increase the photostability of the electrodes [Günes *et al.* (2006)].

One of the photonics devices based on the hybrid nanomaterials was presented by [Günes *et al.* (2006)]. This device represents a combination of the concepts of a nanocrystal/polymer-blend solar cell and a solid state nanocrystal-sensitized solar cell, and denoted as an *ASOS* device, contains both types of nanocrystals, HgTe-AS and HgTe-OS. The ASOS device refers to aqueous soluble HgTe-AS nanocrystals coated onto nanoporous TiO_2 (NP-TiO_2) electrodes, which were further coated with organic-solvent-soluble HgTe-OS in a poly(3-hexylthiophene) (P3HT) matrix.

It has been demonstrated few decades ago that doping of conjugated polymers may result in high (near metallic) conductivity of these materials. The applications of conjugated polymers range from making conducting transparent plastics to exploiting the nonlinear (both second-order and third-order) optical response and the semiconducting properties, for instance in thin-film field-effect transistors, light-emitting diodes, solid-state lasers, and photovoltaic devices (see e.g. [Bredas *et al.* (1999)] and references therein). In this specific class of polymers, small variations in chemical structure play an essential role and the properties of interest directly depend on the electronic structure. This takes the chemical nature fully into account, which helps in forging a fundamental understanding of the electronic and optical characteristics of the conjugated materials and in guiding the experimental efforts toward novel compounds with enhanced characteristics [Bredas *et al.* (1999)].

In order to understand nature of the optical properties of conjugated polymers it is instructive to consider first one of the simplest typical unit of these materials having the π-conjugated system. By π-conjugated system, it is meant a (macro)molecule along the backbone of which there occurs a continuous path of carbon atoms or heteroatoms, each carrying both σ- and π-atomic orbitals. When the backbone is fully planar, there exists a strict distinction between π-orbitals (antisymmetric with respect to the backbone plane) and σ-orbitals (symmetric).

The smallest conjugated molecule is ethylene, $CH_2 = CH_2$. The occupied valence molecular orbitals (MOs) are characterized by there one σ-type C-C MO, four σ-type C-H MOs, and one π-type MO. The latter is formed by the bonding combination of the two π-atomic orbitals and constitutes the highest occupied MO (HOMO). The antibonding combination of the π-atomic orbitals leads to the lowest unoccupied MO (LUMO).

One of the examples of the conjugated system is the *octatetraene* which has the following structure:

$$CH_2 - CH = CH - CH = CH - CH = CH - CH_2 \qquad (10.4)$$

Electronic structure of octatetraene is presented in Fig. 10.8(a). The Hückel model is widely used for quantum chemical interpretation of molecular electronic structure [Pope and Swenberg (1982)]. This model correctly reproduces ground state of various molecular system, however it fails in description of conjugated polymers optics (e.g. luminescence, electron-hole separation, nonlinear optical response, etc.). The model requires a proper description of electronic excited states and interchain interactions, for which the role of many-body effects is large.

In this context, the theoretical treatment of π-conjugated systems often becomes very elaborate because of the need (a) to incorporate electron correlation effects and (b) to take account of the strong connection between, and mutual influence of, the electronic and geometric structures.

Important contribution of the electron correlation effects could be seen from an analysis of the ordering of the lowest singlet excited states in octatetraene, Fig. 10.8. A one-electron (e.g., Hückel or Hartree-Fock) treatment produces eight π MOs whose symmetries alternate between symmetric [g: (*gerade*, German)] and asymmetric [u: (*ungerade*, German)] and energies increase with the number of nodes in the wave function [Bredas *et al.* (1999)].

The eight π electrons distribute among the eight π MOs; each of their possible repartitions defines a so-called electronic configuration whose individual wave function can be cast in the form of one Slater determinant. The lowest energetic configuration corresponds to the situation where the π electrons occupy two by two the four lowest π MOs and defines the singlet ground-state S_0 (of A_g symmetry). At the one electron level, the lowest (one-photon allowed) excited state (of B_u symmetry) is described by promotion of a single electron from the HOMO to the LUMO; any (one photon forbidden) Ag excited state lies higher in energy because it requires promotion of a single electron from the HOMO to the LUMO + 1 or the HOMO - 1 to the LUMO or promotion of two electrons from the HOMO to the LUMO, and these processes nominally cost a larger energy (see Fig. 10.8(a)). However, an entirely different picture emerges when electron correlation is switched on.

Various methods have been developed to take account of electron correlation effects; for instance, in the often-exploited configuration interaction (CI) approach, each state is cast into a

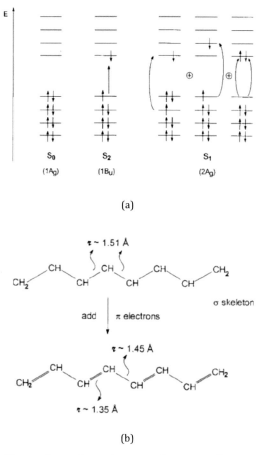

Figure 10.8 Electronic configuration contributing to the ground state S_0 and the lowest singlet excited states S_1 and S_2 of octatetraene (a); the dimerization of the geometric structure due to the uneven distribution of the π-electrons over the bonds (b). (Adapted from [Bredas et al. (1999)].)

linear combination of electronic configurations (the wave function is then described by a linear combination of the corresponding Slater determinants). In polyenes, it so happens that the singly excited HOMO to LUMO + 1 and HOMO − 1 to LUMO configurations and the doubly excited HOMO to LUMO configuration (illustrated in Fig. 10.8(a)) strongly mix and result in the $2A_g$ state being located below the $1B_u$ state; in other words, in polyenes (at least those longer than butadiene), the lowest excited singlet state, S_1, is one-photon forbidden versus the ground state. The consequence is that polyenes and polyacetylene do not luminesce.

Indeed, according to Kasha's rule [Kasha (1959)] (see Section 10.1), luminescence takes place from the lowest excited state; in order to observe strong fluorescence, a large one-photon coupling between S_0 and S_1 is thus required. A major feature that is specific of conjugated compounds is the interconnection between electronic structure and geometric structure. This connection has been beautifully exemplified in the field of conducting polymers with the emergence of concepts such as solitons, polarons, and bipolarons.

Consider again octatetraene, Fig. 10.8(b), and suppose that, in a first step, only the σ backbone is taken into account. The geometry would then be such that all carbon-carbon bond lengths are roughly the same and equal to ca. 1.51 Å (the typical single bond length between two sp^2 carbons). When the π electrons are thrown in, the main aspect is that they distribute unevenly over the bonds and in such a way that alternating larger and smaller π bond densities appear when starting from one end of the molecule; this results in alternating (physicists would say *dimerized*) double-like carbon-carbon bonds (ca. 1.35 Å long) and single-like bonds (ca. 1.45 Å long). A notable feature is that this geometry is reflected into the bonding-antibonding pattern of the HOMO wave function while the LUMO wave function displays the exactly opposite bonding-antibonding pattern; see Fig. 10.8(a).

It can then be easily understood that, in the $1B_u$ excited state (which mainly involves promotion of one electron from the HOMO to the LUMO), the π-bond densities are strongly modified; as a consequence, the equilibrium geometry in the $1B_u$ state is markedly different from that in the ground state and is characterized by a significant reduction in C-C bond alternation. All other π excited states also have equilibrium geometries determined by their π-bond density distributions. Note also that, as the molecules become longer, the geometry relaxations no longer affect the whole chain but become localized (in the $1B_u$ state of long polyenes, the optimal relaxation is calculated to extend over 20–25 Å).

The optical spectra of PPV or oligomers and their derivatives present clear signs of strong vibronic coupling. A very good theoretical simulation of the absorption and emission spectra of

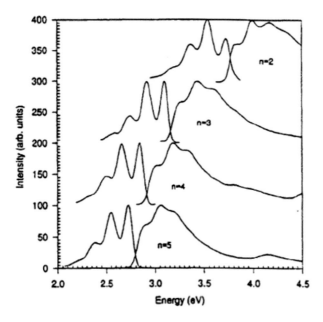

Figure 10.9 Experimental absorption and photoluminescence spectra measured at 77 K of PPV oligomers containing from two to five rings. (Adapted from [Bredas et al. (1999)].)

oligomers containing from two to five rings is obtained (Fig. 10.9) by considering the coupling of the S_0 to S_1 or S_1 to S_0 electronic transition with two Raman modes (at 0.16 and 0.21 eV) corresponding to the C-C bond stretching and phenylene breathing modes, respectively [Bredas et al. (1999)].

From the evolution of the vibronic progression, the relaxation energy in the long chains can be estimated to be on the order of 0.15–0.2 eV. The very presence of a vibronic progression in the polymers indicates that the electronic excitation produces a localized geometry relaxation around it (which extends over 20 Å) and as such can hardly be related to a pure band gap transition. It was found that the lowest B_u-type excited state has a dominant intrachain excitonic character.

Interchain interaction may substantially modify electronic structure and optical spectra of conjugated polymers [Gavrilenko et al. (2008)]. In the case of photoluminescence, a number of experimental studies have highlighted the influence of interchain interactions when going from dilute solutions to the solid state.

Optical absorption and emission spectra of conjugated polymers exhibit well pronounced peak attributed to the excitations of $\pi - -\pi^*$ electron transitions. Vibronic excitations in well-ordered polymers could be seen as additional fine structures.

However, not all components of the fine structure in the optical absorption and emission spectra can be attributed to exciton–phonon coupling. It has been demonstrated earlier that the long-wavelength shoulder in optical absorption spectra of poly(3-hexylthiophene) and poly(thienylene vinylene) (PTV)-conjugated polymers has a different nature, and it could be interpreted as the effect of interchain interaction [Gavrilenko *et al.* (2008)].

Previous works on PPV (see [Gavrilenko *et al.* (2008)] and references therein) showed that crystalline arrangement crucially affects the optical properties of the polymer films and interchain interactions can be viewed as a tunable parameter for the design of efficient electronic devices based on organic materials. 3D arrangement is also a crucial element for the design of materials with efficient transport properties. The effect of the aggregation in PTV, which could be considered as an intermediate phase between liquid and solid has been studied by [Gavrilenko *et al.* (2008)].

Optical absorption spectra of poly(thienylene vinylene) (PTV) samples (the chemical structure is given in Fig. 10.10(a)) measured at room temperature are shown in Fig. 10.10(b).

Figure 10.10(b) presents effects of heating and sonification treatments of regioregular PTV solutions on their optical absorption spectra in visible and near ultraviolet regions [Gavrilenko *et al.* (2008)]. The solid (black) line is the absorption of the PTV heated for 1 min without sonification. The dashed (red) line and the dotted (blue) line show the absorption spectrum of the PTV heated for 1 and 3 min, respectively, followed by sonification. Sonification was performed at a frequency of 40 kHz for 60 min.

A dominant absorption peak located at 577 nm accompanied by a prominent shoulder located at 619 nm (A shoulder) were measured [Gavrilenko *et al.* (2008)]. The A-shoulder was observed on all samples studied and it showed relatively weak dependence on concentration and heat-sonication treatment.

Another low intensity shoulder is observed around 685 nm (see B-shoulder in Fig. 10.10(b)). Its strong dependence on external

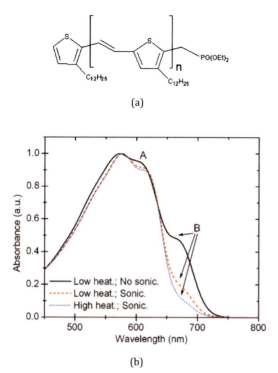

Figure 10.10 (a) Chemical structure of PTV. (b) Optical absorption spectra dependence on heating and sonification treatment of PTV polymers at the concentration of 1 mM. (Adapted from [Gavrilenko et al. (2008)].)

treatment is essentially different from A-shoulder as demonstrated in Fig. 10.10(b): the intensity of the B-shoulder decreases with more intense treatment until becoming undetectable.

The low heating caused slight red shift of optical absorption from its initial spectral location as shown in Figure 10.10(b). Further external treatment did not affect the spectral location of the optical absorption [Gavrilenko et al. (2008)].

Ab initio pseudopotential method within the DFT and the supercell (sc) formalism has been applied to study the effects of interchain interaction of the PTV optical spectra. The system was modeled as an infinite chain as shown in Fig. 10.11a.

The unit cell geometries were optimized using the DFT-LDA method, see [Gavrilenko et al. (2008)]. Only Coulombic interaction providing the most important contribution to the interchain

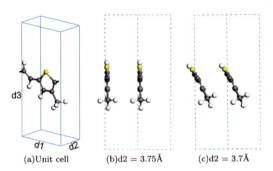

(a) Unit cell   (b) d2 = 3.75Å   (c) d2 = 3.7Å

Figure 10.11 3D view of the PTV unit cell and equilibrium geometries of straight and tilted chain configurations. Optimized unit cell dimensions are shown by numbers. (Adapted from [Gavrilenko *et al.* (2008)].)

interaction was included. The unit cell was replicated in all three dimensions; the height, d_3, was chosen to be 15 Å in order to quench the interaction between the thiophene ring and the methyl group. A very weak effect of the d_3 on optics was observed. The equilibrium intermolecular distance, $d_1 = 6.55$ Å, was determined by cluster calculation of a single polymer chain of 3 units.

A strong dependence of the PTV polymer ground-state on the interchain distance has been reported [Gavrilenko *et al.* (2008)]. The equilibrium interchain distance was studied by a series of unit cell length d_2 optimizations varying between 10 and 3 Å in steps of 0.05 Å. Two equilibrium geometry configurations characterized by different total energy relaxation paths but having almost the same unit cell length: $d_2 = 3.75$ and 3.7 Å; were obtained.

These geometries correspond to the different metastable phases characterized by the total energy relaxation curves shown in Fig. 10.12. Equilibrium geometry analysis of PTV polymers clearly predicted two geometrical phases with straight (Fig. 10.11b) and tilted (Fig. 10.11c) chain geometries [Gavrilenko *et al.* (2008)].

The predicted equilibrium value of $d_2 = 3.75$ Å (that is equal to the interchain distance, d_i) corresponds to absolute energy minimum of straight geometry (see Fig. 10.11b). Another energetically favourable geometrical phase of the PTV polymer (see Fig. 10.12) had a predicted equilibrium value of $d_2 = 3.70$ Å but much smaller corresponding interchain distance value of $d_i = 3.29$ Å. The equilibrium PTV configuration in this case was characterized

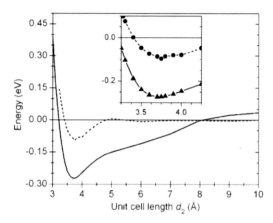

Figure 10.12 Total energy of the system vs PTV unit cell length d_2. Dashed and solid lines correspond to straight and tilted chain geometries shown in Parts (b) and (c) of Fig. 10.11, respectively. (Adapted from [Gavrilenko et al. (2008)].)

by substantial atomic reconstruction: the out-of-plane rotation of the thiophene ring by 27°, relative displacement of neighbouring chains, and additional in-plane atomic distortion.

Predicted optical absorption spectrum of the tilted phase shows pitting of the dominant peak due to the strong interchain interaction (see Fig. 10.13(a)).

Accordingly to the total energy minimization study, the tilted geometry is the most favourable one (see Fig. 10.12). However, the predicted energy difference of 0.16 eV between two phases minima allows for their coexistence at room temperature. For comparison with experiment in Fig. 10.13(b) the theoretical absorption spectrum was generated from the data of both predicted phases.

All things considered, the results reviewed in present section highlight the ways of the polymer optics engineering that include a controllable change of the interchain distances in BCP which substantially modify optical absorption spectra.

10.4 Biological Materials

The optical properties of a tissue or a biomaterial are described in terms of the optical functions and coefficients discussed in previous

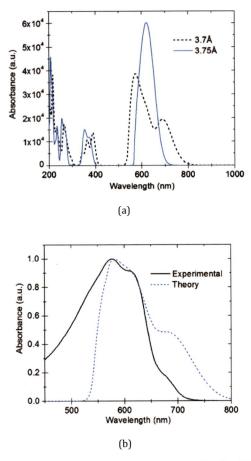

Figure 10.13 Calculated optical absorption spectra for the PTV configurations (a) shown in Parts (b) and (c) of Fig. 10.11. Comparison of experimental spectrum coresponding to low-heating and sonification treatment of the PTV polymer, solid (black) line, and the calculated optical absorption spectra including contributions of the two (tilted and untilted) predicted phases shown in (a), dashed (blue) line (b). (Adapted from [Gavrilenko et al. (2008)].)

chapters, i.e. absorption coefficient, the scattering efficiency, refractive index, etc. Methods for determining the optical parameters of tissues and biomaterials can be divided into two large groups: direct and indirect methods.

Direct methods include those based on the fundamental concepts and rules in optics reviewed in this book. These methods are advantageous in that they use very simple analytic expressions for data processing and reconstruction of optical parameters. Their disadvantages are related to the necessity to strictly fulfill experimental conditions dictated by the selected model, single scattering in thin samples, exclusion of the effects of light polarization, etc. in the case of slabs with multiple scattering, the recording detector must be placed far from both the light source and the medium boundaries.

Indirect methods obtain the solution of the inverse scattering problem using a theoretical model of light propagation in a simulation of light propagation in a medium. These methods are out of scope of this book.

In this section two experimental methods widely used for characterization of biological materials are considered: the surface plasmon resonance (SPR) and the luminescence (or fluorescence)[1] of materials is considered.

10.4.1 Surface Plasmon Resonance for Biosensing

A key element in optical characterization of biological materials are *biosensors*. The biosensors could be fabricated as hybrid materials that include biomolecules, as well as metallic, and/or semiconductor nanoparticles. In modern biological nanotechnology the ordered nanoparticle arrays are applied in order to fabricate the biomaterials (see [Gavrilenko (2020)] for more details). These materials could be used as biosensors in which the biological target is bounded to the plasmon-active nanoparticles. By appropriately functionalizing surfaces of nanostructures, the surface plasmon biosensors can be designed to detect a variety of biological and pathogenic molecules. Such biosensors will make them a valuable diagnostics tool for the biomedicine. Within the recent decade a remarkable success of SPR biosensors has been achieved in a wide range of fields from

---

[1] As noted in Section 4.6, *luminescence* is any emission of light from a material that does not arise from heating. Under photoluminescence one understands the emission of light from a material following the absorption of light. Term *fluorescence* is most commonly used to refer to photoluminescence from molecular systems.

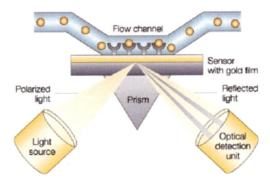

Figure 10.14 Typical set-up of a biosensor based on the surface plasmon resonance (SPR). Changes in the refractive index in the area near the surface layer of a sensor result in variations of spectral characteristics of the SPR. The SPR phenomenon is observed as a negative peak in the angular dependence of reflected light from the surface at an angle of incidence that is dependent on the amount of material at the surface. (Redrawn from [Cooper (2002)]).

fundamental biological studies to clinical diagnosis applications [Hoa et al. (2007)]. Physics behind the SPR is described in Chapter 7.

Practical applications of the SPR biosensors for a number of biological systems has been demonstrated within the recent decades (see e.g. [Hoa et al. (2007)] and references therein). The SPR phenomenon is widely used in different commercial biosensing tools [Tudos and Schasfoort (2008)].

Optical biosensors exploit the evanescent-wave phenomenon to characterize interactions between *receptors* that are attached to the biosensor surface and *ligands* that are in solution above the surface. Typical experimental set-up of the SPR-based biosensor is shown in Fig. 10.14.

The biological nanostructures could be fabricated as hybrid materials including biomolecules, metallic, and/or semiconductor nanoparticles. The biological nanotechnology can also include the ordered nanoparticle arrays to fabricate the biomaterials. These materials could be used as biosensors in which the biological target could be bounded to the plasmon-active nanoparticles. This phenomena has been quantitatively studied in biomaterials bounded to metallic nanoparticles (see (Gavrilenko, 2020) for details). By appropriately functionalizing QD surfaces, the surface plasmon

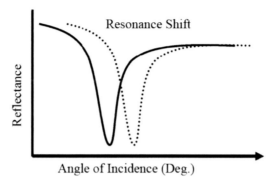

Figure 10.15 Resonance shift in the reflected light spectrum related to the excitation of surface plasmon polaritons. (Redrawn from [Hoa *et al.* (2007)].)

biosensors can be designed to detect a variety of biological and pathogenic molecules. Such biosensors will make them a valuable diagnostics tool for the biomedicine. Within the recent decade a remarkable success of surface plasmon resonance (SPR) biosensors has been achieved in a wide range of fields from fundamental biological studies to clinical diagnosis applications [Hoa *et al.* (2007)]. Physics behind the SPR is described in Section 7.4.

Practical applications of the SPR biosensors for a number of biological systems has been demonstrated within the recent decades (see e.g. review by [Hoa *et al.* (2007)] and references therein). This principle is widely used in different commercial biosensing tools [Tudos and Schasfoort (2008)].

The excitation of surface plasma waves was first discovered by Wood at the beginning of the twentieth century [Wood (1902)] and described as an anomalous diffraction on diffraction gratings [Wood (1935)]. The surface plasmons and surface plasmon polariton phenomena are addressed in detail in Section 7.4. The SPR-based optical biosensors can be designed to detect a variety of biological and pathogenic molecules, which makes them as a valuable diagnostic tool for biology and medicine.

Application of the SPR for optical biosensing is based on the phenomena that the binding of a biological target in most cases induces a measurable shift of the optical spectrum on the angle of incidence (or wavelength) scale related to the SPR [Liedberg *et al.* (1995); Hoa *et al.* (2007)] as demonstrated in Fig. 10.15.

In SPR biosensing, the adsorption of a targeted analyte by a surface bioreceptor is measured by tracking the change in the conditions of the resonance coupling of incident light to the propagating surface electromagnetic wave. The generation of this wave is dictated by the electronic properties of a metal, typically gold or silver, and a dielectric (sample medium) interface (see Section 7.4). The resonance coupling appears as a dip in the light reflection spectrum, which is traditionally tracked by measuring the intensity of the reflected light as a function of the angle of incidence (or of the wavelength) as schematically shown in Fig. 10.15. The coupling of the light to the electromagnetic wave penetrating along the interface requires a high-index prism or a periodic grating surface (see Fig. 10.14).

The sensitivity of the SPR diagnostics is caused by a strong enhancement of the surface electromagnetic wave. Commercial SPR biosensors are generally capable of detecting at least 1 pg/mm^2 of absorbed analytes. This sensitivity is strongly dependent on many parameters, but is particularly dependent on surface functionalization. In comparing sensitivity between reported SPR biosensors, one must be cautious as the sensitivity values are often described independently of the surface functionalization chemistry or for a specific application [Hoa *et al.* (2007)].

Since early 80s pharmaceutical industry became interested in the possibility of biosensor technology and a decision was made to investigate the possibility of developing methods for the direct detection (without labels) of biomolecular interactions [Liedberg *et al.* (1995)]. Practical application of biomolecule sensing by SPR has been demonstrated within recent decades. For examples, in [Liedberg *et al.* (1995)] a silver film was evaporated onto a microscope slide and used as the sensing surface, and a goniometer arrangement with a photodiode as a light detector which was used to measure the position of the resonance angle. An antigen (an immunoglobulin in this case) was spontaneously adsorbed on the silver surface. The subsequent binding of an antibody (a-IgG) was detected as shown in Fig. 10.16.

Experimental results shown in Fig. 10.16 demonstrate that SPR is a promising candidate for the basis of a practical immunosensor [Liedberg *et al.* (1995)].

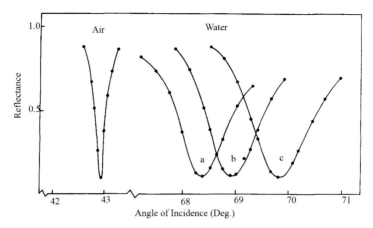

Figure 10.16 SPR resonance angle shift in the reflected light spectrum due to an adsorption of a protein molecule in air or in water (a), or due to an adsorption of an antigen (IgG) (b), and the subsequent binding of its antibody (anti-IgG) (c). (Redrawn from [Liedberg et al. (1995)].)

10.4.2 Fluorescence of Biomaterials

The fluorescence (or luminescence) phenomena, the photoluminescence in particular, has had a major impact on different scientific and technological disciplines, including chemistry, biology, medicine, physics, materials science, and nanotechnology [Lakowicz (2006); Drummen (2012); Bains et al. (2011); Gavrilenko (2020)]. Physical mechanisms that govern fluorescence are addressed in Chapter 4.

Fluorescence is widely used to characterize biological materials for decades. Among biopolymers, proteins are unique in displaying useful intrinsic fluorescence. However, the intrinsic fluorescence of DNA is too weak to be useful and many proteins show low-contrast fluorescence. Thus different molecules with excellent radiative properties sensitive to an ambient are used as labels in fluorescence studies of biomaterials [Lakowicz (2006)]. Good examples for such labels are molecular complexes based on pyrene molecule.

Pyrene is a polycyclic aromatic hydrocarbon (PAH) consisting of four fused benzene rings, resulting in a flat aromatic system. The chemical formula is $C_{16}H_{10}$ and its molecular structure is shown in Fig. 10.17.

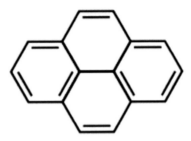

Figure 10.17 Molecular structure of pyrene molecule $C_{16}H_{10}$.

Pyrene was the first molecule for which *excimer*[2] behavior was discovered.

Excimers are molecular complexes that exist only in excited electronic states. They could be detected only in emission spectra, e.g. in fluorescence spectra. Excimers have a very short lifetimes, but they are responsible for many photophysical and photochemical effects [Förster (1969)]. The fluorescence emission spectrum of pyrene is characterized by five major vibronic band designated with well defined peaks at 375, 379, 385, 395, and 410 nm and it is very sensitive to solvent polarity, so pyrene has been used as a probe to determine solvent environments (see Fig. 10.18).

The observed spectral sensitivity of pyrene is due to its excited state having a different, non-planar structure than the ground state. Certain emission bands are unaffected, but others vary in intensity due to the strength of interaction with a solvent. The fluorescent of pyrene molecule and its spectral properties are highly sensitive to changes in the probe's environment that is shown in Fig. 10.18. Spectral characteristics of pyrene monomer emission bands (in the range 350 to 400 nm) are used for polarity measurements.

The formation of a broad excimer-excited state dimer of two interacting pyrene molecules emission peak at near 460 nm can be utilized to study protein conformation, conformational changes, protein folding and unfolding, and protein–protein, protein–lipid,

---

[2] Excimer (or excited dimer) is a dimeric molecule formed from two species, at least one of which has completely filled valence shell by electrons. In this case, formation of molecules is possible only if such atom is in an electronic excited state.

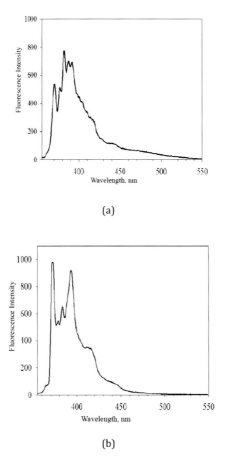

Figure 10.18 Sensitivity of pyrene fluorescence to the solvent: hexane (a) and dimethylsulfoxide (b). (Redrawn from [Bains *et al.* (2011)].)

and protein–membrane interactions [Bains *et al.* (2011); Drummen (2012)].

Increase of the normalized fluorescence intensity with increase of the pyrene concentration in egg yolk phosphatidylcholine (egg-PC) liposomes is depicted in Fig. 10.19 [Bains *et al.* (2011); Drummen (2012)].

Another example relates to the fluorescence study of pyrene interacting with DNA molecules was reported by [Okamoto *et al.* (2004)] where a DNA hybridization was detected by pyrene excimer

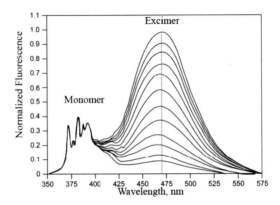

Figure 10.19 Emission spectra of pyrene eximer molecules in egg yolk phosphatidylcholine (egg-PC) liposomes. Intensity of eximer related radiation increases with increasing pyrene concentration. (Redrawn from [Drummen (2012)].)

formation. The excited pyrene molecule can form an excited-state complex with another ground-state pyrene, forming an excimer. This complex displays a broad emission peak located near 500 nm. In the absence of target DNA the emission was observed near 400 nm that characterized the pyrene monomer (see Fig. 10.18). Titration with increasing amounts of target DNA results in increasing emission from the eximer near 500 nm similar to the spectrum depicted in Fig. 10.19 [Okamoto et al. (2004)].

Physical mechanisms that govern the observed high sensitivity of molecular complexes to the environment have been reviewed by [Gemeda (2017)]. It has been shown that molecular complexes (like dimers) arise due to a long-range electrostatic intermolecular interactions (such as Coulomb, ion–dipole, dipole–dipole, etc.). In most cases the interaction electric field penetrates over a substantial range involving many neighboring molecules (or substantial parts of the attached large molecules such as protein, DNA) of the environment. Thus any changes in the structure, configuration as well as in electronic processes of the environment will result in electron energy structure and radiative dipole changes of the molecular system and consequently in the observed variations of the fluorescence spectra.

Chapter 11

Optics of Moving Media

11.1 Introduction

Basics of the special theory of relativity was developed by A. Einstein in his article *On the Electrodynamics of Moving Bodies* based on the English translation of his original German-language paper published in 1905 (see [Einstein (1905)]). This theory explains any electromagnetic and optical effects in moving environment. Two postulates govern the special theory of relativity: Postulate 1. Physical laws in all inertial systems are identical, however, physical quantities differ by their values. Postulate 2. Light velocity does not depend on the velocity of the light source. This velocity is the same in all inertial systems of reference.

Consider two systems of reference (K, K') where $\mathbf{v}$ denotes a velocity in system K'.

Einstein showed that space coordinates and times in systems (K, K') are linked to each other by the Lorentz relationships:

$$x = \frac{x' + vt'}{\sqrt{1 - \beta^2}}, \qquad y = y', \quad z = z', \qquad t = \frac{t' + vx'/c^2}{\sqrt{1 - \beta^2}}, \quad (11.1)$$

Fundamentals of the Optics of Materials: Tutorial and Problem Solving
Vladimir I. Gavrilenko and Volodymyr S. Ovechko
Copyright © 2024 Jenny Stanford Publishing Pte. Ltd.
ISBN 978-981-4877-93-0 (Hardcover), 978-1-003-25694-6 (eBook)
www.jennystanford.com

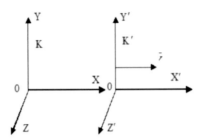

Figure 11.1 Two inertial systems of reference K, K', **v** defines a velocity in system K'.

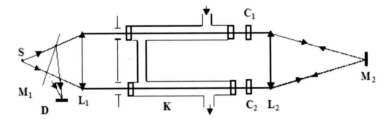

Figure 11.2 Fizeau experimental setup for light propagation measurements in moving medium (water): C_1, C_2 denote compensators, S is a light source, K is a cuvette, M_1, M_2 are mirrors, L_1, L_2 are lenses, and D is a detector.

where $\beta = v/c$, and c is the light velocity in vacuum. From Eq. (11.1) follows that a velocity of material objects does not exceed the velocity of light in vacuum. The two well-known light relativistic effects are considered in the next paragraph: the Fizeau effect (light propagation in moving medium) and the Sagnac effect (light interference of opposing waves in circular resonator).

11.2 Relativistic Effects in Optics

It is instructive to discuses experiment which was performed by Fizeau[1] in 1851. The experimental setup is shown in Fig. 11.2.

---

[1] Armand-Hippolyte-Louis Fizeau was a French physicist, best known for measuring the speed of light in the namesake Fizeau experiment.

Fundamentals of the Optics of Materials: Tutorial and Problem Solving | **285**

If water is not moving in K, the light velocity is defined by $v = c/n$. Relativistic contribution to the velocity is given by

$$v' = \frac{v + V}{1 + vV/c^2},$$ (11.2)

where V is the velocity of the water flow. Within the approximation of $V \ll c$ equation (11.2) results in the following

$$\begin{aligned} v' &= \frac{v\left(1 + \frac{V}{v}\right)}{v\left(\frac{1}{v} + \frac{V}{c^2}\right)} \\ &\simeq \frac{c}{n}\left(1 + \frac{V}{c}n\right)\left(1 - \frac{V}{cn}\right) \\ &\simeq \frac{c}{n}\left[1 + \frac{V}{c}n\left(1 - \frac{1}{n^2}\right)\right] \end{aligned}$$ (11.3)

Initially the interference stripes (D) were observed when the water did not move. If $V \neq 0$ the stripes displaced by δx and the strip width changed to Δx. Thus

$$\frac{\delta x}{\Delta x} = \frac{c\Delta t}{\lambda} = \frac{c}{\lambda}\left(\frac{2l}{v - V} - \frac{2l}{v + V}\right) \simeq \frac{4ln^2}{\lambda}\frac{V}{c},$$ (11.4)

where l is the cuvette length and $V \ll v$. Using calculated phase difference, the light velocity is given by

$$v' - \frac{c}{n} = V\left(1 - \frac{1}{n^2}\right)$$ (11.5)

Another well-known relativistic effect is the *Sagnac* effect (also known as Sagnac interference) that was named after experiments completed by French physicist *Georges Sagnac* in 1913. Sagnac effect manifests itself as a displacement of interference stripes in a ring interferometer by its rotation.

A beam of light is split into two beams that are made to follow the same path but in opposite directions. On return to the point of entry the two light beams are directed to exit the ring and undergo interference. Both the relative phases and thus the position of the interference fringes of the two exiting beams are shifted according to the angular velocity of the apparatus. In other words, when the

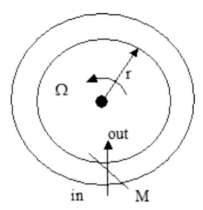

Figure 11.3 Schematic of the optical fibre Sagnac interferometer: M is a mirror.

interferometer is at rest with respect to the earth, the light travels at a constant speed. However, when the interferometer system spins, one beam of light will slow down with respect to the other beam of light.

A ring interferometer resonator setup is depicted in Fig. 11.3. Notations in Fig. 11.3 define the following: M is a mirror, Ω stands for a rotation frequency.

For a fundamental mode of light in fiber the time of light propagation is given by

$$t = \frac{2\pi r m}{v}, \tag{11.6}$$

where m is a number of fiber rings, v is a light velocity.

If interferometer rotates the mirror M displaces to distance l according to,

$$l = \Omega r t = \Omega r \frac{2\pi r m}{v} = \frac{2\Omega S}{v} m, \tag{11.7}$$

where S is a ring area.

Equation (11.7) explains a physical reason for the phase difference $\Delta\Phi$ given by

$$\Delta\Phi = \frac{2\pi}{\lambda} 2l = \frac{8\pi}{\lambda v} S m \Omega, \tag{11.8}$$

Fundamentals of the Optics of Materials: Tutorial and Problem Solving | **287**

where λ is a wavelength.

Accuracy of the rotation frequency measurements in terms of the phase measurement precision $(\Delta\Phi_0)$ is given by

$$\Delta\Omega = \frac{\Delta\Phi_0 \lambda v}{8\pi Sm} \tag{11.9}$$

Assume that $\Delta\Phi_0 = \pi/10$, $\lambda = 500$ nm, $v = 2 \cdot 10^8 m/s$, $S = 0.1m^2$, $m = 100$. In this case $\Delta\Omega \simeq 0.1 c^{-1}$.

Gyroscopes operating on principles of the Sagnac effect and using the optical fibres as light transducer are widely used in airplanes navigation.

11.3 Light Propagation in Ether: Michelson-Morley Experiment

In 1881 year Michelson and Morley carried out experiment with two-beam interferometer (the so-called *Michelson* interferometer). Configuration of the Michelson interferometer is depicted in Fig. 11.4

The Michelson interferometer consists of mirrors M_1, M_2 and a beam splitter M. In Fig. 11.4, a light source emits light that hits the beam splitter which is partially reflective, so part of the light is transmitted through to M_1 while some is reflected in the direction of M_2. Both beams after reflection from M_1 and M_2 recombine to produce an interference pattern incident on the detector.

Michelson and Morley completed their experiments in order to answer the question whether the so called *ether* has an influence upon the light propagation. In other words whether the ether exists in nature or not? Initially the interferometer shown in Fig. 11.4 was configured in such a way that the MM_1 path was oriented parallel to the vector velocity $(\mathbf{v})$ of the Earth. The velocity $\mathbf{v}$ (where $|\mathbf{v}| = 30 \ km/s$) is an orbital velocity of the Earth rotating around the Sun. Consider times t_1 and t_2 required for light to travel along l_1 and l_2, respectively. The time difference $\Delta t = t_2 - t_1$ determines the interference strips displacement.

After rotation of the configuration shown in Fig. 11.4 by $90°$ the time difference is increased twice. In this case the time delay t_1 is

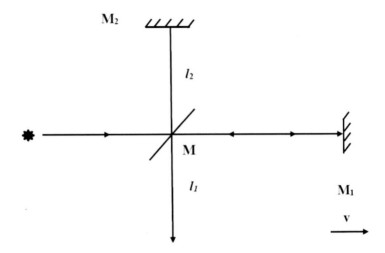

Figure 11.4 Configuration of Michelson interferometer: M_1, M_2 are mirrors, M is a beam-splitter; l_1, l_2 are lengths of optical paths.

given by

$$t_1 = \frac{l_1}{c-v} + \frac{l_1}{c+v} = \frac{2l_1}{c}\frac{1}{1-\beta^2}, \qquad (11.10)$$

where $\beta = v/c$.

The time delay t_2 depends upon a displacement of the mirror M during the light travel (see Fig. 11.5)
Therefore

$$t_{MM_2} = \frac{\sqrt{l_2^2 + (vt_{MM_2})^2}}{c}. \qquad (11.11)$$

The value of t_{MM_2} is given by

$$t_{MM_2} = \frac{l_2}{\sqrt{c^2 - v^2}} = \frac{l_2}{c}\frac{1}{\sqrt{1-\beta^2}}. \qquad (11.12)$$

Obviously $t_2 = 2t_{MM_2}$. Consequently,

$$t_2 = \frac{2l_2}{c}\frac{1}{\sqrt{1-\beta^2}}. \qquad (11.13)$$

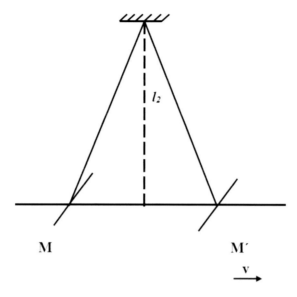

Figure 11.5 Configuration of the experiment for measurements of time delay, t_2.

Using equations (11.10) and (11.13) one can derive the following:

$$\Delta t = t_1 - t_2 = (\text{assume } l_1 = l_2 = l) = \frac{2l}{c}\left(\frac{1}{1-\beta^2} - \frac{1}{\sqrt{1-\beta^2}}\right). \tag{11.14}$$

After taking into account that $\beta = 30 km/s/3 \cdot 10^5 km/s = 10^{-4} \ll 1$, follows:

$$\Delta t \simeq \frac{l}{c}\beta^2. \tag{11.15}$$

Thus the final result is given by:

$$\Delta\varphi = 2\frac{\Delta t}{T} = \frac{2l}{\lambda}\beta^2, \tag{11.16}$$

where $\Delta\varphi$ is the interference strip displacement.

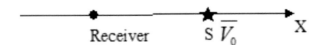

Figure 11.6 Trace diagram indicating light source movement with respect to the light receiver.

In early Michelson's experiments the following setup parameters were used: $l = 1\ m$, $\lambda = 500\ nm$, then $\Delta\varphi \simeq 0.04$ strip. The last value was within the limits of experimental error. Later Michelson repeated the experiments with $l = 11\ m$. Thus an error of experiment (0.02 strip) was much less then the calculated value of 0.4 strip. From these experiments follows that no *ether* exist as an absolute coordinate system.

11.4 Practical Examples

For tutorial purposes this section offers several typical problems with solutions.

11.4.1 Frequency of Moving Light Source

Problem 1. Light source is moving with velocity $v_x = v_0$, light frequency is $\nu = \nu_0$. What frequency does detect a receiver which is placed in a medium with a refraction index n?

Solution. Phase velocity of light in a medium is given by

$$v = c/n \tag{11.17}$$

A distance between two successive maxima of the light wave (wave length) propagating in a forward direction (see Fig. 11.6) is equal to

$$\lambda_1 = (c/n - v_0)T = (c/n - v_0)/\nu, \tag{11.18}$$

and the same for a propagation in an opposite direction:

$$\lambda_2 = (c/n + v_0)T = (c/n + v_0)/\nu \tag{11.19}$$

where

$$\nu = \nu_0/\gamma, \; \gamma = 1/\sqrt{1 - \beta^2}, \; \beta = v_0/c. \tag{11.20}$$

The frequency of light in the R-point is given by

$$\nu_1 = \frac{c/n}{\lambda_1} = \nu_0 \frac{1}{\gamma(1 - n\beta)}, \tag{11.21}$$

$$\nu_2 = \frac{c/n}{\lambda_2} = \nu_0 \frac{1}{\gamma(1 + n\beta)} \tag{11.22}$$

Equations (11.21) and (11.21) represent *Doppler effect* in optics. This effect determines nonhomogeneous spectral broadening of moving molecules and atoms in gas phase.

11.4.2 Doppler Effect in Gas

Problem 2. Calculate a spectral contour area of a molecule absorption for a nonhomogeneous Doppler broadening in a gas. The homogeneous Lorentz spectral distribution function is given by

$$\sigma_h = \frac{\sigma_0}{\left(\frac{2(\omega - \omega_0)}{\Delta\omega_h}\right)^2 + 1} \tag{11.23}$$

Solution. Frequency change of a moving atom due to the Doppler effect is given by

$$\omega' = \omega\left(1 + \frac{v_z}{c}\right) \tag{11.24}$$

The atomic density distribution function $n_v(v_z)$ is given by

$$n(v_z)dv_z = n_0 \sqrt{\frac{M}{2\pi k_B T}} e^{-\frac{M v_z^2}{2k_B T}} dv_z \tag{11.25}$$

292 | *Optics of Moving Media*

where n_0 is a concentration of atoms, M is an atomic mass. In the case of $\Delta\omega_h \ll \Delta\omega_{nh}$, equation (11.25) has a form of

$$\sigma_h(\Delta) = \sigma_{nh} e^{-(\Delta/\Gamma_{nh})^2}, \tag{11.26}$$

where according to Eqs. (11.24) and (11.25), $\Gamma_{nh} = \frac{\omega_0}{c}\sqrt{\frac{2k_BT}{M}}$.

The frequency variation of the absorption spectrum due to the Doppler effect does not change an area of the spectral contour i.e.

$$\int \sigma_h(\Delta)d\Delta = \int \sigma_{nh}(\Delta)d\Delta. \tag{11.27}$$

Solution of Eq. (11.27) results in the following

$$\frac{\pi}{2}\sigma_0\Delta\omega_h = \sqrt{\pi}\sigma_{nh}\Gamma_{nh}$$

$$\sigma_{nh} = \frac{\sqrt{\pi}}{2}\frac{\Delta\omega_h}{\Gamma_{nh}}\sigma_0. \tag{11.28}$$

11.4.3 Light Propagation in a Moving Glass Bar

Problem 3. Glass bar of thickness d is moving along the x-axis with a velocity of $\bar{v}$. Bottom face of the glass bar is a perfect mirror (see Fig. 11.7). Calculate the time the light striking a moving bar from the top will travel inside the bar?

Solution. The law of light refraction for moving glass in a relativistic form can be written as

$$\sin\alpha' = n\sin\varphi' = \frac{v}{c} = \beta. \tag{11.29}$$

where α' is an incidence angle of an incoming beam in bar-coordinate system. Distance and time for a light propagation in one direction within the bar are given by

$$x' = -d\tan\varphi' = -\frac{d}{\sqrt{(\frac{n}{\sin\alpha'})^2 - 1}} \tag{11.30}$$

$$t' = \frac{d}{\cos\varphi'}\frac{n}{c} = \frac{dn}{c\sqrt{1 - \frac{\sin^2\alpha'}{n^2}}} \tag{11.31}$$

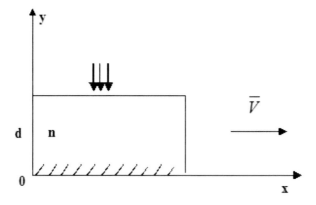

Figure 11.7 Light propagation through the moving glass bar: n is a refraction index, $\bar{v}$ is a velocity of the bar.

In the laboratory coordinate system the time is given by

$$t = \frac{t' + vx'/c^2}{\sqrt{1-\beta^2}} = \frac{d}{c\sqrt{1-\beta^2}}\sqrt{n^2 - \beta^2} \tag{11.32}$$

And the total propagation time in both directions is given by

$$\Delta t = 2t = \frac{2d}{c}\left(\frac{n^2 - \beta^2}{1-\beta^2}\right)^{1/2} \tag{11.33}$$

11.5 Review Questions and Exercises

11.5.1 Optical Doppler Effect

Consider an observer located in a fixed coordinate system. Derive expression for the optical Doppler effect for two cases:

1. optical source is moving in the direction of observation,
2. optical source is moving perpendicular to the direction of observation.

11.5.2 Speed of Light in a Moving Medium

In a stationary medium the light speed is given by: $v = c/n(\lambda)$, where λ is a wavelength of light. Calculate speed of light in a medium that moves relative to the source at a speed of v for two cases

- take into account dispersion of a medium, i.e. $n = n(\lambda)$;
- consider a ratio of v/c as a small parameter.

11.5.3 Optical Radar

Optical radar operates at a frequency of ν_0 and irradiates an object that moves at a speed of v in an arbitrary direction relative to the direction to the radar. Taking into account the optical Doppler effect, find a frequency of the signal received by the radar.

11.5.4 Emission of a Rarefied Gas

Calculate spectral distribution of light emitted by a rarefied gas, which is in thermodynamic equilibrium at a temperature T. Frequency of atomic radiation in the coordinate system of an atom is ν_0.

11.5.5 Gyroscope

Show that a gyroscope whose operation principle is based on the Sagnac effect[2] for waves in a medium (in contrast to the light waves in free space) can have a sensitivity greater by a factor of $(Mc^2/\hbar\omega)$ relative to the angular rotation.

---

[2] The Sagnac effect appears in a ring interferometer. A beam of light is split and the two beams are directed to follow the same path but in opposite directions. On return to the point of entry the two light beams are allowed to exit the ring and undergo interference. Consequently, the relative phases of the two exiting beams (and therefore locations of the interference fringes) are shifted according to the angular velocity of the apparatus (see Section 11.2).

11.5.6 Dispersion in Moving Medium

In one dimensional case obtain relationship between frequency (ω) and wave number (k) in moving medium. Take into account that optical wave phase, ($\omega t - kx$) is an invariant with respect to another inertial system.

11.5.7 Optical Doppler Effect Example

A light source with a frequency ν_0 moves with a velocity v in a stationary medium with a refractive index n and passes to a stationary observer. What frequency will the observer record when the light source approaches and when it moves away?

Chapter 12

Applied Optics

12.1 Introduction

Applied optics is a branch of optics and photonics that focuses on application of light for practical purposes, in other words, it is a study focusing on questions of how optical elements can be applied to a design and construction of optical instruments, and their application to practical engineering problems. This chapter provides an introduction to selected areas of modern applied optics and optical engineering, covering related fundamental concepts as well as practical techniques and applications.

12.2 Thermal Optics

The refractive index of materials such as crystals, semiconductors, and glasses is not a constant parameter but depends on many factors including pressure and temperature. The variation of the refractive index with the temperature at a constant pressure is called the *thermo-optic coefficient*. Measurements of the refractive index changes as a function of temperature for different optical materials are reported by various groups. In ionic materials with a low melting

---

Fundamentals of the Optics of Materials: Tutorial and Problem Solving
Vladimir I. Gavrilenko and Volodymyr S. Ovechko
Copyright © 2024 Jenny Stanford Publishing Pte. Ltd.
ISBN 978-981-4877-93-0 (Hardcover), 978-1-003-25694-6 (eBook)
www.jennystanford.com

point, thermal expansion is high and the thermo-optic coefficient is negative. However, in materials with the small thermal expansion and a high melting point, the thermo-optic coefficient is positive and dominated by the volume change in polarizability [Bosman and Havanga (1963)].

It is instructive to consider a derivation of the thermo-optic coefficient (dn/dT) within a classical picture from the Clausius-Mossotti equation (see Section 1.4). Consider an isotropic matter where a local polarizability is represented by spheres of volume V and polarizability α_m. In this case the equation (1.94) can be written as

$$\frac{\varepsilon - 1}{\varepsilon + 2} = \frac{4\pi}{3}\left(\frac{\alpha_m}{V}\right). \tag{12.1}$$

Generally the function of $\alpha_m(\omega)$ depends on the number of particles (or crystal unit cells) and within the microscopic picture is a complicated function which incorporated all local polarization processes of material (see Section 1.3). In order to obtain a thermo-optic coefficient the equation (12.1) has to be differentiated with respect to temperature at constant pressure that results in the following [Bosman and Havanga (1963)]:

$$\frac{1}{(\varepsilon - 1)(\varepsilon + 2)}\left(\frac{d\varepsilon}{dT}\right)_P = -\frac{1}{3V}\left(\frac{dV}{dT}\right)_P\left[-1 + \frac{V}{\alpha_m}\left(\frac{d\alpha_m}{dV}\right)_P\right]$$
$$+ \frac{1}{3\alpha_m}\left(\frac{d\alpha_m}{dT}\right)_V \tag{12.2}$$

The thermo-optic coefficient follows from the relationship $\varepsilon(\omega) = n^2$. The value of $K_T = -(1/V)(dV/dT)_P$ is the *thermal expansion* coefficient.

The first two terms are the principal contributors in ionic materials: a positive thermal expansion coefficient K_T results in a negative thermo-optic coefficient. In ionic materials with a low melting point, thermal expansion is high and the thermo-optic coefficient is negative; when thermal expansion is small for some nonlinear crystals and has a high melting point, hardness, and high elastic moduli optical materials, the thermo-optic coefficient is positive, dominating by the volume change in polarizability.

Consider a thermal sensing based on the heating of an absorbing medium by a high power laser radiation. Following physical mechanisms govern this phenomenon, i.e.

inhomogeneous heating of an optical element along a direction orthogonal to the optical beam propagation direction, which causes an inhomogeneous variation of refractive index. This phenomenon is called a *thermo-optical effect*;

thermally induced mechanical stress that is called a *photoelastic effect* and results in refractive index change;

pressure and mechanical stresses lead to deformations of the optical element outer faces which results in *lensing*.

12.2.1 Thermal Lensing

Under the acting of powerful laser beam any optical sample is heating. It causes change of refractive index

$$n(\Delta T) = n(0) + \frac{dn}{dT}\Delta T, \tag{12.3}$$

where dn/dT is a *thermal refraction coefficient*.

A laser beam has an axial symmetry thus a temperature in the center of a sample $T(0)$ is higher than that on it's faces $T(r_0)$, i.e.

$$T(0) > T(r_0) \tag{12.4}$$

Thus dn/dT 0 and an induced optical lens has a positive *optical power* $D = 1/f$ 0. Consider this issue more in detail. By taking info account the *photo-tension* effect one can arrive in the following

$$\Delta n(\Delta T) = \Delta n_T + \Delta n_\sigma = \left(\frac{\partial n}{\partial T}\right)_\sigma \Delta T + \Delta n_\sigma. \tag{12.5}$$

For example, Δn_T and Δn_σ have following values for the $LiNbO_3$ crystals

$$\Delta n_T = 1.4 \cdot 10^{-5}\Delta T, \tag{12.6}$$

$$\Delta n_\sigma = 0.66 \cdot 10^{-5}\Delta T. \tag{12.7}$$

300 | *Applied Optics*

In some cases the refractive index variations Δn_T, and Δn_σ may have different signs thus $\Delta n_T + \Delta n_\sigma \simeq 0$.

Several problems with solutions are considered here for tutorial purposes.

Problem 1. Consider an optical plate irradiated by a homogeneous laser beam of intensity I_0. In many problems of the thermo-optics a spacial dependence of the refractive index is given by

$$n(r) = n_0 + ar^2 \tag{12.8}$$

Consequently the plate will act as a thermal lens. Calculate a focal length of this thermal lens.

Solution. With respect to the focus F the propagation time is independent on the transverse coordinate r. Equity condition for the optical paths can be written as

$$n(r)d + \sqrt{r^2 + f_0^2} = n_0 d + f_0, \tag{12.9}$$

where $n_0 = n(0)$. Withing the *paraxial approximation*[1] equation (12.9) can be written in the following form

$$n(r) = n_0 - \frac{r^2}{2df_0}, \tag{12.10}$$

where $(r/f_0)^2 \ll 1$. Solving Eqs. (12.8) and (12.10) simultaneously gives the following

$$n_0 + ar^2 = n_0 - \frac{r^2}{2df_0} \tag{12.11}$$

Consequently

$$f_0 = -\frac{1}{2da} \tag{12.12}$$

If $a < 0$ (i.e. if $dn/dT > 0$) the thermal lens is positive (i.e. the focusing one). If $a > 0$ (i.e. if $dn/dT < 0$) the thermal lens is negative (i.e.

---

[1] In geometrical optics the paraxial approximation is called the *small-angle* approximation. A *paraxial* ray is a ray that makes a small angle to the optical axis of the system, and is close to the axis throughout the system.

Fundamentals of the Optics of Materials: Tutorial and Problem Solving 301

defocusing one). For example, in a gaseous medium $\alpha \simeq 10^{-4} cm^{-2}$ thus for $d = 25$ cm the focal length is negative, i.e. $f_0 \simeq -200$ cm.

Problem 2. Consider a plate with a thermal sources located at the bottom of the plate. The temperature distribution function along the r-axis is $dT/dr = a_T$ The refractive index variation is given by $dn/dr = a_n$. Calculate optical beam tracing.

Solution. Beam propagation is described by the following equation:

$$\frac{1}{n}\frac{dn}{dr} = \frac{d^2 r}{dz^2} \tag{12.13}$$

where $n(r) = n_0 + a_n r$; and $r(z = 0) = r_0$, $dr/dz(z = 0) = 0$ are the boundary conditions. Equation (12.13) can be derived from Fermat's principle by varying optical path (i.e. $\int n(r)(1 + (dr/dz)^2)^{1/2} dz$) within the small angle approximation, $|dr/dz| \ll 1$.
Withing $a_n r \ll n_0$ approximation equation (12.13) is written as

$$\frac{a_n}{n_0} = \frac{d^2 r}{dz^2}. \tag{12.14}$$

The solution of equation (12.14) is given by

$$r(z) = r_0 + \frac{a_n}{2n_0} z^2. \tag{12.15}$$

Refraction angle φ on the outer surface of the plate is defined by

$$\varphi \mid_{z=d} \simeq n_0 \frac{dr}{dz} \mid_{z=d} = a_n d. \tag{12.16}$$

Angular dispersion D_λ is given by

$$D_\lambda = \frac{\partial \varphi}{\partial \lambda} = \frac{\partial^2 n}{\partial \lambda \partial r} d \tag{12.17}$$

Therefore such a plate acts as a *dispersive prism*.

Problem 3. Specify requirements for temperature stabilization by measurements using Fabry–Perot interferometer. Linear expansion coefficient of fused quartz by heating is $\alpha = 10^{-6}$.

Solution. Assume that variable distance between mirrors of Fabry–Perot interferometer is δh. The condition for maximum intensity on the output of interferometer is given by

$$m\lambda = 2nh\cos\beta, \tag{12.18}$$

where $m = 1, 2, ...$, h is an interferometer base, n is an index of refraction.

Angular variation $\delta\beta$ with a variation of h has to be calculated from equation (12.18)

$$2n\delta h\cos\beta - 2nh\sin\beta\delta\beta = 0 \tag{12.19}$$

Thus

$$\delta\beta = \frac{\delta h}{htg\beta} \tag{12.20}$$

The value of δh changes with temperature variation δT, i.e.

$$\delta h = \alpha h\delta T. \tag{12.21}$$

Angular width of an intensity maximum $d\beta$ is given by

$$d\beta = \frac{\lambda}{2h\sin\beta N_e}, \tag{12.22}$$

where $N_e = \pi\sqrt{\rho}(1-\rho)$ is an *effective number* of light rays, ρ is a reflection coefficient. From Eqs. (12.20), (12.21), and (12.22) follows

$$\frac{\alpha h\delta T}{htg\beta} \preceq \frac{\lambda}{2h\sin\beta N_e}. \tag{12.23}$$

Solving (12.23) one can get

$$\delta T \preceq \frac{\lambda}{2h\alpha\cos\beta Ne}. \tag{12.24}$$

Plugging of: $\lambda = 500$ nm, $h = 3$ cm, $\cos\beta \simeq 1$, and $N_e = 25$ by ($\rho = 0.88$) one can get the following value for an allowed temperature variation by Fabry–Perot experiments: $\delta T \preceq 0.33°C$.

12.2.2 Laser Heat Treatment of Metals

Laser heat treating is a surface modification process designed to change the microstructure of metals through controlled heating. An advantage that lasers offer in this process is the ability to heat treat localized areas without affecting the entire work piece. A unique property of the pulsed laser beam is the possibility to concentrate high thermal power within a small volume (i.e. within an area $\sim \lambda^2$) and during a short pulse time ($10^{-12} - -10^{-14}$ sec). As a result the short time processes of heating, smelting,[2] or hardening are observed. Consider numerical description of a heat flow.

Equation of a heat transfer is given by

$$\rho T \frac{dS}{dt} = -\operatorname{div}\bar{g} + \left(\frac{\partial Q}{\partial t}\right)_{ext}, \tag{12.25}$$

where ρ is the density of a substance, S is the entropy of the mass unit, $\bar{g}$ is the averaged density of a heat flow.

Equation (12.25) is a single-stream equation. Assume that the pressure is not changing. Thus $T\partial S/\partial t = C_p \partial T/\partial t$, where C_p is a heat capacitance. In this case $\bar{g} = -\kappa\nabla T$, where κ is a heat conduction coefficient. Taking into account that values of ρ, C_p, and κ are temperature independent, one can get the following equation:

$$\frac{\partial T}{\partial t} = \chi\Delta T + (\rho C_p)^{-1}\left(\frac{\partial Q}{\partial t}\right)_{ext}, \tag{12.26}$$

---

[2] Smelting is a process of applying heat to ore in order to extract out a base metal. It is a form of extractive metallurgy. It is used to extract many metals from their ores, including silver, iron, copper, and other base metals.

304 | *Applied Optics*

where $\chi = \kappa/\rho C_p$ is the heat conductivity. The source of heat in volume is characterized by

$$\left(\frac{\partial Q}{\partial t}\right)_{ext} = (1-R)\alpha I_0 \exp(-\alpha z)\exp\left[-r_\perp^2/r_0^2\right] f\left(\frac{t}{\tau_p}\right) \quad (12.27)$$

where α is the absorption coefficient, R is the reflection coefficient, r_0 is the light beam radius, τ_p is the duration of the laser pulse. Boundary conditions are given by:

$$q_z|_{z=0} = -K\left.\frac{\partial T}{\partial z}\right|_{z=0} = 0 \quad (12.28)$$

The boundary conditions follow from the assumption of no heat exchange. Generally speaking, the boundary conditions stated in Eq. (12.28) are not realistic, because the cooling happens partly due to a convective exchange with the surrounding medium. However, in many cases, where a powerful source of cooling was introduced, one can ignore this convective exchange, thus justifying the choice of boundary conditions.

Problem 4. Assume that a source of heat is a point source. In this case equation (12.27) reads:

$$\frac{1}{\rho C_p}\left(\frac{\partial Q}{\partial t}\right)_{ext} \simeq Q_0\delta(\mathbf{r}-\mathbf{r}')\delta(t-t'). \quad (12.29)$$

Using equation (12.26) calculate coordinate and time dependence of the temperature.

Solution. Equation for $T(\bar{r}, t)$ can be written in the following form

$$\frac{\partial T}{\partial t} = \chi\Delta T + Q_0\delta(\mathbf{r})\delta t. \quad (12.30)$$

Equivalent to equation (12.30) is the following homogeneous equation

$$\frac{\partial T}{\partial t} = \chi\Delta T, \quad (12.31)$$

with the non-zero boundary condition, i.e.

$$T|_{t=0} = Q_0\delta(\mathbf{r}). \quad (12.32)$$

Representation of the desired function $T(\mathbf{r}, t)$ by Fourier integral reads

$$T(\mathbf{r}, t) = \frac{1}{(2\pi)^3} \int T(\mathbf{k}, t) exp(i\mathbf{kr}) d^3\mathbf{k}, \tag{12.33}$$

where $T(\mathbf{k}, t) = \int T(\mathbf{r}, t) exp(-i\mathbf{kr}) d^3\mathbf{r}$. Thus the problem (12.31), (12.32) reduces to the following

$$\frac{\partial T(k, t)}{\partial t} + \chi k^2 T(k, t) = 0, \tag{12.34}$$

with $T(k, t)|_{t=0} = Q_0$. Therefore the Fourier transform is given by: $T(k, t) = Q_0 exp(-\chi k^2 t)$. The inverse Fourier transform represents the solution:

$$T(r, t) = \frac{Q_0}{(\chi t)^{3/2}} \exp\left(-\frac{r^2}{4\chi t}\right) \tag{12.35}$$

Following conclusions can be made from an analysis of the solution of Eq. (12.35):

1. temperature at the center of the laser beam decreases in time according to $\sim (t)^{-3/2}$;
2. area of the heating region increases in time according to $\sim \chi t$;

For different materials coefficient χ (cm^2/sec) has the following numerical values: 0.74 (Al), 0.52 (Cu), 0.054 (steel). Within one second values of the heating areas are: 0.74 cm^2 (Al), 0.52 cm^2 (Cu), 0.054 cm^2 (steel).

Problem 5. If the surface temperature of solid exceeds the smelting temperature $(T > T_{sm})$, the aggregation state is changing. Calculate velocity of the smelting phase boundary v_{fr} and the time of smelting. **Solution**. Neglect a heat transfer during the pulse τ_p, i.e. $(r_0 \succeq \sqrt{\chi \tau_p})$. Thus the temperature can be calculate from the local energy balance:

$$\rho C_p(T_{sm} - T_0) + L_{sm} = at(1 - R)I_0 exp(-\alpha z_{fr}), \tag{12.36}$$

where L_{sm} is heat of smelting of the volume unit, T_{sm} is temperature of smelting, and T_0 is initial temperature. Calculate a derivative of

Applied Optics

(12.36) with respect to t that gives

$$v_{fr} = \frac{dz_{fr}}{dt} = \frac{1}{\alpha \tau_p} = \frac{(1-R)I_0 \exp(-\alpha z)}{\rho C_p(T_{sm} - T_0) + L_{sm}} \qquad (12.37)$$

If $T_0 \to T_{sm}$ one gets

$$v_{fr} \simeq \frac{1-R}{L_{sm}} I_0 \bigg|_{z \to 0} \qquad (12.38)$$

The time of smelting t_{sm} is given by

$$t_{sm} \sim \frac{d}{v_{fr}} \simeq \frac{1}{\alpha v_{fr}} = \frac{L_{sm}}{\alpha(1-R)I_0}. \qquad (12.39)$$

Numerical estimation for a metal sample gives: $L_{sm} \simeq 2000\ v/cm^3$, $\alpha \simeq 10^4\ cm^{-1}, R \simeq 0.9, I_0 = 10^8\ J/(sec \cdot cm^2)$. Therefore

$$t_{sm} = 2 \cdot 10^{-8}\ sec = 20\ ns. \qquad (12.40)$$

12.3 Waveguides: Fiber-optics

Development of the lightwave communication since late 70s of the last century gave rise to optical waveguides. An *optical waveguide* could be defined as a spatially inhomogeneous structure for guiding light, i.e. for restricting the spatial region in which light can propagate. There are different types of optical waveguides and different techniques for fabricating dielectric waveguides. Here are some examples:

- *Planar waveguide* can be fabricated on various crystal and glass materials with epitaxy or with polishing methods;
- *Channel waveguide* can be made on semiconductor, crystal, or glass materials with lithographic methods in combination with, e.g. epitaxy, ion exchange, or thermal indiffusion;
- *Optical fibers* can be fabricated by drawing from a preform, which is a large glass rod with a built-in refractive index profile. Fibers can again be drawn into waveguides with reduced dimensions, e.g. resulting in *nanofibers*.

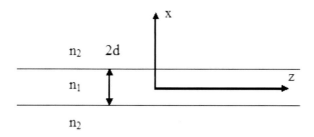

Figure 12.1 Planar waveguide guides light only in the vertical direction; the waveguide thickness (2d) is indicated by arrow.

Most waveguides exhibit two-dimensional guidance, thus restricting the extension of guided light in two dimensions and permitting propagation essentially only in one dimension, i.e. the *channel waveguide*.

Planar waveguide is one of the simplest types of optical waveguides having basic properties of guided structures and it is instructive to consider first this type of optical waveguides as an example [Dmytruk et al. (1996); Amerov et al. (1989)]. The waveguides can be placed between other solid layers as shown in Fig. 12.1.

Consider TE- and TM-modes propagating in a planar waveguide. The modes do not vary in y-direction therefore ($\frac{\partial}{\partial y} = 0$). For the TE-modes one has $E_z = 0$. By taking into account a representation of the plane waves in the following form:

$$E(x, z) = E_0(x) \exp[i(\omega t - \beta z)]$$
$$H(x, z) = H_0(x) \exp[i(\omega t - \beta z)] \qquad (12.41)$$

From Maxwell's equation follows:

$$E_x = -\frac{i}{\kappa^2}\left(\beta \frac{\partial E_z}{\partial x} + \omega\mu \frac{\partial H_z}{\partial y}\right) \qquad (12.42)$$

$$E_y = -\frac{i}{\kappa^2}\left(\beta \frac{\partial E_z}{\partial y} - \omega\mu \frac{\partial H_z}{\partial x}\right) \qquad (12.43)$$

$$H_x = -\frac{i}{\kappa^2}\left(\beta \frac{\partial H_z}{\partial x} - \omega\varepsilon \frac{\partial E_z}{\partial y}\right) \qquad (12.44)$$

$$H_y = -\frac{i}{\kappa^2}\left(\beta \frac{\partial H_z}{\partial y} + \omega\varepsilon \frac{\partial E_z}{\partial x}\right), \qquad (12.45)$$

Applied Optics

where $\kappa^2 = k^2 - \beta^2$, and $k^2 = \omega^2 \varepsilon \mu / c^2$. The non-zero components are H_z, H_x, and E_y (TE-mode). Consequently:

$$H_x = -\frac{i}{\omega \mu} \frac{\partial E_y}{\partial z} \tag{12.46}$$

$$H_z = \frac{i}{\omega \mu} \frac{\partial E_y}{\partial x}. \tag{12.47}$$

For the E_y amplitude we have the following wave equation:

$$\frac{\partial^2 E_y}{\partial x^2} + \frac{\partial^2 E_y}{\partial z^2} + n^2 k_0^2 E_y = 0, \tag{12.48}$$

where $n^2 = \varepsilon / \varepsilon_0$, and $k_0^2 = \omega^2 \varepsilon_0 \mu_0 / c^2$.
For the plane waves given by Eq. (12.41) one has:

$$\frac{\partial^2 E_y}{\partial x^2} + \left(n^2 k_0^2 - \beta^2 \right) E_y = 0. \tag{12.49}$$

Consider odd and even modes separately. Within the waveguide, i.e. at $(|x| < d)$ one has:

$$E_y = A \cos(\kappa x), \tag{12.50}$$

where $\kappa^2 = n_1^2 k_0^2 - \beta^2$.

Electrical and magnetic field components outside the waveguide (i.e. at $|x| > d$) are given by:

$$E_y = A \cos(\kappa d) e^{-\gamma(|x|-d)}, \tag{12.51}$$

$$H_z = \frac{x}{|x|} \frac{i\gamma}{\omega \mu_0} A \cos(\kappa d) e^{-\gamma(|x|-d)}, \tag{12.52}$$

where $\gamma^2 = \beta^2 - n_2^2 k_0^2$. Note that $\kappa^2, \gamma^2 > 0$ if $n_1 > n_2$. The condition for mode that could exist in the waveguide is given by:

$$\gamma > 0. \tag{12.53}$$

Taking into account the continuity properties of E_y and H_z amplitudes on the slab boundaries, i.e. at $(x = \pm d)$, the equation for the eigenvalues is given by:

$$\tan(\kappa d) = \frac{\gamma}{\kappa} \qquad (12.54)$$

An amplitude A can be calculated from the total field power P that is given by:

$$P = \frac{1}{2} \int (\mathbf{E} \times \mathbf{H}^*)_z \, dx = -\frac{1}{2} \int_0^\infty E_y H_x^* \, dx = \frac{\beta}{\omega\mu_0} \int_0^\infty |E_y|^2 \, dx. \qquad (12.55)$$

Plugging fields given by Eq. (12.52) into Eq. (12.55) and taking into account Eq. (12.54) the amplitude A is given by:

$$A = \left(\frac{2\omega\mu_0}{\beta d + \frac{\beta}{\gamma}} P \right)^{1/2}. \qquad (12.56)$$

The problem below focuses on a calculation of the odd guided TE-modes.

Problem 6. Calculate an odd guided TE-modes in planar waveguide.

Solution. Here the theoretical method applied in this chapter for even guided TE-modes is used. The electric E_y and magnetic H_z field amplitudes of electromagnetic waves within the waveguide (i.e. at $|x| < d$) are given by

$$E_y = A \sin(\kappa x), \qquad (12.57)$$

$$H_z = \frac{i\kappa}{\omega\mu_0} A \cos(\kappa x). \qquad (12.58)$$

Outside the waveguide (i.e. at $|x| > d$) one has

$$E_y = \frac{x}{|x|} A \sin(\kappa d) e^{-\gamma(|x|-d)} \qquad (12.59)$$

$$H_z = \frac{-i\gamma}{\omega\mu_0} A \sin(\kappa d) e^{-\gamma(|x|-d)}. \qquad (12.60)$$

Continuity requirement for the H_z component results in the following equation for the eigenvalues:

$$\tan(\kappa d) = -\frac{\kappa}{\gamma}. \tag{12.61}$$

Amplitude A can be calculated from the following

$$P = -\frac{1}{2} \int_{-\infty}^{\infty} E_y H_x dx. \tag{12.62}$$

After taken into account waves inside and outside the waveguide given by Eqs. (12.57) to (12.60) as well as the eigenvalues given by Eq. (12.61) one can arrive in the following

$$A = \left(\frac{2\omega\mu_0}{\beta d + \frac{\beta}{\gamma}} P \right)^{1/2} \tag{12.63}$$

Equations (12.56) and (12.63) differ by the values of β and γ for even and odd modes, respectively.

Problem 7. Calculate eigenvalues for the TE-modes.

Solution. Eqs. (12.54) and (12.61) can be solved by a graphical method. One has

$$\kappa^2 = n_1^2 k_0^2 - \beta^2, \tag{12.64}$$

$$\gamma^2 = \beta^2 - n_2^2 k_0^2. \tag{12.65}$$

Excluding β in Eqs. (12.64) and (12.65) one gets

$$\gamma = ((n_1^2 - n_2^2)k_0^2 - \kappa^2)^{1/2}. \tag{12.66}$$

Consequently one gets

$$\tan(\kappa d) = \frac{\gamma}{\kappa} = \frac{((n_1^2 - n_2^2)k_0^2 - \kappa^2)^{1/2}}{\kappa} \tag{12.67}$$

$$\tan(\kappa d) = -\frac{\kappa}{\gamma} = -\frac{\kappa}{((n_1^2 - n_2^2)k_0^2 - \kappa^2)^{1/2}} \tag{12.68}$$

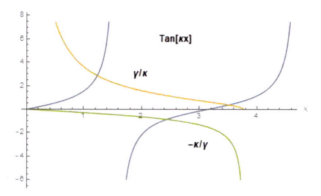

Figure 12.2 Graphical solution of Eqs. (12.67) and (12.68); $X = kd$, $d/\lambda = 2$, with $\sqrt{n_1^2 - n_2^2} = 0.3$.

Equations (12.67) and (12.68) are obtained for even and odd modes, respectively. Results of numerical computations of the left and right parts of Eqs. (12.67) and (12.68) are depicted in Fig. 12.2.
Every intersection point of graphs in Fig. 12.2 corresponds to the fixed waveguide mode.

Problem 8. What is the difference between the odd and even plane modes? Use solution of the previous problem.
Solution. The intersection points in Fig. 12.2 define conditions for modes propagation in planar waveguide. Frequency of an even mode can be very low, i.e. as low as $\kappa \to 0$ that corresponds to the mode existence condition in planar waveguide, ($\gamma > 0$). In contrast, the odd modes do not have such properties.
Cutoff conditions for the even modes are given by

$$\kappa d = \nu \pi, \ \nu = 1, 2, \ldots \qquad (12.69)$$

$$\gamma_c d = 0 \qquad (12.70)$$

One can get

$$\sqrt{n_1^2 - n_2^2} \, k_0 d = \kappa_c d = \nu \pi \qquad (12.71)$$

312 | *Applied Optics*

Cutoff frequency corresponds to the condition of $\beta_c = n_2 k_0$, $\gamma_c = 0$, i.e.

$$\frac{\beta_c}{\kappa_c} = \frac{n_2}{\sqrt{n_1^2 - n_2^2}} \qquad (12.72)$$

The angle of wave propagation α can be calculated from the following

$$\tan\alpha = \pm\frac{\beta}{\kappa}. \qquad (12.73)$$

For the maximum angle of internal reflection (α_i) one has

$$\tan\alpha_i = \frac{\sin\alpha_i}{\sqrt{1 - \sin^2\alpha_i}} = \frac{n_2}{\sqrt{n_1^2 - n_2^2}} \qquad (12.74)$$

Comparison of equations (12.72) and (12.74) indicates that they have the same nature.

12.4 Electro-optics

Electro-optics is a branch of electrical engineering, electronic engineering, and materials science that concerns the interaction between the electromagnetic (optical) and the electrical (electronic) states of materials. The term *electro-optical effect* encompasses a number of distinct phenomena such as *Pockel's* effect (or linear electro-optical effect), *Kerr* effect (or quadratic electro-optical effect), electroabsorption and electroreflectance, *Franz-Keldish* effect, and *Stark* effect.

This section focuses on the phenomena of electro-optics within the basics of crystal physics and crystal optics. The nonlinear electro-optical effect is discussed in conjunction with several practical examples.

12.4.1 Principles of Electro-optics

Electro-optical effect is a change in optical properties (i.e. changes of $a_{ij} = 1/n^2$) of a material in response to an electric field that varies slowly compared to the frequency of light.

Fundamentals of the Optics of Materials: Tutorial and Problem Solving | 313

In external electric field $(\mathbf{E})$ optical functions of most materials become functions of $\mathbf{E}$. In particular, the refractive index, n of an electro-optic medium is a function of electric field $(n(\mathbf{E}))$ and can be expended in a Taylor series:

$$n(\mathbf{E}) = n_0 + a_1 E_k + \frac{1}{2} a_2 E_k E_l + \dots \tag{12.75}$$

The expansion coefficients in Eq. (12.75) characterize the linear and/or quadratic electro-optic effects. However, in electro-optics the introduction of the linear and quadratic electro-optic coefficients is based not on the expansion of $n(E)$ but of an another optical function, the *impermeability*, η, which is inverse of the relative dielectric permittivity function, ε:

$$\eta = \frac{\varepsilon_0}{\varepsilon} = \frac{1}{n^2}. \tag{12.76}$$

Taylor expansion of η is given by:

$$\eta_{ij}(\mathbf{E}) = \eta_{ij}(0) + \Delta \eta_{ij}(\mathbf{E}) \tag{12.77}$$

$$= \eta_{ij}(0) + \left(\frac{\partial \eta_{ij}}{\partial E_k} \right)' E_k + \frac{1}{2} \left(\frac{\partial^2 \eta_{ij}}{\partial E_k \partial E_l} \right)' E_k E_l + \dots$$

Prime in Eq. (12.78) indicates that the expansion coefficients are calculated at zero field (i.e. $\mathbf{E} = 0$). The field induced variation of η, i e the value of $\Delta \eta_{ij}(\mathbf{E})$ is given by:

$$\Delta \eta_{ij}(\mathbf{E}) = \eta_{ij}(\mathbf{E}) - \eta_{ij}(0) = r_{ijk} E_k + R_{ijkl} E_k E_l \tag{12.78}$$

where linear (r) and quadratic (R) electro-optic coefficients describing respectively Pockel's and Kerr electro-optic effects, are given by:

$$r_{ijk} = \left(\frac{\partial \eta_{ij}}{\partial E_k} \right)' \tag{12.79}$$

$$R_{ijkl} = \frac{1}{2} \left(\frac{\partial^2 \eta_{ij}}{\partial E_k \partial E_l} \right)'$$

314 | *Applied Optics*

It is instructive to obtain a similar to the Eq. (12.78) expression for n. Using equation (12.76) the field induced change of the refractive coefficient, i.e. the value of $\Delta n_{ij}(\mathbf{E}) = n_{ij}(\mathbf{E}) - n_{ij}(0)$, is given by:

$$\Delta n_{ij}(\mathbf{E}) = -\frac{1}{2} r_{ijk} n^3 E_k - \frac{1}{2} R_{ijkl} n^3 E_k E_l \qquad (12.80)$$

The third and fourth rank tensor coefficients of r_{ijk} and R_{ijkl} are called Pockel's and Kerr coefficients, respectively.

Polarization constant are defined matching established definitions (see Section 1.2)

$$
\begin{aligned}
\Delta a_{11} &= a_{11} - a_{11}^0, & \Delta a_{23} &= a_{23} - a_{23}^0, \\
\Delta a_{22} &= a_{22} - a_{22}^0, & \Delta a_{31} &= a_{31} - a_{31}^0, \\
\Delta a_{33} &= a_{33} - a_{33}^0, & \Delta a_{12} &= a_{12} - a_{12}^0.
\end{aligned} \qquad (12.81)
$$

Plugging expressions for (a_{ij}) into equation of optical indicatrix (9.10) and in the case when coordinate system (x, y, z) matches axes of refractive indexes ellipse one can get the following:

$$(a_{11}^0 + \Delta a_{11})x^2 + (a_{22}^0 + \Delta a_{22})y^2 + (a_{33}^0 + \Delta a_{33})z^2$$
$$+2\Delta a_{23}yz + 2\Delta a_{31}zx + 2\Delta a_{12}xy = 1, \qquad (12.82)$$

where

$$a_{11}^0 = \frac{1}{n_x^2}, \quad a_{22}^0 = \frac{1}{n_y^2}, \quad a_{33}^0 = \frac{1}{n_z^2}. \qquad (12.83)$$

The Eq. (12.82) describes the change of n^{-2}, rather than directly the change of refractive index n. One distinguishes between linear (*Pockel's*) and quadratic (*Kerr*) electro-optical effects.

Variation of optical functions at linear electro-optical effect is proportional to the first order of applied electric field E_j, i.e.

$$\Delta a_{ij} = r_{ijk} E_k, \qquad (12.84)$$

The electro-optic coefficient r_{ijk} is a third rank tensor as it relates a second rank tensor $(\Delta a_{ij} = 1/N_{ij}^2 = 1/\varepsilon_{ij})$ to a vector $(\mathbf{E})$. The third

Table 12.1 Contracted notation of the linear electro-optical coefficients

	E_1	E_2	E_3
Δa_{11}	r_{11}	r_{12}	r_{13}
Δa_{22}	r_{21}	r_{22}	r_{23}
Δa_{33}	r_{31}	r_{32}	r_{33}
Δa_{23}	r_{41}	r_{42}	r_{43}
Δa_{31}	r_{51}	r_{52}	r_{53}
Δa_{12}	r_{61}	r_{62}	r_{63}

rank tensor has 27 elements but only 18 are independent due to the dielectric tensor symmetry (i.e. $\varepsilon_{ij} = \varepsilon_{ji}$) and therefore $r_{ijk} = r_{jik}$. This symmetry allows to write the electro-optic tensor in a contracted notation where the i,j subscripts are replaced by: $xx = 1$, $yy = 2$, $zz = 3$, $yz = zy = 4$, $xz = zx = 5$ and $xy = yx = 6$. The k subscripts denotes $x = 1, y = 2$ and $z = 3$.

Presentation of the electro-optical r_{ijk} tensor in contracted notation as 6×3 matrix is given in Table 12.1.

Symmetry considerations provide an information as to which electro-optic coefficient tensor elements are non-zero and independent. For example, due to the inversion symmetry all r_{ijk} must be zero due to the following consideration. Because under application of the inversion operator (i.e. when $E_k \rightarrow -E_k$) the material parameters (ε_{ji} and r_{ijk}) remain unchanged. Hence $r_{ijk}E_k = -r_{ijk}E_k$ for any specified **E** which requires that all the electro-optical coefficients are zero, i.e. $r_{ijk} = 0$. For other symmetry classes, group theory can be employed to get the form of the electro-optic tensor. For example, the GaAs semiconductor crystals have the zinc-blende (cubic) symmetry, they belong to the class $\bar{4}3m$ crystals and do not have inversion symmetry. Consequently the electro-optical tensor has only one non-zero component: $r_{41} = r_{52} = r_{63}$.

In crystals with the quadratic electro-optical effect, changes of optical functions are proportional to the second order of applied fields, i.e.

$$\Delta a_{ij} = R_{ijkl}E_kE_l, \tag{12.85}$$

Applied Optics

Table 12.2 Contracted notation of the quadratic electro-optical coefficients

	E_x^2	E_y^2	E_z^2	$E_y E_z$	$E_x E_z$	$E_x E_y$
Δa_{11}	R_{11}	R_{12}	R_{13}	R_{14}	R_{15}	R_{16}
Δa_{22}	R_{21}	R_{22}	R_{23}	R_{24}	R_{25}	R_{26}
Δa_{33}	R_{31}	R_{32}	R_{33}	R_{34}	R_{35}	R_{36}
Δa_{23}	R_{41}	R_{42}	R_{43}	R_{44}	R_{45}	R_{46}
Δa_{31}	R_{51}	R_{52}	R_{53}	R_{54}	R_{55}	R_{56}
Δa_{12}	R_{61}	R_{62}	R_{63}	R_{64}	R_{65}	R_{66}

where R_{ijkl} are coefficients of quadratic electro-optical Kerr effect. Due to the dielectric tensor symmetry the fourth rank R_{ijkl} tensor has only 36 independent component which in contracted form are given in Table 12.2.

Actual number of independent nonzero constants is defined by the crystallographic class.

12.4.2 Primary and Secondary Electro-optical Effect

As stated above the linear electro-optical effect (or the primary electro-optical effect) contributes to a quadratic nonlinear optical response only in non-centrosymmetrical crystals, i.e. in the crystals without the inversion symmetry. This property poses piezoelectric crystals. If piezoelectric crystal is not exposed to any mechanical stress the external electric field induces an elastic strain, this phenomena is called the *reverse piezoelectrical* effect. The last effect induces changes in refractive index as a result of crystal photoelasticity:

$$\Delta a_{ij} = P_{ijkl} S_{kl}, \tag{12.86}$$

where P_{ijkl} is the photoelasticity tensor components, S_{kl} is the elastic strain.

Elastic strain in piezoelectric crystals are determined by *piezoelectric* coefficients d_{klm} defined according to

$$S_{kl} = d_{klm}E_m \tag{12.87}$$

Secondary linear electro-optical effect is attributed to the reverse piezoelectrical effect.

Relationship between primary and secondary electro-optical effects is described by the following equation:

$$\Delta a_{ij} = r_{ijm}^*E_m + P_{ijkl}S_{kl} = (r_{ijm}^* + P_{ijkl}d_{klm})E_m, \tag{12.88}$$

where r_{ijm}^* is the coefficient of primary linear electro-optical effect, values of P_{ijkl} represent photoelastic coefficients.

12.4.3 Practical Examples

Example 1. Calculate relationship between electro-optical coefficient r and the second order nonlinear optical susceptibility $\chi^{(2)}$.

Solution. Consider the electro-optic effect as a nonlinear parametric sum frequency generation process of interaction between an optical wave (at a frequency ω), i.e. $E_1(\omega)$) and an electric field of zero frequency ($E_2(0)$).

The wave equation for a slowly varying wave is given by [Shen (2003)]: $E_1(\omega) = E_{10}e^{-i\omega t}e^{i\Delta\phi(z)}$, $E_2(0) = E_{20}$, $E_3(\omega) = -iE_1(\omega)$

$$\frac{dE_3(\omega)}{dz} = -i\frac{2\pi\omega^2}{kc^2}\chi^{(2)}(\omega, 0, \omega)E_1(\omega)E_2(0). \tag{12.89}$$

This equation has to be solved for the following fields:

$$E_1(\omega) = E_{10}e^{-i\omega t}, \quad E_2(0) = E_{20}, \quad E_3(\omega) = E_{30}e^{-i(\omega t-\varphi)}, \tag{12.90}$$

Plugging (12.90) into (12.89) results in the following

$$E_{30}e^{i\varphi}\frac{d\Delta\varphi(z)}{dz} = -\frac{2\pi\omega^2}{kc^2}\chi^{(2)}E_{10}E_{20} \tag{12.91}$$

Taking into account a relationship of $-1 = \exp(i\pi)$ one can get

$$\frac{d\Delta\varphi(z)}{dz} = \frac{2\pi\omega^2}{kc^2}\chi^{(2)}E_{20},$$
(12.92)

Analyzing equation (12.92) one can conclude that electric field at a crystal entrance has a phase shift of $\varphi = \pi$ and that nonlinear interaction changes phase of optic waves. Assume that $E_1 = E_3 = E_0 e^{i\Delta\varphi(z)}$. Combining this relationship and equation (12.89) one can get the following

$$\Delta\varphi(L) = -\frac{2\pi\omega^2}{kc^2}\chi^{(2)}E_{20}L,$$
(12.93)

where L represents a length of the interaction.
On the other hand

$$\Delta a = \Delta\left(\frac{1}{n^2}\right) = -\frac{2\Delta n}{n^3} = rE_{20}$$
(12.94)

and the phase variation is given by

$$\Delta\varphi(L) = \frac{\omega}{c}\Delta nL = -\frac{\omega}{c}L\frac{rE_{20}n^3}{2}.$$
(12.95)

From equations (12.93) and (12.95) one can get the following

$$\frac{2\pi\omega^2}{kc^2}\chi^{(2)}(\omega, 0, \omega)E_{20}L = \frac{\omega}{c}L\frac{rE_{20}n^3}{2}.$$
(12.96)

Finally the relationship between electro-optic coefficient r and the nonlinear optical 2nd order susceptibility $\chi^{(2)}(\omega, 0, \omega)$ is given by

$$|r| = \frac{4\pi}{n^3}|\chi^{(2)}(\omega, 0, \omega)|.$$
(12.97)

As long as

$$\chi^{(2)}_{ijk}(0, \omega, -\omega) = \chi^{(2)}_{jki}(\omega, -\omega, 0),$$
(12.98)

i.e. the nonlinear optical susceptibility for "optical detection" (the left part of equation (12.98)) is equal to the susceptibility of the electro-optical effect (the right part of equation (12.98)) the coefficient

Fundamentals of the Optics of Materials: Tutorial and Problem Solving | 319

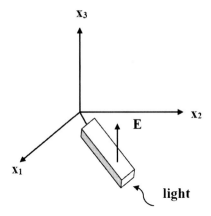

Figure 12.3 Diagram to measure electro-optical coefficient in a sample bar of the KDP-crystal.

r_{ij} can be calculated using equation (12.98) by a simultaneous rearrangement of matrix rows and columns, i.e.

$$r_{ij} = -\frac{4\pi}{n^3}\chi_{ji} \qquad (12.99)$$

where index i ranges from 1 to 6 and index j ranges from 1 to 3.

Example 2. The electro-optical coefficient r_{ij} can be measured using the longitudinal electro-optical effect in the sample bar of the KDP-crystal (KH_2PO_4). Calculate electro-optical coefficient r_{ij} for the following conditions: the phase difference $\Delta\varphi = \pi$ is observed for the optical wave with $\lambda = 546.0\ nm$ at an electric voltage of $V_\pi = 7.9\ kV$ applied as shown in Fig. 12.3.

Solution. KDP-crystal belong to the tetragonal crystal system of 42 m crystallographic class. Consequently it's matrices of the dielectric susceptibility function (ε_{ij}) and that of the linear electro-optical coefficient (r_{ij}) are given by

$$\varepsilon = \begin{bmatrix} \varepsilon_{11} & 0 & 0 \\ 0 & \varepsilon_{22} & 0 \\ 0 & 0 & \varepsilon_{33} \end{bmatrix} \qquad (12.100)$$

320 | *Applied Optics*

$$r_{ij} = \begin{bmatrix} 0 & 0 & 0 \\ 0 & 0 & 0 \\ 0 & 0 & 0 \\ r_{41} & 0 & 0 \\ 0 & r_{52} & 0 \\ 0 & 0 & r_{61} \end{bmatrix} \tag{12.101}$$

Optical indicatrix of KDP-crystal without applied electric field is determined by

$$a_1^0(x_1^2 + x_2^2) + a_3^0 x_3^2 = 1, \tag{12.102}$$

where

$$a_1^0 = \frac{1}{n_0^2}, \qquad a_3^0 = \frac{1}{n_e^2} \tag{12.103}$$

n_0, n_e stand for ordinary and extraordinary refractive indices, respectively.

Electric field E is applied to the plate along z-axis (see Fig. 12.3). Thus $E_1 = E_2 = 0$, and $E_3 \neq 0$.

$$\Delta a_6 = r_{63} E_3 \tag{12.104}$$

Optical indicatrix is defined by the following

$$a_1^0 \left(x_1^2 + x_2^2\right) + a_3^0 x_3^2 + 2 r_{63} E_3 x_1 x_3 = 1 \tag{12.105}$$

Light propagates along the axis x_3.

Refractive indexes ellipse (with and without applied field) is shown in Fig. 12.4

Ellipse of the optical indicatrix is described by the following equation:

$$\frac{x_1^2 + x_2^2}{n_0^2} + 2 r_{63} x_1 x_2 E_3 = 1. \tag{12.106}$$

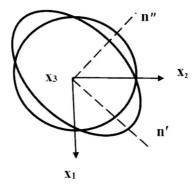

Figure 12.4 Optical indicatrix with (ellipse) and in the absence of the applied field (circle).

Calculate new basic refractive indices within the $x_1 x_2$ plane transforming equation (12.106) to the basic ellipse axes, i.e.

$$\left(\frac{1}{n_0^2} + r_{63} E_3\right) x_1'^2 + \left(\frac{1}{n_0^2} - r_{63} E_3\right) x_2'^2 = 1. \qquad (12.107)$$

The refractive indices are given by

$$\frac{1}{n'^2} = \frac{1}{n_0^2} + r_{63} E_3, \quad \frac{1}{n''^2} = \frac{1}{n_0^2} - r_{63} E_3 \qquad (12.108)$$

It follows

$$n'^2 = \frac{n_0^2}{1 + r_{63} E_3 n_0^2}, \quad n''^2 = \frac{n_0^2}{1 - r_{63} E_3 n_0^2}, \qquad (12.109)$$

Typically $|r_{63} E_3 n_0^2| \ll 1$. Therefore equation (12.109) can be rewritten as

$$n' \simeq n_0 - \frac{1}{2} r_{63} E_3 n_0^3, \quad n'' \simeq n_0 + \frac{1}{2} r_{63} E_3 n_0^3 \qquad (12.110)$$

322 | *Applied Optics*

Finally one gets a two-axes crystal with the basic refractive indices given by

$$n' = n_0 - \frac{1}{2}r_{63}E_3n_0^3, \quad n'' = n_0 + \frac{1}{2}r_{63}E_3n_0^3, \quad n''' = n_0 \quad (12.111)$$

The birefringence for light beam, propagating through the z-cut plate orthogonal to the facet direction is given by

$$n'' - n' = r_{63}E_3n_0^3. \tag{12.112}$$

Hence

$$\frac{2\pi}{\lambda}r_{63}E_3n_0^3L = \pi \tag{12.113}$$

Plugging $\lambda = 546$ nm, $n_0 = 1,5115$, $V_\pi = E_3L = 7,9KV$ one gets

$$\mid r_{63} \mid = \frac{\lambda}{2V_\pi n_0^3} = 3 \cdot 10^{-7}(SGSE) = 9 \cdot 10^{-2}\frac{m^2}{C} \tag{12.114}$$

Example 3. Calculate the '$\lambda/2$-voltage' for KDP-crystal (see Example 2) (the voltage converting a sample bar to the half-wave plate (i.e. that changes the phase by π)) if electric field is applied to orthogonal faces of the crystal (see Fig. 12.3). Crystal dimensions are $3 \times 3 \times 100\ mm^3$.

Solution. Given $E_3 = \mid \mathbf{E} \mid$. Optical indicatrix is given by

$$a_1(x_1^2 + x_2^2) + a_3x_3^2 + 2r_{63}E_3x_2x_1 = 1. \tag{12.115}$$

One can transform the coordinate system to basic one by rotating around x_3 axis by $45°$ angle. In this case the [110] axis matches the light beam propagation direction. Sectional drawing of refractive index ellipse within the $x_1' = 0$ plane is given by

$$(a_1 + r_{63}E)x_2'^2 + a_3x_3'^2 = 1 \tag{12.116}$$

Hence

$$n_2' = n_0(1 - \frac{1}{2}n_0^2r_{63}E), \quad n_3' = n_e \tag{12.117}$$

Light beams polarized along x'_2 and $x'_3(x_3)$ axes propagate with the phase velocities given by

$$V'_2 \simeq \frac{c}{n_0}(1 + \frac{1}{2}n_0^2 r_{63}E), \quad V'_3 = \frac{c}{n_e} \tag{12.118}$$

A total phase difference between these light beams is given by

$$\Delta\varphi = \frac{2\pi L}{\lambda}(n_e - n_0) + \frac{\pi n_0^3 r_{63}EL}{\lambda} \tag{12.119}$$

The first term is responsible for the natural crystal anisotropy. Second term appeared due to the induced anisotropy. Electric field converting the sample bar to the half-wave plate (i.e. that changes the phase by π) is given by

$$V_\pi = \frac{\lambda}{n_0^3 r_{63}} \frac{d}{L} \tag{12.120}$$

The V_π voltage is much less than that given by equation (12.114). The exact ratio is $\eta = L/2d = 100(mm)/6(mm) \sim 16.7$. Thus $V_\pi = 7.9kV/\eta = 470V$. Another advantage is that the lateral configuration of electro-optic effect does not require a deposition of electrodes on the facial faces of a crystal.

12.5 Magneto-optics

A magneto-optical effect can be any one of a number of phenomena in which an electromagnetic wave propagates through a medium that has been altered by the presence of a quasistatic magnetic field. When light is transmitted through a layer of magneto-optic material, the result is called the *Faraday effect*—the plane of polarization can be rotated, forming a *Faraday rotator*. The results of reflection from a magneto-optic material are known as the magneto-optic Kerr effect (not to be confused with the nonlinear Kerr effect considered in Section 12.4), the *Cotton-Mouton* effect (optical birefringence

324 | *Applied Optics*

induced in the presence of a constant transverse magnetic field),[3] see [Ovechko and Kharchenko (2013)] and references therein.

12.5.1 Principles of Magneto-optics

Quadratic approximation of nonlinear polarization takes into account effect of both electric $E_j(\mathbf{r}, t)$ and magnetic $H_k(\mathbf{r}, t)$ fields, i.e.

$$P_i^{(2)}(\mathbf{r}, t) = \eta_{ijk}^{em} E_j(\mathbf{r}, t) H_k(\mathbf{r}, t), \tag{12.121}$$

where η_{ijk}^{em} is a pseudotensor of electro-magnetic susceptibility, that does not vanish even for isotropic substances. Assuming that magnetic field in (12.121) is a stationary one, i.e. $H_k(\mathbf{r}, t) = H_k(0)$ that results in the following

$$P_i^{(2)}(\mathbf{r}, t) = \eta_{ijk}^{em} E_j(\mathbf{r}, t) H_k(0) \tag{12.122}$$

In this case a variation of dielectric permeability tensor under linear action of magnetic field is given by

$$\Delta\varepsilon_{i,j} = 4\pi\eta_{ijk}^{em} H_k(0). \tag{12.123}$$

For isotropic substance it reads

$$\varepsilon_{i,j} - \varepsilon_0\delta_{ij} = 4\pi\eta_{ijk}^{em} H_k(0). \tag{12.124}$$

Assume that magnetic field is directed along the z-axis. The nonzero components in equation (12.124) are

$$\varepsilon_{xx} = \varepsilon_{yy} = \varepsilon_{zz} = \varepsilon, \quad \varepsilon_{xy} = -\varepsilon_{yx} = 4\pi\eta^{em} H_z(0). \tag{12.125}$$

For nonmagnetic substance (i.e. for $(\mu = 1)$) and for $|\varepsilon_{xy}| \ll \varepsilon$, the refractive index for circularly polarized optical wave is given by

$$n_{\mp} \simeq n \mp \frac{1}{2}i\frac{\varepsilon_{xy}}{n}. \tag{12.126}$$

---

[3] The Cotton-Mouton effect was discovered in 1907 by Aimé Cotton and Henri Mouton.

In this case the circular birefringence can be written as

$$n_+ - n_- = i\frac{\varepsilon_{xy}}{n} = i\frac{4\pi}{n}\eta^{em}H_z(0).$$ (12.127)

Equation (12.127) can be interpreted as a plane rotation of light polarization under magnetic field in isotropic substance, that represents the *Faraday effect*. Angle (θ) of optical wave vector **E** rotation is given by

$$\theta = \frac{\pi}{\lambda}(n_+ - n_-)L = \frac{4\pi^2}{n\lambda}\eta^{em}H_z(0)L.$$ (12.128)

One can define the *Verdet constant* [4] by equation $\theta = V_\lambda H_z(0)L$. One has

$$V_\lambda = \frac{4\pi^2}{n\lambda}\eta^{em}.$$ (12.129)

Using equations (12.129) and (12.121) one can get the following

$$V_\lambda(\omega) = \frac{4\pi}{n(\omega)}\left(\frac{\omega}{c}\right)\chi^{em}_{xyz}(-\omega, \omega, 0).$$ (12.130)

Therefore the Faraday effect is linear with respect to the magnetic field $H_z(0)$ and to the electric optical field $\mathbf{E}(\mathbf{r}, t)$. Selected problems related to the Faraday effect are considered here.

12.5.2 Practical Examples

Example 4. Calculate optical wave propagation through the optical system depicted in Fig. 12.5. Calculate Verdet constant V_λ for a dense flint glass, if $T = I/I_0 = 0.014$ for $H = 4 \cdot 10^5 A/m$, $L = 10^{-1}m$.

Solution. According to the *Malus law* , when completely plane polarized light is incident on the analyzer, the intensity I of the light transmitted by the analyzer is directly proportional to the square of

---

[4] *Verdet constant* (named after the French physicist Émile Verdet) describes the strength of the Faraday effect for a particular material. The Verdet constant for most materials is extremely small and is wavelength dependent.

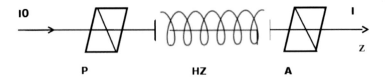

Figure 12.5 Schematics of experimental set-up for the measurements of the Faraday's effect. P - polarizer, A - analyser, H_z - magnetic field.

the cosine of angle between the transmission axes of the analyzer and the polarizer, i.e.

$$I = I_0 \cos^2(\varphi_0 + \varphi_F), \tag{12.131}$$

where $\varphi_0 = \pi/2$ for crossed polarizer (P) and analyzer (A), $\varphi_F = V_\lambda H \cdot L$ is the *Faraday angle*. For small angle $\varphi_F \ll 1$ equation (12.131) reads

$$I = (V_\lambda H \cdot L)^2 I_0. \tag{12.132}$$

Thus

$$V_\lambda = \left(\frac{I}{I_0}\right)^{1/2} \frac{1}{HL} = (0.014)^{1/2} \frac{1}{4 \cdot 10^5 \frac{A}{m} \cdot 0.1 m}$$
$$= 3 \cdot 10^{-6} \frac{rad}{A} = 0.01 \frac{min}{A} \tag{12.133}$$

Example 5. Within the *Lorentz model* discussed in Section 1.4, calculate dispersion of Verdet constant $V_\lambda(\omega)$.
Solution. Assume that magnetic field is applied along z-axis, i.e. $H = H_z$ and the optical beam propagates along z-direction. Equating of electron motion is given by

$$m\ddot{\mathbf{r}} + f\mathbf{r} = e\left(\mathbf{E} + \frac{1}{c}[\dot{\mathbf{r}} \times \mathbf{H}]\right). \tag{12.134}$$

Dissipation is neglected in equation (12.134). Using notation $f/m = \omega_0^2$, where ω_0 denotes the resonance frequency of oscillation, one gets the following

$$\ddot{\mathbf{r}} + \omega_0^2 \mathbf{r} = \frac{e}{m}\left(\mathbf{E} + \frac{1}{c}[\dot{\mathbf{r}} \times \mathbf{H}]\right). \tag{12.135}$$

From equation (12.135) one can get the following scalar relationship

$$\ddot{x} - \frac{e}{mc}\dot{y}H + \omega_0^2 x = \frac{e}{m}E_x, \ddot{y} + \frac{e}{mc}\dot{x}H + \omega_0^2 y = \frac{e}{m}E_y. \tag{12.136}$$

Consider solution of equations (12.136) for two circularly polarized optical waves:

1. $E_x = E_0 \cos \omega t, \quad E_y = E_0 \sin \omega t;$
2. $E_x = E_0 \cos \omega t, \quad E_y = -E_0 \sin \omega t;$

Coordinate dependencies are given by

1. $x = r_0 \cos \omega t, \quad y = r_0 \sin \omega t;$
2. $x = r_0 \cos \omega t, \quad y = -r_0 \sin \omega t.$

Consequently the two solutions of equation (12.136) are given by

$$r_0^{\pm} = \frac{(e/m)E_0}{(\omega_0^2 - \omega^2) \pm \left(\frac{eH}{mc}\right)\omega} \tag{12.137}$$

Following the same procedure as in Section 1.4, consider polarization function in the form of $\mathbf{P} - Ne\mathbf{r}$, where N is a concentration of oscillators. Thus combining

$$n^2 = \varepsilon = 1 + 4\pi P/E. \tag{12.138}$$

with equation (12.138) one can get

$$n_{\pm}^2 = 1 + \frac{4\pi Ne^2/m}{(\omega_0^2 - \omega^2) \pm \left(\frac{eH}{mc}\right)\omega} \tag{12.139}$$

328 | *Applied Optics*

For a very low value of the oscillator density (N) equation (12.139) can be simplified to the following

$$n_\pm \simeq 1 + \frac{2\pi Ne^2/m}{(\omega_0^2 - \omega^2) \pm \left(\frac{eH}{mc}\right)\omega} \qquad (12.140)$$

Accordingly to equation (12.138) Faraday angle θ_F is given by

$$\theta_F = \frac{\omega}{2c}(n_+ - n_-)L = -\frac{2\pi Ne^2 L}{mc}\frac{eH}{mc}\frac{\omega^2}{(\omega_0^2 - \omega^2)^2 - \left(\frac{eH}{mc}\omega\right)^2} \qquad (12.141)$$

Far away from the resonance and under approximation of $\omega_0 \gg \omega \gg \frac{eH}{mc}$ one can further simplify equation (12.141) to the following

$$\theta_F(\omega) = -\frac{2\pi Ne^2 L}{mc}\frac{eH}{mc}\frac{\omega^2}{\omega_0^4} \qquad (12.142)$$

Calculated spectral dependence of $\theta_F(\lambda) \sim \lambda^{-2}$ given by equation (12.142) corresponds to the results observed experimentally. The Verdet constant is given by

$$V_\lambda = -\frac{2\pi Ne^3}{m^2c^2}\frac{\omega^2}{\omega_0^4} \qquad (12.143)$$

12.6 Acousto-optics

The term *acousto-optics* can be defined as an interaction of light with sound (or ultrasound) waves in materials. Acousto-optic interaction occurs in all optical mediums when an acoustic wave and a light beam (e.g. laser) are present in the medium. Thus acousto-optics is a branch of physics that studies the interactions between sound waves and light waves, especially the diffraction of laser light by ultrasound (or sound in general) through an ultrasonic grating. When an acoustic wave is launched into the optical medium, it generates a refractive index wave that behaves like a sinusoidal grating. An acoustic wave creates a perturbation of the refractive index in the form of a wave. The medium becomes a dynamic graded-index medium—an inhomogeneous medium with a time-varying stratified refractive index. An incident laser light beam passing through this grating will diffract the laser beam into several orders.

Since optical frequencies are much greater than the acoustic frequencies, the variations of the refractive index in a medium perturbed by sound are usually very slow in comparison with an optical period. There are therefore two significantly different time scales for light and sound. As a consequence, it is possible to use an adiabatic approach. In this case the optical wave propagation problem is solved separately at every instant of time during the relatively slow course of the acoustic cycle, always treating the material as if it were a static inhomogeneous medium. Within this quasi-stationary approximation, acousto-optics becomes the optics of an inhomogeneous medium.

The acousto-optic effect is extensively used in the measurement and study of ultrasonic waves. Acousto-optic devices are used in laser equipment for electronic control of the intensity and positioning of the laser beam.

12.6.1 Basics of Acousto-optics

The theory of acousto-optics deals with the perturbation of the refractive index. In this section the basics of the optical and acoustic plane waves interaction is considered based on Maxwell's differential equations. For more detailed description see [Born and Wolf (1999)].

Sound is a dynamic strain involving molecular vibrations that take the form of waves which travel at a velocity characteristic of the medium, i.e. the velocity of sound. For example, in gas a harmonic plane wave of compressions and rarefactions creates regions where medium is compressed (the density is higher and the refractive index is larger) followed by regions where medium is rarefied (the density is lower and the refractive index is smaller). In solids, the sound induces vibrations of the molecules around their equilibrium positions, which change the optical polarizability and consequently the refractive index.

Therefore an acoustic wave creates a periodic (in space) pertur-bation of the refractive index in the form of a wave. The medium becomes a dynamic graded-index medium, i.e. an inhomogeneous medium with a time-varying stratified refractive index. The theory

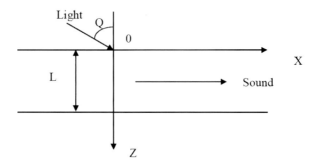

Figure 12.6 Schematics of optical and acoustic waves interaction.

of acousto-optics deals with the perturbation of the refractive index caused by sound, and with the propagation of light through this perturbed time-varying inhomogeneous medium.

Consider a diffraction of an optical plane wave on a grating induced in a medium by an acoustic plane wave. Varied refractive index in medium causes nonlinear optical effects. Consequently the electric induction vector ($D = (D_x, D_y, D_z)$) (see Eq. (1.2) in Chapter 1) will contain nonlinear (cross) terms and can be written as:

$$D_i = (\hat{\varepsilon}_{ij}^0 + \hat{f}_{ijkl} U_{kl}) E_j + \hat{e}_{ijk} \hat{U}_{jk}, \tag{12.144}$$

where $\hat{\varepsilon}_{ij}^0$ – tensor of dielectric permittivity, $\hat{f}_{ijkl}$ – tensor of electrostriction, $\hat{e}_{ijk}$ – tensor of piezomodules, and $\hat{U}_{jk}$– tensor of tension. Here we consider an isotropic solid or liquid. In this case only diagonal elements of tensors $\hat{f}_{ijkl}$ and $\hat{e}_{ijk}$ are not equal to zero. As widely accepted the repeated indexes in Eq. (12.144) imply summation. It is also assumed that the acoustic wave does not change in amplitude. Schematics of the optical and acoustic waves interaction is depicted in Fig. 12.6.

Wave equation for the optical wave amplitude E is given by:

$$\Delta E - \frac{1}{c^2} \frac{\partial^2}{\partial t^2} (\varepsilon E) = 0, \tag{12.145}$$

where
$\varepsilon(x, t) = \varepsilon_0 + \delta\varepsilon \sin(\Omega t - Kx + \delta p)$, Ω, K – frequency and wave vector of sound respectively, and δp – phase shift. Consider a solution of the

equation (12.145) within the Flocke theorem, i.e.:

$$E = \exp[i(\omega t - n_0 k(z \cos \theta + x \sin \theta)] \sum_{n=-\infty}^{n=\infty} F_n(z) \exp[i(\Omega t - Kx)],$$

$$(12.146)$$

where ω, k are frequency and wave vector of the optical wave, respectively, $n_0 = \sqrt{\varepsilon_0}$ stands for the unperturbed refractive index, θ is the angle of incidence, n stands for a diffraction order, and F_n is a slowly varying amplitudes.

One can see that the frequency of the diffracted optical wave has a frequency shift of $n\Omega$ which is caused by the Doppler effect. From Eqs. (12.145) and (12.146) follow n-differential equations for slowly varying amplitudes $F_n(z)$

$$\frac{\partial F_n}{\partial z} + \frac{v}{2L}[F_{n-1} - F_{n+1}] = i\frac{nQ}{2L}(n - 2\alpha)F_n, \qquad (12.147)$$

where
$$v = (1/2)(\delta\varepsilon/\varepsilon_0)kn_0 L, \alpha = -n_0(k/K) \sin \theta, Q = K^2 L/n_0 k.$$

Equations (12.147) were first derived by Raman and Nath in 1935. Consider now two problems with solutions of the equations (12.146) for the following two approximations: 1. $Q \ll 1$ and 2. $Q \gg 1$.

12.6.2 Practical Examples: Optical Difraction Wave Amplitude for $Q \ll 1$

Problem 9. Calculate optical amplitude of diffracted waves of n-th order in $Q \ll 1$.
Solution. The equation (12.147) for $Q \ll 1$ can be written in the following form:

$$\frac{\partial F_n}{\partial z} + \frac{v}{2L}[F_{n-1} - F_{n+1}] = -i\frac{nQ\alpha}{L}F_n, \qquad (12.148)$$

332 | Applied Optics

Here we take into account that $n \ll 2\alpha$, because of $k/K \gg 1$. This approximation is valid for the acoustic frequencies $\Omega/2\pi \leq 100$ MHz.

Exact solution of the equation (12.148) is given by:

$$F_n = \exp(-i\frac{nQaz}{2L})J_n[\frac{2v}{Qa}\sin\frac{Qaz}{2L}], \qquad (12.149)$$

where J_n is the *Bessel* function of n-th order. Assume $z = L$, then the intensity $I_n = |F_n|^2$ is given by:

$$I_n(L) = J_n^2[v\frac{\sin(Qa/2)}{(Qa/2)}], \qquad (12.150)$$

If an angle of incidence is equal to zero (as shown in Fig. 12.7), we have:

$$I_n(L) = J_n^2[v], \qquad (12.151)$$

Recall that $v = (1/2)(\delta\varepsilon/\varepsilon_0)kn_0L$ and $J_n(v)$ is slowly varying function of the order n. Therefore amplitudes of the n-th order of diffraction change slowly.

12.6.3 Practical Examples: Optical Diffraction Wave Amplitude for $Q \gg 1$

Problem 10. Calculate optical amplitudes of diffracted waves corresponding to the n-th order of diffraction, if $Q \gg 1$.

Solution. Consider equation (12.147). Right part of the equation turns zero, if $n - 2\alpha = 0$. For the first order of diffraction $n = \pm 1$, then $\alpha = \pm 1/2$. We have the following set of three equations:

$$\frac{\partial F_1}{\partial z} + \frac{v}{2L}[F_0 - F_2] = 0,$$

$$\frac{\partial F_0}{\partial z} + \frac{v}{2L}[F_{-1} - F_1] = 0, \qquad (12.152)$$

$$\frac{\partial F_{-1}}{\partial z} + \frac{v}{2L}[F_{-2} - F_0] = 0$$

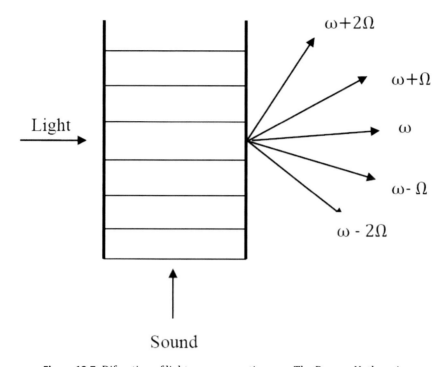

Figure 12.7 Difraction of light on an acoustic wave. The Raman-Nath regime.

By neglect of the higher order of diffraction i.e. $F_2, F_{-2} \ll F_0, F_1, F_{-1}$, the growing amplitudes (for $n = -1$) are given by:

$$\frac{\partial F_{-1}}{\partial z} - \frac{v}{2L} F_0 = 0,$$
$$\frac{\partial F_0}{\partial z} + \frac{v}{2L} F_1 = 0, \quad (12.153)$$

Eqs. (12.153) can be written in the following form:

$$\frac{\partial^2 F_{-1}}{\partial z^2} + \left(\frac{v}{2L}\right)^2 F_{-1} = 0,$$
$$\frac{\partial^2 F_0}{\partial z^2} + \left(\frac{v}{2L}\right)^2 F_0 = 0, \quad (12.154)$$

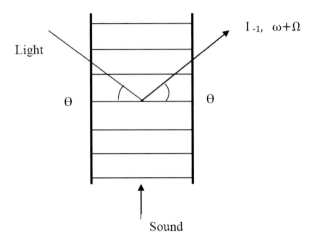

Figure 12.8 Diffraction of an optical beam from an acoustic plane wave. There is only one plane wave that satisfies the Bragg condition. Bragg regime.

with the following boundary conditions $F_{-1}(0) = 0$, $F_0(0) = E_0$. Solutions of equations (12.154) are given by:

$$I_{-1}(L) = I_0 \sin^2(v/2),$$
$$I_0(L) = I_0 \cos^2(v/2), \tag{12.155}$$

where $I_0 = |E_0|^2$, $I_{-1} = |E_{-1}|^2$. Applying $v = \pi$, results in $I_{-1} = I_0$, $I_0(L) = 0$. Therefore the incident light completely passes to the first order of diffraction maximum (see Fig. 12.8).

This type of diffraction is called the *Bragg diffraction*. It occurs if $\lambda L \gg \Lambda^2$, where λ is the optical wavelength, and Λ is the acoustic wavelength. The condition $\alpha = \pm 1/2$ is equivalent to the following expression:

$$K = 2n_0 k \sin\theta. \tag{12.156}$$

Formula (12.156) represents the momentum and energy conservation laws for the interaction of the optical and acoustical plane waves within the classic electrodynamics.

According to the quantum theory of light, an optical wave of angular frequency ω and wave vector $\mathbf{k}$ is viewed as a flux of *photons* with energy $\hbar\omega$ and momentum $\hbar\mathbf{k}$ each. An acoustic wave of angular frequency Ω and wave vector $\mathbf{K}$ is regarded as a stream of *phonons* of energy $\hbar\Omega$ and momentum $\hbar\mathbf{K}$ each. Within the quantum theory the momentum and energy conservation laws for the interaction between the optical and acoustical plane waves are given by:

$$\hbar\omega \pm \hbar\Omega = \hbar\omega',$$

$$\hbar\mathbf{k} \pm \hbar\mathbf{K} = \hbar\mathbf{k}'. \tag{12.157}$$

Appendix A

Selected Physical Constants

Table A.1 Selected physical constants

Quantity	Symbol	Value
Avogadro number	N_A	$6.02214 \times 10^{23}\,\mathrm{mol}^{-1}$
Bohr radius	a_B	$0.52917\,\text{Å}$
Boltzmann constant	k	$1.38066 \times 10^{-23}\,\mathrm{J/K}$
Elementary charge	q	$1.60218 \times 10^{-19}\,\mathrm{C}$
Hartree energy	Hrt	$27.2\,\mathrm{eV}$
Mass of electron at rest	m_0	$0.91094 \times 10^{-30}\,\mathrm{kg}$
Mass of proton at rest	M_p	$1.67262 \times 10^{-27}\,\mathrm{kg}$
Permeability in vacuum	μ_0	$1.25664 \times 10^{-8}\,\mathrm{H/cm}\ (4\pi \times 10^{-9})$
Permittivity in vacuum	c_0	$8.85418 \times 10^{-14}\,\mathrm{F/cm}\ (1/\mu_0 c^2)$
Planck constant	h	$6.62607 \times 10^{-34}\,\mathrm{J\cdot s}$
Planck constant (reduced)	$\hbar$	$1.05457 \times 10^{-34}\,\mathrm{J\cdot s}\ (h/2\pi)$
Rydberg energy constant	Ry	$13.6\,\mathrm{eV}$
Speed of light in vacuum	c	$2.99792 \times 10^{10}\,\mathrm{cm/s}$
Universal gas constant	R	$1.98719\,\mathrm{cal/mol\cdot K}$

Appendix B

Optical Units Conversions

Optical quantities are mostly measured in SI or CGS (Gaussian) systems. In the Gaussian system the polarization (**P**) is given by:

$$\mathbf{P}(t) = \chi^{(1)}\mathbf{E}(t) + \chi^{(2)}\mathbf{E}^2(t) + \chi^{(3)}\mathbf{E}^3(t) + \cdots \tag{B.1}$$

The units of **P** in Gaussian system (the same as of other fields **E, D, B, H**) are [Boyd (1992)]:

$$[\mathbf{P}] = [\mathbf{E}] = \frac{\text{statvolt}}{\text{cm}} = \left(\frac{\text{erg}}{\text{cm}^3}\right)^{1/2}. \tag{B.2}$$

Dimensions for optical susceptibilities are given by:

$$\left[\chi^{(1)}\right] = \text{dimensionless} \tag{B.3}$$

$$\left[\chi^{(2)}\right] = \left[\frac{1}{\mathbf{E}}\right] = \frac{\text{cm}}{\text{statvolt}} = \left(\frac{\text{cm}^3}{\text{erg}}\right)^{1/2} \tag{B.4}$$

$$\left[\chi^{(3)}\right] = \left[\frac{1}{\mathbf{E}^2}\right] = \left(\frac{\text{cm}}{\text{statvolt}}\right)^2 = \frac{\text{cm}^3}{\text{erg}} \tag{B.5}$$

340 | *Optical Units Conversions*

In literature the Gaussian units are frequently referred as *esu* units—the electrostatic units. In SI system, equation (B.1) is written in the following form:

$$\mathbf{P}(t) = \varepsilon_0 \left[\chi^{(1)} \mathbf{E}(t) + \chi^{(2)} \mathbf{E}^2(t) + \chi^{(3)} \mathbf{E}^3(t) + \cdots \right] \tag{B.6}$$

where ε_0 denotes the permittivity of vacuum, $\varepsilon_0 = 8.85 \times 10^{-12}$ F/m (with F=C/V). The units of $\mathbf{P}$ and $\mathbf{E}$ in SI are given by:

$$[\mathbf{P}] = \frac{C}{m^2} \tag{B.7}$$

$$[\mathbf{E}] = \frac{V}{m} \tag{B.8}$$

Dimensions of optical susceptibilities in SI system are given by [Boyd (1992)]:

$$\left[\chi^{(1)} \right] = \text{dimensionless} \tag{B.9}$$

$$\left[\chi^{(2)} \right] = \left[\frac{1}{\mathbf{E}} \right] = \frac{m}{V} \tag{B.10}$$

$$\left[\chi^{(3)} \right] = \left[\frac{1}{\mathbf{E}^2} \right] = \left(\frac{m}{V} \right)^2 \tag{B.11}$$

The relationship for potential units in both systems is: 1 statvolt = 300 V. Consequently, values of the electrical field strength in both systems relate as follows:

$$\mathbf{E}(\text{SI}) = 3 \times 10^4 \, \mathbf{E}(\text{Gaussian}) \tag{B.12}$$

In linear medium the displacement in Gaussian system is given by:

$$\mathbf{D} = \mathbf{E} + 4\pi\mathbf{P} = \mathbf{E} \left(1 + 4\pi\chi^{(1)} \right), \tag{B.13}$$

and in SI system it is given by

$$\mathbf{D} = \varepsilon_0 \mathbf{E} + \mathbf{P} = \varepsilon_0 \mathbf{E} \left(1 + \chi^{(1)} \right). \tag{B.14}$$

Fundamentals of the Optics of Materials: Tutorial and Problem Solving | 341

Table B.1 Units of selected optical quantities

Quantity	Symbol	SI	CGS
Electric field strength	E	V/m	$10^{-4}/3$ esu
Density	d	kg/m^3	10^{-3} g/cm^3
Magnetic induction	B	T = W/m^2	10^4 G
Magnetic field strength	H	A/m	$12\pi 10^7$ esu
Susceptibility, linear	$\chi^{(1)}$	dimensionless	dimensionless
Susceptibility, 2^d order	$\chi^{(2)}$	m/V	4.19×10^{-4} esu
Susceptibility, 3^d order	$\chi^{(3)}$	(m/V)^2	1.4×10^{-8} esu

Therefore

$$\chi^{(1)}(\text{SI}) = 4\pi\chi^{(1)}(\text{Gaussian}) \tag{B.15}$$

$$\chi^{(2)}(\text{SI}) = \frac{4\pi}{3 \times 10^4}\chi^{(2)}(\text{Gaussian})$$

$$= 4.19 \times 10^{-4}\chi^{(2)}(\text{Gaussian}) \tag{B.16}$$

$$\chi^{(3)}(\text{SI}) = \frac{4\pi}{(3 \times 10^4)^2}\chi^{(3)}(\text{Gaussian})$$

$$= 1.40 \times 10^{-8}\chi^{(3)}(\text{Gaussian}) \tag{B.17}$$

Selected linear and nonlinear optical units conversions are summarized in Table B.1.

The widely used units of wavelength are nanometers (nm) and micrometers (μm). [1]

$$1 \text{ nm} = 10^{-3} \text{ } \mu m = 10^{-9} \text{ m,}$$

$$1 \text{ } \mu m = 10^3 \text{ nm} = 10^{-6} \text{ m,}$$

$$1 \text{ A} = 10^{-1} \text{ nm} = 10^{-4} \text{ } \mu m = 10^{-10} \text{m.}$$

Commonly used units (spectral parameters) in optical spectroscopy are wavelength (λ), wavenumber ($\tilde{\nu}$), frequency (ν), and

---

[1] Note that the common (but incorrect) version of the μm units, is microns.

Optical Units Conversions

Table B.2 Spectral parameter conversion factors

	λ (nm)	$\tilde{\nu}$ (cm^{-1})	ν (Hz)	$h\nu$ (eV)
λ (nm)	1	$10^7/\tilde{\nu}$	$3 \times 10^{17}/\nu$	$1240/h\nu$
$\tilde{\nu}$ (cm^{-1})	$10^7/\lambda$	1	$3.3 \times 10^{-11}\nu$	$8065.54 \times h\nu$
ν (Hz)	$3 \times 10^{17}/\lambda$	$3 \times 10^{10}\tilde{\nu}$	1	$2.42 \times 10^{14}h\nu$
$h\nu$ (eV)	$1240/\lambda$	$1.24 \times 10^{-4}\tilde{\nu}$	$4.1 \times 10^{-15}\nu$	1

photon energy ($h\nu$). Spectral parameters conversion factors are given in Table B.2.

From Table B.2 follow the useful units conversions widely used in optical spectroscopy:

$$1 \text{ eV } (\hbar\omega) = 1.60219 \times 10^{-19} \text{ J} = 8.065 \times 10^3 \ cm^{-1};$$

the conversion between photon energy (in eV) and wavelength (in µm) is given by

$$\lambda(\mu\text{m}) = \frac{1.24}{\hbar\omega(\ eV)}$$

The wavenumbers ($\tilde{\nu}$) and wavelength (λ) are related to each other by the following formula, $\tilde{\nu}(\text{cm}^{-1}) = 1/\lambda(\text{cm})$. The frequently used relation between wavelength (in µm) and wavenumbers (in cm^{-1}) is given by:

$$\lambda(\mu\text{m}) = \frac{10^4}{\tilde{\nu}(\text{cm}^{-1})}$$

The conversion factors of wavenumbers to other energy units are given by:

$$1 \text{ cm}^{-1}(\tilde{\nu}) = 1.98646 \times 10^{-23} \text{ J} = 1.24 \times 10^{-4} \text{ eV} = 3 \times 10^{10} \text{ Hz}$$
(1/s).

Appendix C

The Vector Triple Product

Consider vector $\mathbf{g}$ defined as $\mathbf{g} = \mathbf{b} \times \mathbf{c}$. Vector $\mathbf{g}$ is perpendicular to the plane in which vectors $\mathbf{b}$ and $\mathbf{c}$ are lying. Another vector (e.g. $\mathbf{d}$) defined as a vector product of any third vector (e.g. $\mathbf{a}$) and vector $\mathbf{g}$ (i.e. $\mathbf{d} = \mathbf{a} \times \mathbf{g}$) will be oriented perpendicular to $\mathbf{g}$ and therefore lying in the plane of vectors $\mathbf{b}$ and $\mathbf{c}$. Therefore, the vector $\mathbf{d}$ can be expressed as a linear combination of the vectors $\mathbf{b}$ and $\mathbf{c}$, i.e.

$$\mathbf{d} = m\mathbf{b} + n\mathbf{c} \qquad (C.1)$$

Vector $\mathbf{d}$ is perpendicular to $\mathbf{a}$ thus $(\mathbf{a} \cdot \mathbf{d}) = 0$. Therefore taking the scalar product of the both sides of equation (C.1) one obtains

$$m(\mathbf{a} \cdot \mathbf{b}) + n(\mathbf{a} \cdot \mathbf{c}) = 0. \qquad (C.2)$$

Requirement for the equation (C.2) to be valid for any vectors $\mathbf{a}$, $\mathbf{b}$, and $\mathbf{c}$ results in the following condition:

$$m = \lambda(\mathbf{a} \cdot \mathbf{c}),$$
$$n = -\lambda(\mathbf{a} \cdot \mathbf{b}), \qquad (C.3)$$

The Vector Triple Product

where λ is a proportionality constant.
Therefore one has the following

$$\mathbf{d} = \mathbf{a} \times \mathbf{g}$$

$$= \mathbf{a} \times \mathbf{b} \times \mathbf{c} \tag{C.4}$$

$$= \lambda \left[(\mathbf{a} \cdot \mathbf{c}) \, \mathbf{b} - (\mathbf{a} \cdot \mathbf{b}) \, \mathbf{c} \right]$$

Vectors $\mathbf{a}$, $\mathbf{b}$, and $\mathbf{c}$ can be selected arbitrarily, e.g. $\mathbf{a} = \mathbf{x}$, $\mathbf{b} = \mathbf{y}$, and $\mathbf{c} = \mathbf{x}$ (where $\mathbf{x}$ and $\mathbf{y}$ are mutually perpendicular unit vectors) that immediately results in $\lambda = 1$. Consequently, one eventually obtains the following vector identity

$$\mathbf{a} \times \mathbf{b} \times \mathbf{c} = (\mathbf{a} \cdot \mathbf{c}) \, \mathbf{b} - (\mathbf{a} \cdot \mathbf{b}) \, \mathbf{c} \tag{C.5}$$

Appendix D

Mueller versus Jones Vectors and Matrices

In Table D.1 Jones and Stokes vectors are given for different states of polarization for comparison.

346 | *Mueller versus Jones Vectors and Matrices*

Table D.1 Jones and Stokes vectors for selected states of polarization

State of polarization	Jones vector	Stokes vector
Horizontal	$\begin{bmatrix} 1 \\ 0 \end{bmatrix}$	$\begin{bmatrix} 1 \\ 1 \\ 0 \\ 0 \end{bmatrix}$
Vertical	$\begin{bmatrix} 0 \\ 1 \end{bmatrix}$	$\begin{bmatrix} 1 \\ -1 \\ 0 \\ 0 \end{bmatrix}$
$P-$state at $+45°$	$\frac{1}{\sqrt{2}} \begin{bmatrix} 1 \\ 1 \end{bmatrix}$	$\begin{bmatrix} 1 \\ 0 \\ 1 \\ 0 \end{bmatrix}$
$P-$state at $-45°$	$\frac{1}{\sqrt{2}} \begin{bmatrix} 1 \\ -1 \end{bmatrix}$	$\begin{bmatrix} 1 \\ 0 \\ -1 \\ 0 \end{bmatrix}$
Right circular	$\frac{1}{\sqrt{2}} \begin{bmatrix} 1 \\ -i \end{bmatrix}$	$\begin{bmatrix} 1 \\ 0 \\ 0 \\ 1 \end{bmatrix}$
Left circular	$\frac{1}{\sqrt{2}} \begin{bmatrix} 1 \\ i \end{bmatrix}$	$\begin{bmatrix} 1 \\ 0 \\ 0 \\ -1 \end{bmatrix}$

In Table D.2 Jones and Mueller matrices are given for different linear optical elements for comparison. With Stokes vectors and Mueller matrices, one can describe light with arbitrarily combinations of polarized and unpolarized light.

Table D.2 Jones and Mueller matrices for selected optical elements

Optical element	Jones matix	Mueller matrix
Horizontal linear polarizer	$\begin{bmatrix} 1 & 0 \\ 0 & 0 \end{bmatrix}$	$\frac{1}{2}\begin{bmatrix} 1 & 1 & 0 & 0 \\ 1 & 1 & 0 & 0 \\ 0 & 0 & 0 & 0 \\ 0 & 0 & 0 & 0 \end{bmatrix}$
Vertical linear polarizer	$\begin{bmatrix} 0 & 0 \\ 0 & 1 \end{bmatrix}$	$\frac{1}{2}\begin{bmatrix} 1 & -1 & 0 & 0 \\ -1 & 1 & 0 & 0 \\ 0 & 0 & 0 & 0 \\ 0 & 0 & 0 & 0 \end{bmatrix}$
Linear polarizer at $+45°$	$\frac{1}{2}\begin{bmatrix} 1 & 1 \\ 1 & 1 \end{bmatrix}$	$\frac{1}{2}\begin{bmatrix} 1 & 0 & 1 & 0 \\ 0 & 0 & 0 & 0 \\ 1 & 0 & 1 & 0 \\ 0 & 0 & 0 & 0 \end{bmatrix}$
Linear polarizer at $-45°$	$\frac{1}{2}\begin{bmatrix} 1 & -1 \\ -1 & 1 \end{bmatrix}$	$\frac{1}{2}\begin{bmatrix} 1 & 0 & -1 & 0 \\ 0 & 0 & 0 & 0 \\ -1 & 0 & 1 & 0 \\ 0 & 0 & 0 & 0 \end{bmatrix}$
Quater-wave plate with vertical fast axis	$e^{i\pi/4}\begin{bmatrix} 1 & 0 \\ 0 & -i \end{bmatrix}$	$\begin{bmatrix} 1 & 0 & 0 & 0 \\ 0 & 1 & 0 & 0 \\ 0 & 0 & 0 & -1 \\ 0 & 0 & 1 & 0 \end{bmatrix}$
Quatcr wave plate with horizontal fast axis	$e^{i\pi/4}\begin{bmatrix} 1 & 0 \\ 0 & i \end{bmatrix}$	$\begin{bmatrix} 1 & 0 & 0 & 0 \\ 0 & 1 & 0 & 0 \\ 0 & 0 & 0 & 1 \\ 0 & 0 & -1 & 0 \end{bmatrix}$
Homogeneous right circular polarizer	$\frac{1}{2}\begin{bmatrix} 1 & i \\ -i & 1 \end{bmatrix}$	$\frac{1}{2}\begin{bmatrix} 1 & 0 & 0 & 1 \\ 0 & 0 & 0 & 0 \\ 0 & 0 & 0 & 0 \\ 1 & 0 & 0 & 1 \end{bmatrix}$
Homogeneous left circular polarizer	$\frac{1}{2}\begin{bmatrix} 1 & -i \\ i & 1 \end{bmatrix}$	$\frac{1}{2}\begin{bmatrix} 1 & 0 & 0 & -1 \\ 0 & 0 & 0 & 0 \\ 0 & 0 & 0 & 0 \\ -1 & 0 & 0 & 1 \end{bmatrix}$

Bibliography

Adachi, S. (1999). *Optical Constants of Crystalline and Amorphous Semiconductors: Numerical Data and Graphical Information* (Kluwer Academic Publishers, Norwell, MA).

Agranovich, V. M., Shen, Y. R., Baughman, R. H. and Zakhidov, A. A. (2004). Linear and nonlinear wave propagation in negative refraction metamaterials, *Phys. Rev. B* **69**, p. 165112.

Akhmanov, S. A. and Khokhlov, R. V. (1972). *Problems of nonlinear optics: Electromagnetic Waves in Nonlinear Dispersive Media* (Gordon and Breach Science Publishers, New York).

Alivisatos, A. P. (1996). Semiconductor clusters, nanocrystals, and quantum dots, *Science* **271**, p. 933.

Alivisatos, A. P., Gu, W. and Larabell, C. (2005). Quantum dots as cellular probes. *Annu. Rev. Biomed. Eng.* **7**, p. 55.

Alu, A. and Engheta, N. (2005). Achieving transparency with plasmonic and metamaterial coatings, *Phy. Rev. E* **72**, p. 016623.

Amerov, A., Ovechko, V. and Strizhevski, V. (1989). Optical damage plasma in liquids under influence of picocesond pulses as an laser active environment, *Acad. Sci. USSR Izvestiya* **53**, p. 714.

Arici, E., Sariciftci, N. S. and Meissner, D. (2003). Hybrid solar cells based on nanoparticles of $CuInS_2$ on organic matrices, *Adv. Func. Mat.* **13**, p. 165.

Ashcroft, N. W. and Mermin, N. D. (1976). *Solid State Physics* (Thomson Learning).

Aspnes, D. E. and Studna, A. A. (1983). Dielectric functions and optical parameters of Si, Ge, GaP, GaAs, GaSb, InP, InAs, and InSb from 1.5 to 6.0 ev, *Phys. Rev. B* **27**, p. 985.

Avramenko, V. G., Dolgova, T. V., Nikulin, A. A., Fedyanin, A. A., Aktsipetrov, O. A., Pudonin, A. F., Sutyrin, A. G., Prohorov, D. Y. and Lomov, A. A. (2006). Subnanometer-scale size effects in electronic spectra of Si/SiO_2 multiple quantum wells: Interferometric second-harmonic generation spectroscopy, *Phys. Rev. B* **73**, p. 155231.

Bains, G., Patel, A. B. and Narayanaswami, V. (2011). Pyrene: A probe to study protein conformation and conformational changes, *Molecules* **16**, p. 7909.

Barford, W. (2005). *Electronic and Optical Properties of Conjugated Polymers* (Clarendon Press, Oxford).

Bassani, F. and Parravicini, G. P. (1967). Band structure and optical properties of graphite and of the layer compounds GaS and GaSe, *Nuovo Cimento* **B50**, p. 95.

Bechstedt, F. (2003). *Principles of Surface Physics* (Springer, Berlin, Heidelberg).

Benten, W., Nilius, N., Ernst, N. and Freud, H.-J. (2005). *Phys. Rev. B* **72**, p. 045403.

Bergmann, M. J. and Casey, H. C. (1998). Optical-field calculations for lossy multiple-layer $Al_xGa_{1-x}N/In_xGa_{1-x}N$ laser diodes, *J. Appl. Phys.* **84**, p. 1196.

Bessonov, E. G. (1992). Conditionally-strange electromagnetic waves, *Quantum Electronics* **19**, p. 35.

Bloembergen, N. (1965). *Nonlinear Optics* (Benjamin, New York).

Boguslawski, P. and Bernholc, J. (2002). Surface segregation of Ge at SiGe(001) by concerted exchange pathways, *Phys. Rev. Lett.* **88**, p. 166101.

Borensztein, Y., Pluchery, O. and Witkowski, N. (2005). Probing the Si-Si dimer breaking of Si(100)2 $\times$ 1 surfaces upon molecule adsorption by optical spectroscopy, *Phys. Rev. Lett* **95**, p. 117402.

Born, M. and Wolf, E. (1999). *The Principles of Optics*, 7th edn. (Cambridge University Press).

Bosman, A. J. and Havanga, E. E. (1963). Temperature dependence of dielectric constant of cubic ionic compounds, *Phys. Rev.* **129**, p. 1593.

Boyd, R. W. (1992). *Nonlinear Optics* (Academic Press).

Bredas, J.-L., Cornil, J., Beljonne, D., dos Santos, D. A. and Shuai, Z. (1999). Excited-state electronic structure of conjugated oligomers and polymers: A quantum-chemical approach to optical phenomena, *Acc. Chem. Res.* **32**, p. 267.

Cai, W., Chettiar, U. K., Kildishev, A. V. and Shalaev, V. M. (2007). Optical cloaking with metamaterials, *Nature Photonics* **1**, p. 224.

Cardona, M. (1982). Resonance phenomena, in M. Cardona and G. Güntherodt (eds.), *Light Scattering in Solids II*, Part of the Topics in *Applied Physics* book series, Vol. 50 (Springer, Berlin, Heidelberg), p. 19.

Cardona, M., Grimsditch, M. and Olego, D. (1979). Theoretical and experimental determination of Raman scattering cross sections in simple solids, in J. L. Birman, H. Z. Cummins and K. K. Rebane (eds.), *Light Scattering in Solids* (Plenum, New York), p. 249.

Ceperley, D. M. and Adler, B. J. (1980). Ground state of electron gas by a stochastic method, *Phys. Rev. Lett.* **45**, pp. 566–569.

Chen, C., Liu, J., Yu, B. and Dai, Q. (2007). Determination of boron concentration in heavily doped p-type $Si_{1-x}Ge_x/Si$ heterostructure by infrared ellipsometric spectroscopy, *Microelectr. J.* **38**, p. 392.

Chen, W., Westhoff, R. and Reif, R. (1997). Determination of optical constants of strained $Si_{1-x}Ge_x$ epitaxial layers in the spectral range 0.75-2.75 ev, *Appl. Phys. Lett.* **71**, p. 1525.

Cobet, C., Esser, N., Zettler, J. T., Richter, W., Waltereit, P., Brandt, O., Ploog, K. H., Peters, S., Edwards, N. V., Lindquist, O. P. A. and Cardona, M. (2001). Optical properties of wurtzite $Al_xGa_{1-x}N$ (x<.1) parallel and perpendicular to the c axis, *Phys. Rev. B* **64**, p. 165203.

Cooper, M. (2002). Optical biosensors in drug discovery, *Nature Review. Drug Discovery* **515**, p. 1.

Cörekci, S., Öztürk, M. K., Akaoglu, B., Cakmak, M., Özcelik, S. and Özbay, E. (2007). Structural, morphological, and optical properties of AlGaN/GaN heterostructures with aln buffer and interlayer, *J. Appl. Phys.* **101**, p. 123502.

Davydov, A. S. (1976). *Quantum mechanics*, 2nd edn. (Pergamon Press, New York).

Davydov, A. S. (1980). *Solid State Theory* (Academic Press, New York).

Del Sole, R. (1995). Reflectance spectroscopy - Theory, in P. Halevi (ed.), *Photonic probes of surfaces* (Elsevier, Amsterdam), p. 131.

Depine, R. A. and Lakhtakia, A. (2004) A new condition to identify isotropic dielectric-magnetic materials displaying negative phase velocity. *Microwave Opt. Technol. Lett.* **41**, p. 315.

Dmytruk, A., Fursenko, O., Lepeshkina, O. and Ovechko, V. (2001). Ellipsometery and spectroscopy of porous glass surface, *Vacuum* **61**, p. 123.

Dmytruk, A., Mulenko, S., Ovechko, V. and Pogorelyi, A. (1996). Laser deposition of thin metal film from gas phase under optical control, *Met. Phys. Adv. Techn.* **18**, p. 32.

Dmytruk, A., Mygashko, V., Mulenko, S. and Ovechko, V. (2000). Near IR spectroscopy of porous glass, *Vibr. Spec.* **22**, p. 87.

Dmytruk, A. and Ovechko, V. (1995). Nonlinear and interferometric optical method of diagnostics of semiconductors, *Proc. SPIE* **2648**, p. 239.

Bibliography

Dmytruk, A. and Ovechko, V. (2003). Spectroscopic study of physical adsorption in porous glass, *Surf. Rev. Let.* **10**, p. 289.

Doni, E. and Parravicini, G. P. (1969). Energy bands and optical properties of hexagonal boron nitride and graphite. *Nuovo Cimento* **64**, p. 117.

Downer, M. C., Mendoza, B. S. and Gavrilenko, V. I. (2001). Optical second harmonic spectroscopy of semiconductor surfaces: advances in microscopic understanding, *Surf. Interf. Anal.* **31**, pp. 966–986.

Dresselhaus, M. S., Dresselhaus, G., Saito, R. and Jorio, A. (2007). Exciton photophysics of carbon nanotubes, *Annu. Rev. Phys. Chem.* **58**, p. 719.

Drummen, G. P. C. (2012). Fluorescent probes and fluorescence (microscopy) techniques—illuminating biological and biomedical research, *Molecules* **17**, p. 14067.

Einstein, A. (1905). Zur Elektrodynamik bewegter Körper, *Ann. Physik* **17**, p. 891.

Emory, S. R., Haskins, W. E. and Nie, S. (1998). Direct observation of size-dependent optical enhancement in single metal nanoparticles, *J. Am. Chem. Soc.* **120**, p. 8009.

Erley, G. and Daum, W. (1998). Silicon interband transitions observed at $Si(100) - SiO_2$ interfaces, *Phys. Rev. B* **58**, p. R1734.

Feynman, R. P., Metropolis, N. and Teller, E. (1949). Equations of state of elements based on the generalized Thomas-Fermi theory, *Phys. Rev.* **75**, pp. 1561–1573.

Fischetti, M. V. and Laux, S. E. (1996). Band structure, deformation potentials, and carrier mobility in strained Si, Ge, and SiGe alloys, *J. Appl. Phys.* **80**, p. 2234.

Förster, T. (1969). Excimers, *Angew. Chem.* **8**, p. 333.

Foss, C. A., Hornyak, G. L., Stockert, J. A. and Martin, C. R. (1994). Template-synthesized nanoscopic gold particles: Optical spectra and the effects of particle size and shape, *J. Phys. Chem.* **98**, p. 2963.

Fox, M. (2003). *Optical Properties of Solids*, 2nd edn. (Oxford University Press, New York).

Fuchs, F., Schmidt, W. G. and Bechstedt, F. (2005). Understanding the optical anisotropy of oxidized $Si(001)$ surfaces, *Phys. Rev. B* **72**, p. 075353.

Fursenko, O., Bauer, J., Zaumseil, P., Yamamoto, Y. and Tillack, B. (2008). Doping concentration control of SiGe layers by spectroscopic ellipsometry, *Thin Sol. Films* **517**, p. 259.

Gavrilenko, A. V., Black, S. M., Sykes, A. C., Bonner, C. E. and Gavrilenko, V. I. (2008). Computations of ground state and excitation energies of poly(3-methoxy-thiophene) and poly(thienylene vinylene) from first principles, in M. B. et al. (ed.), *Lecture Notes in Computational Science, ICCS LNCS 5102*, Vol. Part II (Springer), p. 396.

Gavrilenko, A. V., McKinney, C. S. and Gavrilenko, V. I. (2010). Effects of molecular adsorption on optical losses of the Ag (111) surface, *Phys. Rev. B* **82**, p. 155426.

Gavrilenko, V. I. (1987). Electronic structure and optical properties of polycrystalline cubic semiconductors, *Phys. Stat. Solidi (b)* **19**, p. 457.

Gavrilenko, V. I. (1993). Adsorption of hydrogen on the (001) surface of diamond, *Phys. Rev. B* **47**, p. 9556.

Gavrilenko, V. I. (2008). Differential reflectance and second-harmonic generation of the Si/SiO_2 interface from first principles, *Phys. Rev. B* **77**, p. 155311.

Gavrilenko, V. I. (2009). Optics of nanostructured materials from first principles, in M. A. Noginov, M. W. McCall, G. Dewar and N. I. Zheludev (eds.), *Tutorials in Complex Photonic Media*, Chap. 15 (SPIE Press, Bellingham), pp. 479–524.

Gavrilenko, V. I. (2020). *Optics of Nanomaterials*, 2nd edn. (Jenny Stanford Publishing, Singapore).

Gavrilenko, V. I. and Bechstedt, F. (1997). Optical functions of semiconductors beyond density functional theory, *Phys. Rev. B* **55**, p. 4343.

Gavrilenko, V. I., Frolov, S. I. and Pidlisnyj, E. V. (1990). Optical properties of graphite-like carbon films, *Thin Solid Films* **190**, p. 255.

Gavrilenko, V. I. and Noginov, M. A. (2006). Ab initio study of optical properties of Rhodamine 6G molecular dimers, *J. Chem. Phys.* **124**, p. 44301.

Gavrilenko, V. I. and Wu, R. Q. (2000). Linear and nonlinear optical properties of group-III nitrides, *Phys. Rev. B* **61**, p. 2632.

Gavrilenko, V. I. and Wu, R. Q. (2002). Second harmonic generation of GaN(0001), *Phys. Rev. B* **65**, p. 035405.

Gavrilenko, V. I., Wu, R. Q., Downer, M. C., Ekerdt, J. G., Lim, D. and Parkinson, P. (2001). Optical second-harmonic spectra of Si(001) with H and Ge adatoms: First-principles theory and experiment, *Phys. Rev. B* **63**, p. 165325.

Gebeyehua, D., Brabec, C. J., Sariciftcia, N. S., Vangeneugdenb, D., Kieboomsb, R., Vanderzandeb, D., Kienbergerc, F. and Schindler, H. (2001). Hybrid solar cells based on dye-sensitized nanoporous TiO_2 electrodes and conjugated polymers as hole transport materials, *Synth. Met.* **125**, p. 279.

Gemeda, F. T. (2017). A review on effect of solvents on fluorescent spectra, *Chem. Sci. Int. J.* **18**, p. 1.

Glinka, Y. D., Lin, S.-H. and Chen, Y.-T. (2002). Time-resolved photoluminescence study of silica nanoparticles as compared to bulk type-III fused silica, *Phys. Rev. B* **66**, p. 035404.

Gray, H. B. (1965). *Electrons and Chemical Bonding* (W. A. Benjamin Inc).

Günes, S., Neugebauer, H., Sariciftci, N. S., Roither, J., Kovalenko, M., Pillwein, G. and Heiss, W. (2006). Hybrid solar cells using *HgTe* nanocrystals and nanoporous TiO_2 electrodes, *Adv. Func. Mat.* **16**, p. 1095.

Harrison, W. A. (1989). *Electronic Structure and the Properties of Solids: The Physics of the Chemical Bond* (World Scientific).

Hoa, X., Kirk, A. and Tabrizian, M. (2007). Towards integrated and sensitive surface plasmon resonance biosensors: A review of recent progress, *Biosens. Bioelectr.* **23**, p. 151.

Hofmann, P. (2008). *Solid State Physics* (RSC Publishing, Berlin).

Hohenberg, P. and Kohn, W. (1964). Inhomogeneous electron gas, *Phys. Rev.* **136**, pp. B864–B871.

Holder, E., Tesslerb, N. and Rogach, A. L. (2008). Hybrid nanocomposite materials with organic and inorganic components for opto-electronic devices, *J. Mater. Chem.* **18**, p. 1064.

Hornyak, G. L., Patrissi, C. J. and Martin, C. R. (1997). Fabrication, characterization, and optical properties of gold nanoparticle/porous alumina composites: The nonscattering Maxwell-Garnett limit, *J. Phys. Chem. B* **101**, p. 1548.

Huang, S.-P., Wu, D.-S., Hu, J.-M., Zhang, H., Xie, Z., Hu, H. and Cheng, W.-D. (2007). First-principles study: size-dependent optical properties for semiconducting silicon carbide nanotubes, *Optics Express* **15**, p. 10947.

Hughes, J. L. P. and Sipe, J. E. (1996). Calculation of second-order optical response in semiconductors, *Phys. Rev. B* **53**, p. 10751.

Humliček, J., Garriga, M., Alonso, M. I. and Cardona, M. (1989). Optical spectra of $Si_{1-x}Ge_x$ alloys, *J. Appl. Phys.* **65**, p. 2827.

Huynh, W. U., Dittmer, J. J. and Alivisatos, A. P. (2002). Hybrid nanorod-polymer solar cells, *Science* **295**, p. 2425.

Ilchenko, L. G., Ilchenko, V. V., Gavrilenko, A. V. and Gavrilenko, V. I. (2013). Realistic electric field modeling of multilayered nanostructures by classic electrodynamics and first principles theory, in A. D. Boardman, N. Engheta, M. A. Noginov and N. I. Zheludev (eds.), *Proceeding of SPIE* (SPIE Press), p. 8806.

Jackson, J. D. (1998). *Classical Electrodynamics*, 3rd edn. (Wiley, New York).

Kasai, H., Oikawa, H. and Nakanishi, H. (2000). Chemistry for the 21st century, in H. Masuhara and F. C. DeSchryver (eds.), *Organic Mesoscopic Chemistry* (Blackwell Science, Oxford), p. 145.

Kasha, M. (1959). Relation between exciton bands and conduction bands in molecular lamellar systems, *Rev. Mod. Phys.* **31**, p. 162.

Kawashima, T., Yoshikawa, H., Adachi, S., Fuke, S. and Ohtsuka, K. (1997). Optical properties of hexagonal GaN, *J. Appl. Phys.* **82**, p. 3528.

Kelso, S. M., Aspnes, D. E., Pollack, M. A. and Nahory, R. E. (1982). Optical properties of $In_{1-x}Ga_xAs_yP_{1-y}$ from 1.5 to 6.0 ev determined by spectroscopic ellipsometry, *Phys. Rev. B* **26**, p. 6669.

Kiefer, W. (2008). Recent advances in linear and nonlinear Raman spectroscopy, *J. Raman Spectr.* **39**, p. 1710.

Kikteva, T., Star, D., Zhao, Z., Baislev, T. L. and Leach, G. W. (1999). Molecular orientation, aggregation, and order in rhodamine films at the fused silica/air interface, *J. Phys. Chem. B* **103**, p. 1124.

Kildishev, A. V. and Shalaev, V. M. (2006). Negative refractive index in optics of metal-dielectric composites. *J. Opt. Soc. Am. B* **23**, p. 423.

Kim, C. C., Garland, J. W., Abad, H. and Raccah, P. M. (1992). Modeling the optical dielectric function of semiconductors: Extension of the critical-point parabolic-band approximation, *Phys. Rev. B* **45**, p. 11749.

Kinoshita, T. (1996). The fine structure constant, *Rep. Prog. Phys.* **59**, p. 1459.

Kiselev, V. F. and Krylov, O. V. (1985). *Adsorption Processes on Semiconductor and Dielectric Surfaces* (Springer, Berlin, Heidelberg).

Kittel, C. (1986). *Introduction to Solid State Physics*, 6th edn. (Wiley, New York).

Kneipp, K., Kneipp, H., Itzkan, I., Dasari, R. R. and Feld, M. S. (1999). Ultrasensitive chemical analysis by Raman spectroscopy, *Chem. Rev.* **99**, p. 2957.

Kneipp, K., Wang, Y., Kneipp, H., Perelman, L. T., Itzkan, I., Dasari, R. R. and Feld, M. S. (1997). Single molecule detection using Surface-Enhanced Raman Scattering (SERS), *Phys. Rev. Lett.* **78**, p. 1667.

Kohn, W. (1999). Electronic structure of matter - wave functions and density functionals, *Rev. Mod. Phys.* **71**, pp. 1253–1266.

Kohn, W. and Sham, L. J. (1965). Self-consistent equations including exchange and correlation effects, *Phys. Rev.* **140**, p. A1133.

Kolasinski, K. W. (2008). *Surface Science: Foundation of Catalysis Nanoscience*, 2nd edn. (Wiley, New York).

Kreibig, U. and Vollmer, M. (1995). *Optical Properties of Metal Clusters* (Springer Series in Material Science 25, Springer, Berlin).

356 | *Bibliography*

Kulzer, F. and Orrit, M. (2004). Single-molecule optics, *Ann. Rev. Phys. Chem.* **55**, p. 585.

Lakowicz, J. R. (2006). *Principles of Fluorescence Spectroscopy*, 3rd edn. (Springer).

Lamb, H. (1904). On group-velocity, *Proc. Lond. Math. Soc.* **1**, p. 473.

Landau, L. D. and Lifshits, E. M. (1980). *Quantum Mechanics* (Academic Press, New York).

Lebedenko, A. N., Guralchuk, G. Y., Sorokin, A. V., Yefimova, S. L. and Malyukin, Y. V. (2006). Pseudoisocyanine J-aggregate to optical waveguiding crystallite transition: Microscopic and microspectroscopic exploration, *J. Phys. Chem. B* **110**, p. 17772.

Levy, O. and Stroud, D. (1997). Maxwell-Garnett theory for mixtures of anisotropic inclusions: Application to conducting polymers, *Phys. Rev. B* **56**, p. 8035.

Li, Y., Cai, W. and Duan, G. (2008). Ordered micro/nanostructured arrays based on the monolayer colloidal crystals, *Chem. Mater.* **20**, p. 615.

Liebsch, A. (1997). *Electronic Excitations at Metal Surfaces* (Plenum Press, New York).

Liedberg, B., Nylander, C. and Lundström, I. (1995). Biosensing with surface plasmon resonance - how it all started, *Biosensors & Bioelectronics* **10**, p. i.

Lim, D., Downer, M. C., Ekerdt, J. G., Arzate, N., Mendoza, B. S., Gavrilenko, V. I. and Wu, R. Q. (2000). Optical second harmonic spectroscopy of boron-reconstructed $si(001)$, *Phys. Rev. Lett.* **84**, p. 3406.

Link, S. and El-Sayed, M. A. (2000). Shape and size dependence of radiative, non-radiative and photothermal properties of gold nanocrystals, *Int. Rev. Phys. Chem.* **19**, p. 409.

Logothetidis, S., Alouani, M., Garriga, M. and Cardona, M. (1990). E_2 interband transitions in $Al_xGa_{1-x}N$ alloys, *Phys. Rev. B* **41**, p. 2959.

Logothetidis, S., Petalas, J., Cardona, M. and Moustakas, T. D. (1994). Optical properties and temperature dependence of the interband transitions of cubic and hexagonal GaN, *Phys. Rev. B* **50**, p. 18017.

Lü, C., Guan, C., Liu, Y., Cheng, Y. and Yang, B. (2005). *PbS*/polymer nanocomposite optical materials with high refractive index, *Chem. Mat.* **17**, p. 2448.

Malinsky, M. D., Kelly, K. L., Schatz, G. C. and Duyne, R. P. V. (2001). Nanosphere lithography: Effect of substrate on the localized surface plasmon resonance spectrum of silver nanoparticles, *J. Chem. Phys. B* **105**, p. 2343.

Mandelstam, L. I. (1945). Group-velocity in a crystal lattice, *Zh. Eksp. Teor. Fiz.* **15**, p. 475.

Manghi, F., Sole, R. D., Selloni, A. and Molonari, E. (1990). Anisotropy of surface optical properties from first-principles calculations, *Phys. Rev. B* **41**, p. 9935.

Martin, R. M. (2004). *Electronic Structure. Basic Theory and Practical Methods* (Cambridge University Press, New York).

Martin, R. M. (2005). *Electronic Structure. Basic theory and practical methods*, 2nd edn. (Cambridge university press).

Martinez, V. M., Arbeloa, F. L., Prieto, J. B., Lopez, T. A. and Arbeloa, I. L. (2004). Characterization of rhodamine 6G aggregates intercalated in solid thin films of laponite clay. 1. Absorption spectroscopy, *J. Phys. Chem. B* **108**, p. 20030.

Maxwell-Garnett, J. C. (1904). Colours in metal glasses and metal films, *Philos. Trans. R. Soc. London* **3**, p. 385.

McCall, M. W., Lakhtakia, A. and Weiglhofer, W. S. (2002). The negative index of refraction demystified. *Eur. J. Phys.* **23**, p. 353.

McDonald, S. A., Konstantatos, G., Zhang, S., Cyr, P. W., Klem, E. J. D., Levina, L. and Sargent, E. H. (2005). Solution-processed *PbS* quantum dot infrared photodetectors and photovoltaics, *Nature Materials* **4**, p. 138.

McQuarrie, D. A. and Simon, J. D. (1997). *Physical Chemistry: A Molecular Approach* (University Science Books).

Molebny, V., Ovechko, V. and Strizhevski, V. (1974). Point monochromatic sources visualization by means of nonlinear optics, *Quant. Electr.* **1**, p. 2328.

Mousseau, N. and Thorpe, M. F. (1993). Structural model for crystalline and amorphous Si-Ge alloys, *Phys. Rev. B* **48**, p. 5142.

Murzina, T. V., Maydykovskiy, A I, Gavrilenko, A. V. and Gavrilenko, V. I. (2012). Optical second harmonic generation in semiconductor nanostructures, *Physics Research International* **2012**, p. 836430.

Nazzeruddin, M. K., Kay, A., Rodicio, I., Humphry-Baker, R., Müller, E., Liska, R., Vlachopoulos, N. and Grätzel, M. (1993). Conversion of light to electricity by $cis - X_2\text{Bis}(2, 2' - \text{bipyridyl} - 4, 4' - \text{dicarboxylate})$ruthenium (II) charge-transfer sensitizers ($X = Cl^-, Br^-, I^-, CN^-$, and SCN^-) on nanocrystalline TiO_2 electrodes, *J. Am. Chem. Soc.* **115**, p. 6382.

Nuss, M. C. and Orenstein, J. (1998). Terahertz time-domain spectroscopy, in G. Grüner (ed.), *Millimeter and Submillimeter Wave Spectroscopy of Solids*, *Topics Appl. Phys.*, Vol. 74 (Springer, Berlin), p. 19.

Okamoto, A., Kanatani, K. and Saito, I. (2004). Pyrene-labeled base-discriminating fluorescent dna probes for homogeneous snp typing, *JACS* **126**, p. 4820.

Onida, G., Reining, L. and Rubio, A. (2002). Electronic excitations: density functional versus many-body Green's-function approach, *Rev. Mod. Phys.* **74**, pp. 601–656.

Onida, G., Schmidt, W. G., Pulci, O., Palummo, M., Marini, A., Hogan, C. and Sole, R. D. (2001). Theory for modeling the optical properties of surfaces, *Phys. Stat. Sol (a)* **188**, p. 1233.

Ovechko, V. (2012). Femtosecond optics – optics of the elementary wave packets, *J. Opt. Soc. Am. B* **29**, p. 799.

Ovechko, V. (2017). Femtosecond optical pulse propagation through the single resonance lorentz model dielectric, *Int. J. Adv. Res. Phys. Sci.* **4**, p. 28.

Ovechko, V. and Kharchenko, N. (2013). *Atomic Physics. Practical work in physics*, 2nd edn. (Kyiv University Publishing).

Ovechko, V., Mygashko, V. and Kornienko, A. (2015). Slowly varying amplitude approximation in optics, in *Proceeding of the XI International Conference: Electronics and Applied Physics* (Kyiv University Publishing, Kyiv, Ukraine), p. 20.

Ovechko, V. and Myhashko, V. (2018). Spectral particularities of femtosecond optical pulses. propagating in dispersive medium, *Ukr. J. Phys.* **63**, p. 479.

Ovechko, V. and Sheka, D. (2006). *Physics of Atoms and Atomic Structures: From Classical to Quantum* (Kyiv University Publishing).

Ovechko, V., Shur, A. and Myagashko, V. (2005). Optical properties of the porous glass composite material, *Optica Applicata* **38**, p. 75.

Ovechko, V. and Shur, O. (2005). Size spectroscopy of porous glass and porous glasses with metal nanoparticles using UV-VIS and X-ray radiation, *Optica Applicata* **35**, p. 735.

Ovechko, V. S. (2020). Kramers-Kronig relations-supplementary technique for the time-domain spectroscopy, *Ukrainian Physical Journal* **65**, p. 1051.

Palik, E. D. (ed.) (1985). *Optical constants of solids* (Academic Press).

Parazzoli, C., Greegor, R. B., Li, K., Koltenbah, B. E. and Tanielian, M. (2003). Experimental verification and simulation of negative index of refraction using Snell's law, *Phys. Rev. Lett.* **90**, p. 107401.

Parr, R. G. and Yang, W. (1989). *Density Functional Theory of Atoms and Molecules* (Oxford University Press).

Patra, A., Hebalkar, N., Sreedhar, B. and Radhakrishnan, T. P. (2007). Formation and growth of molecular nanocrystals probed by their optical properties, *J. Phys. Chem. C* **111**, p. 16184.

Paul, D. J. (1999). Silicon-germanium strained layer materials in microelectronics, *Adv. Mat.* **11**, p. 191.

Peeters, E., Ramos, A. M., Meskers, S. C. J. and Janssen, R. A. J. (2000). Singlet and triplet excitations of chiral dialkoxy-p-phenylene vinylene oligomers, *J. Chem. Phys.* **112**, p. 9445.

Pendry, J. B. (2000). Negative refraction makes a perfect lens, *Phys. Rev. Lett.* **85**, p. 3966.

Pendry, J. B. (2004). Negative refraction, *Cont. Phys.* **45**, p. 191.

Pendry, J. B., Schurig, D. and Smith, D. R. (2006). Controlling electromagnetic fields. *Science* **312**, p. 1780.

Perdew, J. P. and Wang, Y. (1992). Accurate and simple analytic representation of the electron-gas correlation energy, *Phys. Rev. B* **45**, pp. 13244–13249.

Pickering, C. and Carline, R. T. (1994). Dielectric function spectra of strained and relaxed $Si_{1-x}Ge_x$ alloys (x =0-0.25), *J. Appl. Phys.* **75**, p. 4642.

Pitarke, J. M., Silkin, V. M., Chulkov, E. V. and Echenique, P. M. (2007). Theory of surface plasmons and surface-plasmon polaritons, *Rep. Prog. Phys.* **70**, p. 1.

Podgorny, M., Wolfgarten, G. and Pollmann, J. (1986). The band structure of $Si_{1-x}Ge_x$ alloys: the self-consistent virtual-crystal approximation, *J. Phys. C* **19**, p. L141.

Pope, M. and Swenberg, C. E. (1982). *Electronic Processes in Organic Crystals* (Oxford University Press, New York).

Raether, H. (1988). *Surface Plasmons on Smooth and Rough Surfaces and on Gratings* (Springer Verlag, Berlin, Heidelberg, New York).

Ritchie, R. H. (1957). Plasma losses by fast electrons in thin films. *Phys. Rev.* **106**, p. 074.

Ritchie, R. H. (1973). Surface plasmons in solids, *Surf. Sci.* **34**, p. 1.

Rowell, N., Lafontaine, H. and Dion, M. (2002). Photoluminescence of boron-doped $Si_{1-x}Ge_x$ epilayers grown by uhv-cvd, *Mat. Sci. Eng. B* **89**, p. 141.

Rumpel, A., Manschwetus, B., Lilienkamp, G., Schmidt, H. and Daum, W. (2006). Polarity of space charge fields in second-harmonic generation spectra of $Si(100)/SiO_2$ interfaces, *Phys. Rev. B* **74**, p. 081303(R).

Saleh, B. E. A. and Teich, M. C. (2007). *Fundamentals of Photonics*, 2nd edn. (Wiley).

Sarychev, A. K., McPhedran, R. C. and Shalaev, V. M. (2000). Electrodynamics of metal-dielectric composites and electromagnetic crystals, *Phys. Rev. B* **62**, p. 8531.

Sarychev, A. K. and Shalaev, V. M. (2007). *Electrodynamics of Metamaterials* (World Scientific).

Sasai, R., Fujita, T., Iyi, N., Itoh, H. and Takagi, K. (2002). Aggregated structures of Rhodamine 6G intercalated in a fluor-taeniolite thin film, *Langmuir* **18**, p. 6578.

Schatz, G. C., Young, M. A. and Duyne, R. P. V. (2006). Electromagnetic mechanism of SERS, in K. Kneipp, H. Kneipp and M. Moskovits (eds.), *Surface Enhanced Raman Scattering. Physics and Applications, Topics Appl. Phys.*, Vol. 103 (Springer, Berlin, Heidelberg), p. 19.

Schiek, M., Balzer, F., Al-Shamery, K., Brewer, J. R., Lützen, A. and Rubahn, H.-G. (2008). Organic molecular nanotechnology, *Small* **2**, p. 176.

Schmidt, W. G. (1997). (4×2) and (2×4) reconstructions of GaAs and InP(001) surfaces, *Appl. Phys. A* **65**, p. 581.

Schmidt, W. G., Bechstedt, F., Fleischer, K., Cobet, C., Esser, N., Richter, W., Bernholc, J. and Onida, G. (2001). GaAs(001): Surface structure and optical properties, *Phys. Stat. Sol (a)* **188**, p. 1401.

Schmidtling, T., Pohl, U. W., Richter, W. and Peters, S. (2005). In *situ* spectroscopic ellipsometry study of GaN nucleation layer growth and annealing on sapphire in metal-organic vapor-phase epitaxy, *J. Appl. Phys.* **98**, p. 033522.

Schuster, A. (1904). *An Introduction to the Theory of Optics* (Arnold, London).

Shalaev, V. M. (2007). Optical negative-index metamaterials, *Nature Photonics* **1**, p. 41.

Shelby, R., Smith, D. R. and Schultz, S. (2001). Experimental verification of a negative index of refraction, *Science* **292**, p. 77.

Shen, Y. R. (2003). *The Principles of Nonlinear Optics*, 2nd edn. (Wiley).

Shvets, G. and Urzhumov, Y. A. (2006). Negative index meta-materials based on two-dimensional metallic structures. *J. Opt. A* **8**, p. S122.

Sieg, R. M., Alterovitz, S. A., Croke, E. T., Harrell, M. J., Tanner, M., Wang, K. L., Mena, R. A. and Young, P. G. (1993). Characterization of Si_xGe_{1-x}/Si heterostructures for device applications using spectroscopic ellipsometry, *J. Appl. Phys.* **74**, p. 586.

Silinsh, E. A. and Capek, V. (1994). *Organic Molecular Crystals: Interaction, Localization and Transport Phenomena* (AIP Press, New York).

Silveirinha, M. G., Alu, A. and Engheta, N. (2008). Cloaking mechanism with antiphase plasmonic satellites, *Phy. Rev. B* **78**, p. 205109.

Singleton, J. (2004). *Band Theory and Electronic Properties of Solids* (Oxford University Press, New York).

Sipe, J. E. and Boyd, R. W. (2002). Nanocomposite materials for nonlinear optics based on local field effect, in V. M. Shalaev (ed.), *Optical Properties of Nanostructured Random Media* (Springer, Berlin, Hedelberg), p. 1.

Sivukhin, D. V. (1957). The energy of electromagnetic waves in dispersive media. *Opt. Spektrosk* **3**, p. 308.

Slater, J. C. and Koster, G. F. (1954). Simplified LCAO method for the periodic potential problem, *Phys. Rev.* **94**, p. 1498.

Slonczewski, J. C. and Weiss, P. R. (1958). Band structure of graphite. *Phys. Rev.* **109**, p. 109.

Smith, D. R., Padilla, W. J., Vier, D. C., Nemat-Nasser, S. C. and Schultz, S. (2000). Composite medium with simultaneously negative permeability and permittivity, *Phys. Rev. Lett.* **84**, p. 4184.

Stroscio, M. A. and Dutta, M. (2001). *Phonons in Nanostructures* (Cambridge University Press).

Sun, F., Cai, W., Li, Y., Duan, G., Nichols, W. T., Liang, C., Koshizaki, N., Feng, Q. and Boyd, I. W. (2005). Laser morphological manipulation of gold nanoparticles periodically arranged on solid supports, *Appl. Phys. B* **81**, p. 765.

Sutton, A. P. (2004). *Electronic Structure of Materials* (Clarendon Press).

Tannor, D. J. (2007). *Introduction to Quantum Mechanics. A Time-Dependent Perspective* (University Science Books, Sausalito, California).

Tennakone, K., Perera, V. P. S., Kottegoda, I. R. M. and Kumara, G. R. R. A. (1999). Dye-sensitized solid state photovoltaic cell based on composite zinc oxide/tin (IV) oxide films, *J. Phys. D* **32**, p. 374.

Terry, F. L. (1991). *J. Appl. Phys.* **70**, p. 409.

Tserbak, C. and Theodorou, G. (1991). Optical transitions of infinite and finite strained Si/Ge superlattices, *J. Appl. Phys.* **76**, p. 1062.

Tudos, A. J. and Schasfoort, R. B. M. (2008). Introduction to surface plasmon resonance, in R. B. M. Schasfoort and A. J. Tudos (eds.), *Handbook of Surface Plasmon Resonance* (RCS Publishing, Cambridge), p. 1.

van den Brink, J., Brocks, G. and Morpurgo, A. F. (2005). Electronic correlations in oligo-thiophene molecular crystals, *J. Mag. Mag. Mat.* **290-291**, p. 294.

Venger, E. F., Goncharenko, A. V. and Dmitruk, M. L. (1999). *Optics of Small Particles and Disperse Media* (Naukova Dumka, Kiev).

Veselago, V. G. (1968). The electrodynamics of substances with simultaneously negative values of ε and μ, *Sov. Phys. Uspekhi* **10**, p. 509.

Volkov, V. V., Asahi, T., Masuhara, H., Masuhara, A., Kasai, H., Oikawa, H. and Nakanishi, H. (2004). Size-dependent optical properties of polydiacetylene nanocrystal, *J. Phys. Chem. B* **108**, p. 7674.

Vysotskii, V., Dyachenko, S., Karlash, A., Ovechko, V., Prokopenko, O. and Kharchenko, N. (2011). Atomic and nuclear physics by examples and problems, in V. Vysotskii and V. Ovechko (eds.), *VPC Polygraph Publishing Center* (Kyiv University Publishing, Kyiv, Ukraine), p. 511.

Wallace, P. R. (1947). The band theory of graphite, *Phys. Rev.* **71**, p. 622.

Wang, F., Dukovic, G., Brus, L. E. and Heinz, T. F. (2005). The optical resonances in carbon nanotubes arise from excitons, *Science* **308**, p. 838.

Wood, R. W. (1902). On a remarkable case of uneven distribution of light in a diffraction grating spectrum, *Phil. Mag.* **4**, p. 396.

Wood, R. W. (1935). Anomalous diffraction gratings, *Phys. Rev.* **48**, p. 928.

Xu, X. and Goddard, W. A. (2004). The extended Perdew-Burke-Ernzerhof functional with improved accuracy for thermodynamic and electronic properties of molecular systems, *J. Chem. Phys. B* **121**, p. 4068.

Yin, Y. and Alivisatos, P. (2005). Colloidal nanocrystal synthesis and the organic-inorganic interface, *Nature* **437**, p. 665.

Yoffe, A. D. (2001). Semiconductor quantum dots and related systems: electronic, optical, luminescence and related properties of low dimensional systems, *Adv. Phys.* **50**, p. 1.

Yu, P. Y. and Cardona, M. (2010). *Fundamentals of Semiconductors. Physics and Materials Properties*, 4th edn. (Springer-Verlag, Berlin, Heidelberg, New York).

Zahn, M. (1979). *Electromagnetic Field Theory: A Problem Solving Approach* (Wiley).

Index

absorbance 91, 251, 253
  maximum 252–253
  spectrum 252–254
absorption 29–30, 36, 63, 69–71,
  73, 75, 98, 101, 105, 227–228,
  268, 270
  cross section 71, 74
  infrared 143
  maximum 94, 206, 253
  molecular integral 74
  multi-photon 29
  optical 63, 65, 76, 92, 94,
    182, 205, 255, 257, 259,
    270–271
  spectral 74
  spectrum 139, 183–184, 205,
    207–208, 252, 254–255,
    257–258, 270, 273–274, 292
  two-photon 33, 43
acousto-optics 328–330
adsorption 130, 132, 147, 278–279
  molecular 131, 143–144, 147
AFM *see* atomic force
  microscopy
aggregation 144, 182, 250, 259,
  270
alloy 145, 181–216
amplification 29, 38, 105
amplitude 10, 27, 77, 87, 157,
  166, 192, 194, 225, 309–310,
  330–333
  anisotropy 226, 243
  approximation 32, 34, 42
  circular 242

diffraction field 86
  electric 22
  linear 242
  optical 331–332
  optical-field 141
anharmonicity 39, 160
application
  aerospace 208
  biotechnology 93
  clinical diagnosis 276–277
  photonics 259
  technical 91
approach
  adiabatic 329
  cloaking 215
  fabrication 259
  first principles DFT 256
  mathematical 108
  mechanical computational
    modeling 59
  molecular-orbital 47
  relaxation rate 98
  surface-plasmon polariton 172
  theoretical 05
  tight binding 51
  trial-and-error 46
approximation 35, 55, 59, 85, 118,
  285, 328, 331–332
  adiabatic 141, 162
  dense medium 25
  local density 58–59, 61–62, 135
  paraxial 300
  quasi-stationary 329
  random phase 20, 135, 188

364 *Index*

atomic
  configuration 92–93, 138–139, 148–149, 255
  orbital 54–55, 117, 119, 121, 140
  reconstruction 128, 132–133, 137, 139, 147, 273
  structure 12–13, 109, 111, 121–122, 125, 130, 132, 148, 150, 157, 184, 189
atomic force microscopy (AFM) 141
Avogadro's number 23, 337
azimuth 226, 240, 243, 245

band 119, 123–124, 193, 251, 280
  structure 118–125, 138, 184–187, 190, 192, 196, 200–201
  structure theory 107, 112, 124, 133, 137, 185, 194
biomaterial 273–276, 279
biomedicine 275, 277
biomolecule 275–276, 278
biosensor 275–278
birefringence 222, 228, 231, 247, 322
  circular 325
  optical 323
Bloch
  function 14, 16
  state 116, 117
  theorem 116–117
Bouguer–Lambert Law 105
boundary conditions 86, 97, 114, 151, 166, 168, 170, 236, 301, 304
  Bornvon Karman 114, 115
Bragg
  condition 334
  diffraction 334
  reflection 173
Brillouin zone 115, 117, 120, 125, 139, 192

charge
  conservation 3
  density 3, 13, 18, 41, 51, 58, 60–61
  transfer (CT) 143, 205
Clausius-Mossotti equation 182, 298
CNDO *see* completely neglected differential overlaps method
coefficient 10, 25, 53, 70, 99, 115, 175, 179, 302, 304, 316–318
  absorption 37, 64, 71, 139, 153, 174, 233–234, 274, 304
  damping 10
  diffusion 97
  Einstein 69–71, 73–74, 105–106
  electro-optic 313, 318
  extinction 4, 24, 64, 100
  heat conduction 303
  nonlinear 32, 42, 151, 298
  optic 297
  photoelastic 317
  proportionality 100
  reflectance 147
  refraction 6, 174
  tensor 314
  thermal refraction 299
coherent length 35
collinear synchronism 35
collision time 163
completely neglected differential overlaps method (CNDO) 53
conductivity 217, 265
configurational coordinate diagram 92, 95
continuity equation 3, 160, 162, 246, 310
convergence 82, 134–136
correlation
  function 77, 82
  interaction 59
  see electron correlation 262

see electron exchange and
correlation (XC) 60, 61
Cotton-Mouton effect 324
Coulomb gauge 5, 65
crystal 36, 108–110, 112–114,
116–120, 128–129, 133–134,
151, 186, 189, 218–220, 222,
227–229, 231, 233, 236, 247,
250, 306, 315–316, 322–323
  anisotropic 222–223, 231,
233–237
  biaxial 218, 222–223, 228
  calcite 233, 236, 247
  centrosymmetric 155
  covalent 128
  cubic 111, 220
  dichroic 234
  double-refracting 222
  ionic 250
  molecular 108, 249–250, 259,
262
  non-centrosymmetrical 316
  photonic 184
  piezoelectric 316–317
  transparent 217
  uniaxial 222, 227, 229, 234
  virtual 186
crystal lattice 95, 108, 112, 116
CT *see* charge transfer

dangling bond 128, 137–138
de Broglie wavelength 114
delta function 71, 109
density functional theory (DFT)
13, 56, 58–60, 62, 187, 271
density operator 13–16, 41
dependence 9, 207, 240, 262, 270,
272, 300
  algebra spectral 164
  angular 276
  concentration 95
  electron energy 120
  exponential 99

functional 175
device 90, 103, 241, 259, 264
  cloaking 215
  electronic 190, 270
  engineered 188
  medical 208
  optoelectronic 197, 264
  photonic 139, 182, 250, 264
  photovoltaic 263, 265
  solid-state 107
DFT *see* density functional
theory
dielectric
  function spectrum 190, 192,
197, 201, 203
  permittivity 1, 25, 139, 162,
164, 188, 208, 213, 220,
330
  permittivity tensor 246, 330
  tensor 201, 227–228, 231
diffraction 277, 328, 330, 332–334
dimer 130, 251–258, 280, 282
dipole 8–9, 20, 68, 98–99, 123,
163, 252, 256–257, 282
  approximation 3, 67, 69, 106
  dynamic 206
  induced charge 21
  induced magnetic 213
  molecular 258
  moment 68, 183, 251, 256
  monomer transition 253
  static 256
dispersion 6, 12, 19, 23, 169,
171–172, 177, 179, 192, 198, 202
DNA 279, 281–282
Doppler effect 291–292, 295, 331
Drude *see* model
dye 93, 95, 102, 182, 250–251, 259,
264

effective medium approximation
(EMA) 182, 184
Einstein coefficient *see*
coefficient

366 | *Index*

Einstein summation 228
electric field 2, 6, 11, 21, 28, 32, 73–74, 162–163, 167, 312–314, 316–318, 320, 322–323
electromagnetic field 3, 13, 15, 65, 163, 176, 215
electromagnetic radiation 1, 12, 65, 69–70, 107, 125, 139–140, 179, 208
electron
  correlation 262
  density 13, 56–59, 61, 140, 159, 162
  exchange and correlation 60, 61
electronegativity 51
electronic excitation 21, 132, 161, 169, 269
electronic structure 45, 59, 94, 116, 127, 148, 197, 256, 265–266, 268
electron transition 1, 12, 30, 66, 74, 92, 123, 190, 192, 194, 200
electro-optical effect 32, 312, 314–319
electro-optical Kerr effect 316
electro-optical Pockel's effect *see* Pockel's effect
electrostatic units (*esu*) 340
elementary wave packets (EWP) 85, 88, 90
ellipsometry *see* spectro-ellipsometry
EMA see effective medium approximation
emission 29–30, 32, 69–71, 77, 91–92, 94, 100, 103, 106, 204, 282
  cross section 94
  spectrum 91, 94, 204, 268, 270, 280
energy gap 62, 124–125, 192–195
envelope approximation 85
esu see electrostatic units

Euler equation 81
evolution 189–190, 192, 197–198, 208, 259, 269
EWP *see* elementary wave packets
excitation 12–13, 21, 27, 29, 91–92, 94–95, 141, 145, 164, 172–173, 177–178, 270, 277
  infrared laser 262
  optical 12, 16, 20, 41, 91, 94, 139
  optical plasmon 12
  optical two-photon 155
  short light pulse 98
  vibronic 249, 270
excited state 75–76, 92–93, 105, 251–252, 263, 266, 268–269, 280

Fabry–Perot interferometer 301–302
Faraday angle 326, 328
Faraday effect 323, 325
Fermat's principle 301
Fermi energy 56–57, 123
Feynman diagrams 29–33
Fick's law 97
Fizeau effect 284
Flocke theorem 331
fluorescence 63–106, 251–252, 268, 275, 279
  donor 100
  intrinsic 279
  low-contrast 279
  pyrene 281
  red-shifted 252
Fourier reverse transform *see* reverse Fourier transform
Fourier series 113
Fourier transform 17–18, 86, 305
Fourier transform spectrometer 77–78, 82, 85
fractional coordinates 218–220
Franck-Condon principle 92

Franz-Keldish effect 312
Fredholm equation 81
frontier orbitals 55
function 6–7, 9–11, 21, 28, 53,
  55–59, 67–68, 71–72, 76–81,
  83–84, 88, 115–116, 164,
  183–184, 297–298, 313
  autocorrelation 82
  Bessel 332
  Bloch 14, 16
  dielectric 20–21, 147, 179,
    187–188, 190–193, 198,
    203, 206, 227
  dielectric permittivity 2, 164,
    174, 191, 313
  Fermi 16, 41
  Kronecker delta 109
  optical 1, 4, 7, 12, 14, 16, 182,
    184–185, 187–188, 190,
    194, 201, 313–315
  oscillator 191
  periodic 113
  polarization 1, 5, 7–9, 11,
    19–20, 28, 40, 142, 188,
    327
  space-time 78
  time-dependent 76
  trial 40
  wave 13, 17, 46–51, 59–60,
    114–115, 117, 185, 187,
    251, 266–267
  work 136

gauge transformation 4
Gaussian
  orbitals 60
  system 68, 339
  units see electrostatic units
    (esu) 68, 340
graphene 120–121, 123, 134
graphite 120–123, 134, 219–220
grating 173, 328–330
ground state 13, 47, 48, 50, 55, 56,
  58, 59, 61, 62, 69, 75, 92–95,

251, 266–268, 272, 280, 282,
  351, 352
gyroscope 287, 294

Hamiltonian 47, 49, 53–54,
  65–66, 116–117, 251
Hartree energy 337
Hartree–Fock 59, 266
highest occupied molecular
  orbital (HOMO) 50, 55, 62,
  265–268
HOMO see highest occupied
  molecular orbital
Hückel approximation 53, 266
Huygens' construction 247
hybrid orbital 52–54, 121, 148,
  159

index ellipsoid 222, 233, 320, 322
interaction 36, 65–67, 70, 246,
  251, 257–258, 272, 276, 280,
  282, 317–318, 328, 334–335
  biomolecular 278
  chemical 144
  dipolar 250
  elastic 163
  interchain 266, 269–272
  membrane 281
  short-range 255–256
  van der Waals type 250, 255,
    257
interface 127, 132–133, 135, 139,
  148–151, 156, 165–172, 209,
  212, 234, 236, 278
inverse problem 78–80
inverse scattering problem 275
inverse Fourier transform 305
invisibility cloak 215

Jones matrix 224–225, 238, 246
Jones vector 224, 345–346

Kasha's rule 268
Kerr effect 312, 323

368 | *Index*

Kohn–Sham equation 61, 62
Kramers-Kronig relation 228

laser 39, 70, 94, 100–102, 107, 145, 250, 263, 265, 303, 328
laser pulse 103, 207, 304
lattice constant 111, 118–119, 132, 185–186, 188, 220
lattice plane 109, 111, 218
lattice vector 110–111
    reciprocal 17, 108–109, 113, 131, 133, 185
LCAO *see* linear combination of atomic orbitals
LCAO method 51–52, 116, 121
LDA *see* local density approximation
lens 39, 101–102, 284
    induced optical 299
    thermal 300
Lenz's law 214
linear combination of atomic orbitals (LCAO) 47, 51–52, 117
local density approximation (LDA) 58–59, 61–62, 135–136, 186
local field 9, 19, 20, 144
Lorentz formula 24
Lorentz model *see* model
lowest unoccupied molecular orbital (LUMO) 55, 62, 265–268
luminescence 63, 76, 91, 95, 97–101, 106, 263, 266, 268, 275, 279
    cathodoluminescence 91
    chemiluminescence 91
    donor 100
    electroluminescence 91
    kinetics 95, 97–100
    method 101
    signal 98–99, 101, 103
    two-photon 261

LUMO *see* lowest unoccupied molecular orbital

magnetic field 2, 5, 170, 209, 214, 323–326
magnetic permeability 3, 68, 184, 208–209, 213
magnetic response 213–214
Malus law 325
many-body problem 61
material
    absorbing 64
    bulk 128, 205, 208
    conjugated 265
    double negative 210
    hole transporting 204
    hybrid 264, 275–276
    inhomogeneous 182
    inorganic 203, 205
    ionic 297–298
    lossless 228
    luminescent 95
    magneto-optic 323
    mesoporous 259
    metallic 145
    micro-crystalline 125
    nanostructured 95, 207
    negative-index 210–214
    negative refractive 210
    non-homogeneous 181
    nonlinear 31
    nonmagnetic 2, 8
    nonmetallic 181
    non-periodic 107
    optical 36, 100, 228, 297–298
    organic 46, 262, 270
    positive index 210
    transparent 64, 228
Maxwell
    equation 3, 165, 167, 209, 215, 229, 307, 329
    theory 2
Maxwell-Garnett theory 182, 184
MBE *see* molecular beam epitaxy

medium 2–4, 6, 8, 25, 28–29, 38, 85, 290, 294, 323, 328–330
  absorbing 299
  condensed 24
  dynamic graded-index 328–329
  electro-optic 313
  gaseous 301
  homogeneous 28
  linear 340
  magnetic 2
  moving 284, 294–295
  non-cubic crystal 220
  nondispersive 6
  nonlinear 30, 32, 38, 43, 152
  stationary 294–295
metamaterial 208, 210–211, 214–215
method
  lithographic 306
  polishing 306
  pseudopotential 185, 187
  semi-empirical 46
  spectral 91
  super-cell 133
  tight-binding 187
methane 52, 54
Michelson interferometer 82, 287–288
Mie theory 182–183
Miller indices 109–111, 218
MO *see* molecular orbital
model 12, 55, 59, 136–137, 189–190, 192, 194, 197, 252, 256, 266
  computer simulated 155
  Drude 12, 22, 163, 164, 166, 169, 177
  Drude–Lorenz 22
  electromagnetic 143
  fife-oscillator 194
  Lorentz 326
  microscopic 227
  oscillator 42, 187, 192–193, 195–196

  semiempirical 190
  stick-and-ball computer generator 93
modeling 1, 12–13, 45–46, 52, 87, 108, 135
  first-principles 13, 200
  microscopic 13
  optical 12
  theoretical 135
molecular beam epitaxy (MBE) 182
molecular complex 250–251, 279–280, 282
molecular orbital (MO) 50, 54–55, 265–266
momentum operator 21, 66–67, 113–114
monochromator 101, 103
Mueller matrix 240, 243–244, 246, 346–347
Mueller matrix method 244–246

nanocrystal 203, 263–265
nanofiber 259–262, 306
nanoparticle 56, 143, 145, 156, 183–184, 203, 205, 207, 275–276
negative
  absorption 70, 71
  anisotropy 36, 231, 233
  focal length 301
  helicity 5
  index 211, 213, 215, 223
  lens 300
  magnetic response 214
  refraction 208, 209, 212
  refractive materials 210
Neumann's principle 151
nonlinear
  light interaction 28, 30, 34, 36, 40
  medium 6, 43
  optics 69

oligomer 268–269
optical
    constant 38, 107, 200
    density 99, 104
    effect 10, 27, 29, 283, 299, 312, 330
    element 101, 225, 297, 299, 346–347
    field 13, 15, 22, 32, 153, 216, 325
    indicatrix 221, 223, 314, 320–322
    path 288, 300–301
    pulse 85, 90
    rectification 7, 31–32
    response 6–8, 10, 12–13, 16, 21–22, 27, 41, 123, 125, 132, 139, 147, 191–192, 200
    spectroscopy 63, 73, 76, 85, 143, 161, 342
    spectrum 73, 76–78, 94, 139, 148–149, 191, 205–206, 268–269, 271
    transition 32–33, 66, 94, 124, 257
organic molecule 52, 249–250, 259
oscillator 23, 154, 192–194, 196, 327

paraxial approximation *see* approximation
Parseval's theorem 88
periodicity 17, 112, 114, 116, 128, 131–133, 173
permeability 66, 176, 213–214, 337
permittivity 68, 209, 213, 337, 340
phase velocity 4, 171, 211–212, 230–231, 290, 323
phonon 29, 93, 141–142, 335

phosphorescence 91
photodiode 103–104, 278
photoelastic effect 299, 316, 317
photoluminescence 91, 204, 269, 275, 279
photometer 101, 103
photon 7, 10, 27, 29–32, 69–71, 92, 178, 266, 334
photon energy 70, 92, 95, 195, 342
piezoelectric crystal *see* crystal
piezoelectrical effect 316
Planck constant 56, 66, 337
Planck radiation law 70
plasma 11–12, 161, 163, 177, 179, 205
    frequency 12, 162, 164, 175
plasmon 29, 161, 164, 166, 174
Pockel's effect 312–314
polarizability
    dynamic 148
    hyperpolarizability 260
    function 8
    local 298
    molar 23
polarization 3, 5, 32, 34, 37, 141–142, 148, 221, 224–225, 231, 233, 241–242, 345–346
polarizer 225, 233, 239–241, 243, 247, 326, 347
polymer 45, 52, 205, 249, 263, 265, 268–270
    conjugated 204, 263–265, 269–270
    lipophilic 202
    organic 203
    π-conjugated 263
    PTV 271–272, 274
Poynting vector 7, 73, 209, 214–215, 229, 246
primitive vector 108, 114, 131

propagation 77, 85, 165, 229, 231, 291, 307, 330
protein 279–282
pyrene 279–282

quantum
  confinement 135, 151, 166, 204
  dot 166, 202
  well 157–159
  yield 94, 99, 104–106, 204
quarter wave plate 225
quasiparticle 32, 61

radiation 2, 7, 66, 70, 73, 77–78, 95, 100–101, 103, 211, 215
Raman scattering 29–31, 37, 142–145
  cross section 143, 144, 145
random phase approximation (RPA) 20, 135, 188
Rayleigh scattering 263
  cross section 263
RDS see reflectance differential spectroscopy
reciprocal
  lattice 17, 108–111, 113–114, 131, 133, 185
  space 120
red shift 169, 182, 206, 252–253, 258, 271
reflectance differential spectroscopy (RDS) 147–150, 156
refraction 4, 64, 104, 209–210, 213, 218, 221, 234, 247, 302
refractive index 24–26, 36–37, 208–209, 211–213, 221–222, 231, 295, 297, 299–300, 313–314, 316, 320–322, 324, 328–331
  extraordinary 222, 231, 237, 320
  ordinary 222, 231, 237, 320

reverse
  counter 104
  Fourier transform 83
  photoelectrical effect 316
  piezoelectrical effect 316
rhodamine 93, 102, 146, 251, 255
RPA see random phase approximation

Sagnac effect 284–285, 287, 294
scanning tunneling microscopy (STM) 141
Schrödinger equation 15, 47–48, 51, 113, 115–117, 185–186
second harmonic generation (SHG) 31, 34, 36, 42, 150–151, 155–156, 159, 261
semiconductor 62, 106, 124, 192, 195, 204, 249, 263–264, 297, 306
  alloy 181, 182, 188, 197
  band structure 124
  heterostructures 195
  luminescence 106
  nanomaterials 275
  surface 153
  waveguide 306
sensitivity 101, 106, 156, 278, 280–282, 294
SERS see surface-enhanced Raman scattering
SHG see second harmonic generation
SHG response 154–156
signal to noise ratio 83
Snell's law 210, 235–236
spectro-ellipsometry 190, 193, 195–201
split-ring resonator (SRR) 214
spontaneous
  emission 32, 69–71, 106
  Raman scattering 30, 31
  Stokes scattering 38

372 | *Index*

SPP *see* surface plasmon polariton
SPR *see* surface plasmon resonance
SRR *see* split-ring resonator
Stark effect 312
STM *see* scanning tunneling microscopy
Stokes
  parameters 240–242, 245
  scattering 30–31
  vector 240, 242–244, 248, 345–346
surface-enhanced Raman scattering (SERS) 143–145, 147, 173
surface
  atomic structure 129, 131, 133, 141, 149
  free energy 130
  plasmon 161, 164–166, 168–169, 177–178, 276–277
  recombination 160
  super cell 134
surface plasmon polariton (SPP) 161, 166, 171–173, 277
surface plasmon resonance (SPR) 144, 169, 173, 275–278
symmetry 110, 112, 114, 124, 128, 137, 151, 232, 266, 315
  axial 299
  crystal point group 151
  dielectric tensor 315–316
  inversion 151, 154–155, 315–316
  tetrahedral 137
  translational 57
system 9, 11–14, 40–41, 56–57, 59–61, 68, 130–131, 134, 145–146, 178, 189, 239, 283–284, 300, 339–340
  absolute 68
  amorphous 154

bar-coordinate 292
biological 276–277
centrosymmetric 40–41
conjugated 265
coordinate 46, 214–215, 217–218, 221, 227, 238, 290, 293, 294, 314, 322
Gaussian 68, 339–340
inertial 283–284, 295
molecular 23, 72, 91, 250, 259, 266, 275, 282
nanostructured 202
optical 103, 173, 239, 325
orthonormal 218
π-conjugated 265–266

Taylor expansion 14–15, 313
Teflon wedge 210–211
tensor 6–7, 142, 151, 217, 221, 228, 315–316, 324, 330
thermal
  conductivity 217
  expansion coefficient 195
  fluctuation 160
  motion of atoms 105
  optics 297
  oxidation 157
thermo-optical coefficient 297
thiophene 272–273
Thomas–Fermi approximation 56
threshold intensity 37, 38
tight binding method 116, 121
Tikhonov method 80, 82
total
  energy 12, 55, 60, 61, 129, 134, 255, 257, 272
  irradiance 241
  light reflectance 173
  number 57, 98, 114, 140
  power 39, 309
  propagation time 293
  wave function 251

Index | **373**

transition 69–70, 72, 156, 258
    band gap 269
    charge-transfer 143
    dipole-allowed 253
    electron quantum 75
transmittance 184, 239

unit
    cell 58, 108, 112, 119, 121, 129, 132, 137, 140, 148, 149, 218–220, 272
    mass 303
    vector 7, 142, 344
    volume 23
units
    atomic 57, 60, 65
    CGSE system 175
    delay 103
    Gaussian, esu, *see* electrostatic 68, 340
    optical 339, 341, 342

vacuum 3, 66, 68, 177, 212–213, 284, 337, 340
van der Waals force 250
variational
    derivative 58
    principle 59
    minimization procedure 61
VCA *see* virtual crystal approximation
Verdet constant 325, 326, 328

virtual crystal approximation (VCA) 184–186

wave 21, 31, 142, 152, 208–209, 230, 233, 235–236, 278, 284, 328–329
    acoustic 163, 328–330, 333, 335
    amplified 43
    collinear 37
    diffracted 331–332
    electric polarization 151
    electromagnetic 4, 21, 30, 63, 65, 85–86, 152, 175, 177–179, 214, 278
    in-plane 153
    optical 25, 43, 227, 231, 235, 317, 319, 324, 327, 331, 334
    propagation 21, 34, 105, 210, 221, 312, 325
    second harmonic 36, 42
    surface plasma 277
    ultrasonic 329
    vector 114, 163, 166, 170, 172–173, 177–178, 184–185, 209–210, 235, 246, 325, 330–331, 334–335
waveguides 306–310
Wood's notation 131, 132